Alain Meunier
Clays

Alain Meunier

Clays

With 262 Figures

 Springer

PROF. DR. ALAIN MEUNIER
UNIVERSITY OF POITIERS
UMR 6532 CNRS
HYDRASA LABORATORY
40, AVENUE DU RECTEUR PINEAU
86022 POITIERS
FRANCE

E-mail: alain.meunier@hydrasa.univ-poitiers.fr

This book has been translated by Nathalie Fradin (Fradin Traduction sarl) from the original work "Argiles", first published in France in 2003 in the series "Geosciences" by Editions scientifiques GB in collaboration with the Société Géologique de France.

ISBN 978-3-642-06000-7   e-ISBN 978-3-540-27141-3

Springer is a part of Springer Science+Business Media
springeronline.com
© Springer-Verlag Berlin Heidelberg 2010
Printed in Germany

Cover design: E. Kirchner, Heidelberg

Printed on acid-free paper   32/2132/AO 5 4 3 2 1 0

Preface

This book is about the most complete work on the subject of clay minerals thus far conceived. Its scope is one of basics to general principles to use in real geologic situations. In principle any student, advanced student or casual researcher, should be able to find an answer to almost any question posed. The breadth of knowledge presented is truly impressive.

This presentation is especially important at present when the study of the most abundant minerals found near the Earth's surface, clays, appears to find disfavour with students and Universities. In fact we have never before needed such an encyclopaedic work to strengthen the discipline. At present when Earth sciences are slowing in popularity, the need for a precise and general education in the study of clay minerals is greater than ever before. Curiously the study of the environment, those materials found in the sphere of biological activity, is more relevant than ever. The natural progress of human activity through the age of science and then industrial activity up to the present stage of the post-industrial era has been marked by an increasing use of the Earth's surface resources. The steady increase of the human population has called upon the natural resources of the surface in a non-linear manner. In the pre-industrial era agriculture and industry was concerned with the basic subsistence of populations. In good years there was enough to eat and in bad ones not enough. The means of obtaining this production did not draw upon the natural mineral riches of the surface to any great extent. Animals were the major non-human motor force, and iron was only a subsidiary part of the production mechanism. Even in cities most of the means of heating remained as the natural, renewable sources of wood. This of course has changed greatly in the last 150 years.

Today not only the soil of the planet is solicited to produce more than subsistence for a vastly increased population but the desires of these populations to move about has drawn on the buried resources to respond to the high energy needs of motor transportation and total comfort heating. Skiing on the top of a mountain in January is the epitome of the triumph of modern man over the elements where an enormous use of energy and natural resources is used to gratify a desire for exotic experiences. These needs of comfort, abundance and pleasure draw on the natural equilibrium of the Earth's surface.

Clay mineralogy is then the study of the basis of the natural riches found in most surface layers of the Earth. Application of the accumulated knowledge about these fundamental portions of our environment will allow us to search more efficiently for petroleum resources, for metallic concentrations and for an increased efficiency of farming practices. These are the positive aspects of our present human experience. However, as is more and more evident, the negative side of current human activity creates high concentrations of toxic material in the surface environment, and perhaps in the near future at some undetermined depth (toxic, nuclear waste). These must be dealt with intelligently and oft times in great haste. Such problems involve, or should involve, knowledge of clay mineralogy. The natural chemical stability (in chemical – time space) of clays, their physical properties and chemical influences must be known in order to deal efficiently with modern day problems involving the environment. In this sense the book proposed by Alain Meunier is not only timely but also essential. No other work has attempted to unite the dispersed and rich knowledge of clay minerals into a single volume. One can find the relations of chemical stability, X-ray diffraction identification, natural occurrence and abundance, reaction rates and physical properties of the minerals and many other aspects are explained and documented in this text. Virtually all of the major questions concerning clay minerals are treated here.

Thus this book can be used as a resource for teaching clay mineralogy or for students or as a guide for research people from other fields who wish to explore the problems of clay mineralogy on their own. It is sufficiently clearly written so that one can use it as a beginner or an experienced professional working in a field that will need some expertise in clay minerals. This is especially important today when we find a change in the structure of science beginning to modify the classical frontiers of our ancient disciplines. As old boundaries change, new needs arise and a fundamental understanding of clay minerals will always be of greatest help in dealing with the surface of the Earth that is the future of mankind.

Bruce Velde

Foreword

Clays are not the most abundant components in the mineral kingdom when compared to olivines of the Earth's mantle or to feldspars of the continental crusts. However, they hold a special place in scientific research because their environment is criss-crossed constantly by human activity. Indeed, inasmuch as they characterise soils and altered rocks, clays are at the centre of farming activities and civil engineering works; they are formed in diagenetic series prospected for petroleum resources; they crystallise in geothermal fields whose energy and mineral deposits prove valuable. Clays play an important part in everyday life, from the white-coated paper on which we write to the confinement of hazardous waste storage, from cosmetics to pneumatics, from paints to building materials.

Clays alone form an entire world in which geologists, mineralogists, physicists, mechanical engineers, chemists find extraordinary subjects for research. These small, flat minerals actually interface widely with their surrounding environment. They absorb, retain, release, and incorporate into their lattice a great variety of ions or molecules. Their huge external surface area (as compared with their volume) makes them first-class materials for catalysis, retention of toxic substances or future supports for composites. Clays are made up of particles that form stable suspensions in water. These suspensions have long served in drilling applications or tunnel piercing techniques. Suspended clays flow as liquids, thereby both helping to shape manufactured products such as ceramics, but also causing tragic mud flows, lahars or landslides.

In view of the tremendously rich research in the field of clays, this book deals only with the geology, the mineralogy or the chemistry of these minerals. Industrial applications, natural hazards and civil engineering could each be the subject of a separate work, and consequently are not addressed here. This work is heir to two books that made a deep impression on me: "Géologie des argiles" by Georges Millot (1964) and "Clay Minerals. A physico-chemical explanation of their occurrence" by Bruce Velde (1985). I have been fascinated by their comprehensive vision of processes ranging from the scale of the mineral to that of the landscape. They present both a brilliant synthesis and a visionary interpretation of the knowledge of their time. More than ever, I feel like a Lilliputian perched on these Gullivers' shoulders!

When the field pedologist or geologist picks up a lump of earth or breaks a chunk of altered granite or clay sediment, he actually examines clay minerals

analytical techniques	scale of investigation	researched data	expected informations

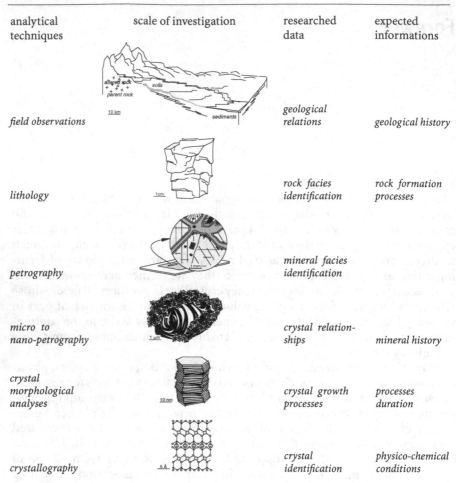

field observations — geological relations — geological history

lithology — rock facies identification — rock formation processes

petrography — mineral facies identification

micro to nano-petrography — crystal relationships — mineral history

crystal morphological analyses — crystal growth processes — processes duration

crystallography — crystal identification — physico-chemical conditions

Fig. 1. Sketch showing the relationships between the different observation scales in a geological study of clay-bearing rocks.

that form complex textures on a scale of one millimetre. In order to understand how such textures have been formed, he must "dismantle the system", meaning he must reach increasingly elementary components: aggregates, particles, crystals and eventually layers. The analysis of each organisational level helps to understand the mechanisms responsible for the genesis of clay minerals and reveals secrets of the rock's formation (Fig. 1).

This book is directed to beginners in clay science. It is aimed at providing clues to "extract the message" underneath the crystallographical characteristics of clay minerals, as well as through their textures in rocks or soils. The book is divided into two parts:

– *Fundamentals (Chap. 1 to 5)*: To deal with the basic concepts of thermodynamics specific to clays, it is first necessary to know the crystal structures and the rules defining the chemical and isotopic compositions of these phyl-losilicates. Surface properties are presented to introduce cation and anion exchange phenomena, as well as their ability for flocculation, aggregation or, on the contrary, dispersion. This is an introduction to the study of the mechanical and rheological properties that characterise the microstructure of natural or artificial clay materials. When necessary, some specific points are developed in separate boxes in the text.

- *Geology of clays (Chaps. 6 to 10)*: The initial acquisition of the fundamentals helps to understand how mineral reactions are determined in the world of clays. Their remarkably high reactivity allows the effect of the passage of time to be measured. This dynamic aspect underlies the presentation of soils and alterites, sediments, diagenetic and hydrothermal formations.

This book ends with a fascinating aspect of these minerals – despite their reputation for fragility, they subsist and form under extreme conditions. In this respect, an astonishing parallel can be drawn between clay and life: both often prove to be much more tenacious than expected.

The main purpose of this book is to share, beyond the facts, my enthusiasm for the study of minerals, undiminished after so many years. This enthusiasm has been kept alive by all the students (more than forty to date) who have done me the honour of working with me. They will be in my heart forever.

The information presented in this book derives in small part from my personal work, from my reading, and for the greater part from countless and sometimes animated discussions I have had with my colleagues from Poitiers (Daniel Beaufort in particular), but also with Bruce Velde, Alain Baronnet, Bruno Lanson and many others. I will never thank them enough for their contribution and am forever indebted to them. I also wish to especially thank Philippe Vieillard who had the courage to review the chapter dedicated to thermodynamics. I am highly indebted with Andreas Bauer who had imprudently promised me to read the English version of the book. He did it courageously!

The references used, although numerous, are notoriously incomplete, as is my knowledge. I hope that those whose work could not be quoted will not hold it against me: they are cited in the articles I reference. Searching the literature is a treasure hunt for which a solid point of departure is absolutely essential. I hope I succeeded in elucidating a starting point for more than a few domains.

Alain Meunier
Poitiers – December 2001

I wish to thank Alain Meunier for his faith in my work. This book has been a deeply enriching experience on both a professional and personal level.

Nathalie Fradin
Neuil – July 2003

Contents

CHAPTER 1

Crystal Structure – Species – Crystallisation

The guiding line of the first part of this book (Chaps. 1 to 5) is to understand some basic principles that repeatedly take place during geological processes. Since the clay scientist is concerned with phyllosilicates, he must first familiarize him- or herself with the fundamentals of the "layer" crystallography. The reconstruction of the silicon sheet and octahedral layer geometries is particularly important before the application of X-ray diffraction (XRD). This is the purpose of Sect. 1.1. Thereafter, this fundamental knowledge of crystallography will be used in Chap. 2 to understand how the chemical elements are distributed inside the layer structure.

The clay scientist's task is not limited to XRD identification of clay species. Indeed, despite their small size, clay minerals are crystals, which means that their dimensions, shape, and number of defects result from specific, discoverable physico-chemical conditions. This is why it is important to understand how crystals were born (nucleation) and how they grow bigger and exhibit crystal faces (crystal growth). This is the purpose of Sect. 1.2.

1.1
The Crystal Structure of Clay Minerals

Clay minerals belong to the phyllosilicate group (from the Greek "phyllon": leaf, and from the Latin "silic": flint). As a distinctive feature, they are very small (a few micrometers maximum) and their preferred formation occurs under surface (alterites, soils, sediments) or subsurface (diagenesis, hydrothermal alterations) conditions. Difficult to observe without using electron microscopy (scanning and transmission), they have been abundantly studied by X-ray diffraction, which is the basic tool for their identification. While the number of their species is relatively small, clay minerals exhibit a great diversity in their composition because of their large compositional ranges of solid solutions and their ability to form polyphased crystals by interstratification. Trying to list them would be gruelling and fruitless work. What is more important is to understand why their crystal structure affords them such a range compositional diversity.

This chapter is aimed at introducing the fundamentals of the crystal structure starting from the layer as the elementary unit and ending with the concepts of crystal, particle and aggregate. These basics are well complemented by re-

ferring to specialised resources in which crystal structures are described from
X-ray diffraction: Brindley and Brown 1980; Moore and Reynolds 1989; Drits
and Tchoubar 1990; Bouchet et al. 2000.

1.1.1
The Elementary Structure Level: the Layer

X-ray diffraction deals with distances between atomic planes. But how are these
distances determined? The following sections will show how they are related
to the length of the chemical bond that links the ions forming the clay mineral.
The scale of investigation is one angström (0.1 nm).

1.1.1.1
Layer: Dimensions and Symmetry

Chemical Bonds and Coordination
As with any mineral species, phyllosilicates are characterized by a "unit cell".
In the first step, the unit cell will be considered at the scale of a single layer
and its dimensions will be determined in 3D space (Fig. 1.1a). The a and
b dimensions are in the x-y plane (the plane in which the largest faces are
oriented). The c dimension along the z-axis corresponds to the "thickness"
of the layer. A simple geometrical calculation provides approximate values
for these three dimensions. One must study the structure of the elementary
organisational level of the layers: the tetrahedral and octahedral "sheets" in
order to make this calculation. The sheet's framework is formed by cation-
anion bonds (bonds intermediate between ionic and covalent bonds) whose

Table 1.1. Effective ionic radii of the main anions and cations contained in phyllosilicates
(Shannon and Prewitt 1976)

Ions	Ionic radius (Å)		
	Coordinence 4	Coordinence 6	Coordinence 12
O^{2-}	1.24	1.26	
OH^-	1.21	1.23	
K^+	1.51	1.52	1.78
Ca^{2+}		1.14	1.48
Na^+	1.13	1.16	1.53
Mg^{2+}	0.71	0.86	
Fe^{2+}	0.77	0.92	
Fe^{3+}	0.63	0.79	
Al^{3+}	0.53	0.68	
Si^{4+}	0.4	0.54	
H_2O^*		1.45	
NH^{4+}*		1.61	

*The shape of H_2O and NH^{4+} here is compared with a sphere

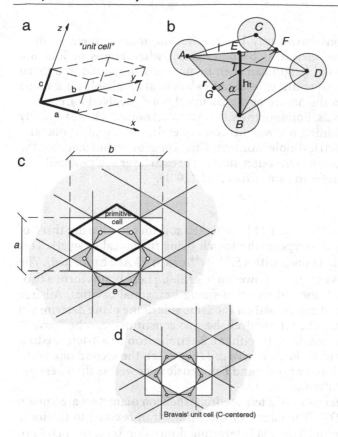

Fig. 1.1a–c. Structure of the tetrahedral sheet. **a)** The a, b and c dimensions of a "unit cell". **b)** The thickness of tetrahedron h_t is 2.12 Å. **c)** The a and b dimensions are 5.36 and 9.27 Å respectively (the primitive unit cell is indicated). **d)** The Bravais' s unit cell is centred on the hexagonal cavity (Méring 1975)

length will be used as a reference in the calculation of the cell dimensions. Ionic diameters will be used for the determination of cation coordination.

Considering the difference in their ionic diameter (Table 1.1), three types of coordination determine the elementary polyhedra that make up the various sheets of the crystal structure:

- 4-fold coordination (SiO_4^{4-} or AlO_4^{5-} tetrahedron).

- 6-fold coordination (octahedron whose centre is occupied by a Al^{3+}, Fe^{3+}, Fe^{2+} or Mg^{2+} cation for the most part – the vertices being formed by O^{2-} or OH^- anions)

- 12-fold coordination (dodecahedron whose centre is occupied by a cation with a wide diameter: K^+, Na^+, Ca^{2+}, and vertices are formed by O^{2-} anions of two opposite tetrahedral sheets).

Size of Atoms and Ions

Because atoms and ions have fuzzy boundaries, the measurement of their radius is difficult. Nevertheless, it becomes easier when atoms or ions are bonded because the bond length can then be precisely measured in a given equilibrium state. Consequently, several radius values are available for a single element, depending on the nature of the chemical bond involved – i.e. ionic, covalent, or van der Waals. Bonds of the silicate crystal framework are typically intermediate between ionic and covalent. Consequently, each bond is polar and characterised by an electric dipole moment. This is of great importance for the electric charge distribution at the outer surfaces of each layer in a phyllosilicate structure. For details, refer to Sainz-Diaz et al. (2001).

The Tetrahedral Sheet

SiO_4^{4-} or AlO_4^{5-} tetrahedra (Fig. 1.1b) are linked together by sharing three of four vertices (three basal oxygens, the fourth being the apical oxygen). This means that one O^{2-} anion bonds with a $Si^{4+}-Si^{4+}$ or a $Si^{4+}-Al^{3+}$ cation pair. The $Al^{3+}-Al^{3+}$ cation pair is excluded (Löwenstein's rule). These bonds form a two-dimensional lattice (tetrahedral sheet) defining hexagonal cavities. All free oxygens (apical oxygens) are located on the same side of the plane determined by the bonded oxygens. The tetrahedral sheet so constituted can be "paved" by translation (without rotation) by either a 2-tetrahedron or a 4-tetrahedron unit. The first one is called *the primitive cell* (Fig. 1.1c), the second one is *the Bravais' unit cell* (Fig. 1.1d) whose *a* and *b* dimensions as well as thickness are easily calculable (C-centred).

Calculating the dimensions of a tetrahedron when ion diameters are known is easy (Jaboyedoff 1999). The edge *r* of the tetrahedron is equal to the ionic diameter of the O^{2-} anion. The first interesting dimension is h_t, the height of a tetrahedron. Figure 1.1b shows the procedure in triangle AFB where AF and BF are the medians of the equilateral triangles ACD and BCD respectively:

$$AE = 2/3AF = 1 \tag{1.1}$$

$$AF^2 + FD^2 = AD^2 \Rightarrow l = \frac{r}{\sqrt{3}}. \tag{1.2}$$

In triangle AEB, $EB = h_t$ can be calculated as follows:

$$h_t = \sqrt{r^2 - l^2} \quad \text{or} \quad h_t = r\sqrt{\frac{2}{3}}. \tag{1.3}$$

In the case of a SiO^{4-} tetrahedron, $r = 2.60$ Å so $h_t = 2.12$ Å.

The relationship between the *a* and *b* unit cell dimensions in the tetrahedral sheet is simple. The distance between one vertex and the centre of the tetrahedron (Si–O bond) determined by the intersection T between EB and FG is given by the relationship:

$$TB = \frac{r}{2\cos\alpha} \quad \text{or} \quad \cos\alpha = \frac{h_t}{r} \quad \text{so,} \quad TB = r\sqrt{\frac{3}{8}} \tag{1.4}$$

The *a* dimension (Fig. 1.1c) is determined as follows: $a = 2r = \frac{4}{3}\sqrt{6}$. TB

Figure 1.1b shows that the b dimension is three times the hexagon side e: $b = 3e$.

And $e = 2l$ hence $e = 2\frac{r}{\sqrt{3}} = \frac{a}{\sqrt{3}}$. Accordingly:

$$b = 3e = a\sqrt{3} = 4\sqrt{2}. \text{ TB} \tag{1.5}$$

The Si–O bond length can be calculated according to the values given in Table 1.1: 1.64 Å. Consequently the theoretical value of the b dimension is 9.27 Å. In reality, the Si–O bond length is 1.618 ± 0.01 Å, so $b = 9.15 \pm 0.6$ Å. The ionic diameter of the 4-fold coordination Al^{3+} cation is greater than that of Si^{4+}; accordingly the theoretical value of the Al–O bond length will be greater than that of the Si–O bond length: 1.77 Å (Table 1.1). In reality, it is 1.748 ± 0.01 Å. Therefore, the b dimension will increase with the substitution rate of Al^{3+} for Si^{4+}. Suppose that this increase is directly proportional to the Al^{3+} content, this leads to the following relation:

$$b_{(Si_{4-x}Al_x)} = 9.15 + 0.74 \times \text{ Å} \tag{1.6}$$

The representation of the distribution of atoms is easier in crystallography (for X-ray diffraction notably) for a four-half-tetrahedron unit cell centred on the hexagonal cavity (Méring 1975).

The Octahedral Sheet

Octahedra are laid on a triangular face (Fig. 1.2a). They are linked together by sharing their six vertices. This means that each anion is bonded to three cations in the trioctahedral type. It is bonded to two cations in the dioctahedral type so that the third site is vacant. These bonds constitute the framework of a continuous sheet in which octahedra form a lattice with hexagonal symmetry. The di- and trioctahedral layers are "paved" by a 6-octahedron unit cell (Fig. 1.2b) whose a and b dimensions as well as thickness can be easily calculated.

The edge of the octahedron corresponds to the ionic diameter of the O^{2-} or OH^- anions considered equivalent here. In triangle OMO', the cation-anion distance MO (or MO') is then given by:

$$MO = \frac{s}{\sqrt{2}} \tag{1.7}$$

The thickness of the sheet is given by the value of RV in rhomb QRTU whose sides equal $s\frac{\sqrt{3}}{2}$. In rectangle triangle QMU, angle α can be determined by its sin: $\sin\alpha = \frac{\frac{s}{2}}{\frac{s\sqrt{3}}{2}} = \frac{1}{\sqrt{3}}$ thus $\alpha = 35.26°$. In triangle RVU, RV is given by: $RV = s\cos\alpha$.

Bond lengths are not always accurately known; as an approximation, they can be considered as the sum of the ionic radii of the cation (Mg^{2+}, Fe^{2+}, Al^{3+}, Fe^{3+} ...) and of the 6-fold coordination O^{2-} anion (Shannon and Prewitt 1976). The values for Mg–O and Al–O bonds are $0.86 + 1.26$ Å and $0.68 + 1.26$ Å

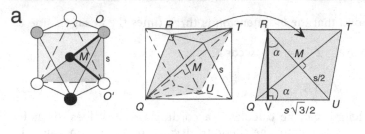

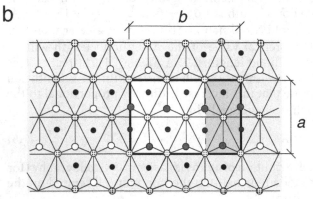

Fig. 1.2a,b. Structure of the octahedral sheet. **a)** Thickness is given by RV $= s\cos\alpha$ (with $\alpha = 35.26°$). **b)** The b dimensions of the trioctahedral (brucite) and dioctahedral (gibbsite) unit cells are 9.43 and 8.64 Å respectively $\left(a = \frac{b}{\sqrt{3}}\right)$

respectively. Therefore, the value of the edge of a "Mg octahedron" will be 3.00 Å and that of an "Al octahedron" will be 2.74 Å while their respective thickness will be 2.45 and 2.24 Å.

Figure 1.2b shows that $b = 3s$, so $b = 3\sqrt{2}. \overline{MO}$. Using the values in Table 1, the theoretical values of this dimension can be calculated for a brucite-like triocta-hedral sheet $[Mg_3(OH)_6]$ and for a gibbsite-like dioctahedral sheet $[Al_2(OH)_6]$. The value of $\overline{MO}$ is 2.12 and 1.94 Å respectively, hence $b_{brucite} = 8.99$ Å and $b_{gibbsite} = 8.23$ Å. These theoretical values differ from the real values 9.43 and 8.64 Å respectively.

Adjustment of the Tetrahedral and Octahedral Sheets

The theoretical b dimensions calculated for the tetrahedral sheet (9.27 Å) and for the tri- and dioctahedral sheets (8.99 and 8.19 Å respectively) differ significantly. It is the same for the a dimension, which depends on the b value. This indicates that the linkage between tetrahedral and octahedral sheets through the free oxygens of the tetrahedra will not take place without deformations. The tetrahedral sheets lose their hexagonal symmetry by rotation of the tetra-hedra about axes perpendicular to the basal plan (Fig. 1.3a). The symmetry

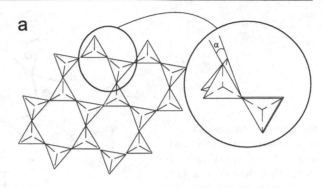

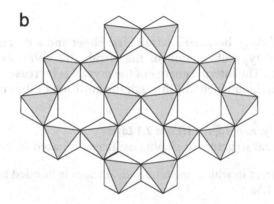

Fig. 1.3a,b. Deformation of sheets. **a)** Rotation of tetrahedra. **b)** Deformation of octahedra in a gibbsite-like structure

becomes ditrigonal and the rotation angle α can be estimated as follows:

$$\cos \alpha = \frac{b_{measured}}{b_{theoretical}} \tag{1.8}$$

Angle α can not exceed 30° for the repulsion that O^{2-} anions exert on each other. More complex distortions occur by rotation about axes that are contained in the plane and by deformation of the tetrahedra. The latter involve changes in the value of angles between the cation and the oxygens of the four vertices (theoretical value 109.47°) as well as in the bond length. Some of them shorten as shown by NMR spectroscopy (see Sect. 2.1.2.2).

Di- and trioctahedral sheets do not undergo identical deformations. The presence of vacant sites (vacancies) in the former alters the geometry of the octahedra (Fig. 1.3b). Indeed, the absence of the cation reduces the attractive forces on anions and leads to the elongation of the edges of the vacant octahedron: from 2.7 to 3.2 Å. Therefore, the occupied octahedra become asymmetrical. Such distortions do not theoretically occur in trioctahedral sheets. In reality they are quite reduced (Bailey 1980). Only the presence of bivalent cations with very different ionic diameter causes local changes in symmetry.

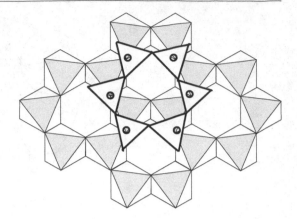

Fig. 1.4. The linkage between a tetrahedral sheet and a dioctahedral-type sheet necessitates their deformation. The hexagonal symmetry transforms into ditrigonal symmetry

The linkage between a tetrahedral sheet and a dioctahedral sheet requires a second type of rotation of the tetrahedra about axes in the basal plane (Fig. 1.4). The internal energy of the crystal is increased by the bond deformations (angles and length) through the addition of the elastic energy.

The Two Types of Layers: 1:1 and 2:1 Layers

The crystal structure of all phyllosilicates is based on two types of layers:

1. 1:1 layers in which one tetrahedral sheet is bonded to one octahedral sheet (Fig. 1.5a);

2. 2:1 layers in which one octahedral sheet is sandwiched between two tetrahedral sheets (Fig. 1.5b).

The theoretical structures (without deformations) of both types of layers depend on the hexagonal symmetry of the tetrahedral and octahedral sheets which are linked to each other. The apical oxygens of the tetrahedra become the vertices of the octahedra. Thus, the six vertices of the octahedra in a 1:1 layer are formed by 4 OH^- radicals and 2 apical oxygens of the tetrahedra. In 2:1 layers, they are formed by 2 OH^- radicals only because the other four vertices are the apical oxygens of the two tetrahedral sheets.

However, the tetrahedral and octahedral sheets exhibit differing a and b dimensions. Thus, their linkage cannot possibly take place without deformation of the angles and lengths of some chemical bonds, as indicated previously. These deformations are significant in dioctahedral layers; they are minor in trioctahedral layers. In all cases, the 6-fold symmetry becomes 3-fold symmetry.

The symmetry group to which phyllosilicates belong depends on the way 1:1 or 2:1 layers are stacked. From the highest to the lowest they are: hexagonal (H), rhombohedral (R), orthorhombic (Or), ditrigonal (T), monoclinic (M) and triclinic (Tc). Therefore, the various layer-stacking modes determine distinct polytypes, the number of which is a function of the mineral species considered

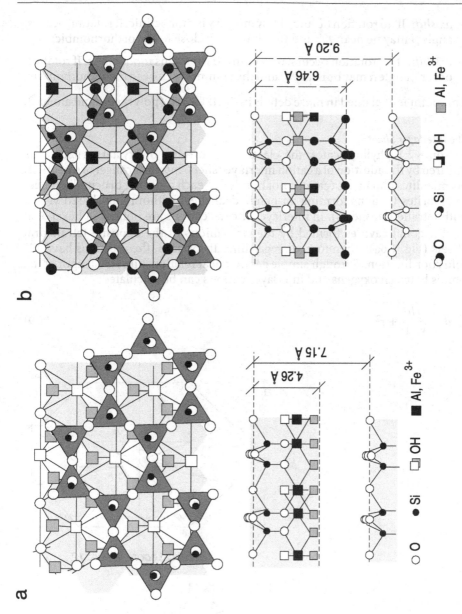

Fig. 1.5a,b. Crystal structure of dioctahedral phyllosilicates. **a** 1:1 layer. **b** 2:1 layer

(cf. 1.1.2.1). Let us consider that there are three possible ways to stack two consecutive layers:

- *no shift*. The symmetry becomes orthorhombic, pseudohexagonal or hexagonal;

- *a:3shift*. If no rotation occurs, the symmetry is monoclinic. If rotation occurs, angle β may be near 90° and the symmetry close to the orthorhombic;

- *b:3shift*. If no rotation occurs, the symmetry remains monoclinic. If rotation occurs, angle α may be near 90° and the symmetry close to the orthorhombic;

This point is explained in more details in Sect. 1.1.2.1. (polytypes) and annex 2.

The Interlayer Sheet

In some 2:1 phyllosilicates, the electrical neutrality of crystal structures is ensured by the addition of a cation interlayer sheet: interlayer sheet of smectites, vermiculites and micas (cations not bonded to each other) or brucite-like sheet of chlorites (cations forming an octahedral sheet without any shared vertex with tetrahedral sheets). In interlayer sheets, cations are housed in hexagonal or ditrigonal cavities formed by the O^{2-} anions of the opposite tetrahedral sheets (Fig. 1.6a). Therefore, disregarding distortions, these cations have 12-fold coordination. Through simple geometrical relationships, the length of the bonds between oxygens and interlayer cations can be calculated:

$$d = \sqrt{\frac{h^2}{4} + r^2} \qquad\qquad (1.9)$$

a

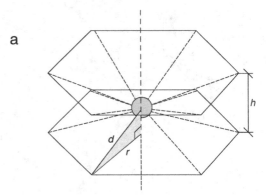

theoretical coordinence: 12

b

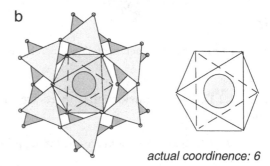

Fig. 1.6a,b. Structure of the interlayer zone. **a)** Thickness h depends on the bond length between the interlayer cation and the O^{2-} anions of the tetrahedra (d). **b)** Distortions of the tetrahedral sheets brings the coordination from 12 down to 6

actual coordinence: 6

where

h: interlayer spacing (shortest distance between the oxygens of the two tetrahedral sheets).

r: distance from the centre of the hexagonal cavity to the oxygens of the plane
in a tetrahedral sheet.

In case of deformation of the tetrahedral sheets by rotation of angle α, r is
shown to vary as follows (Decarreau 1990):

$$r = \frac{b}{6\cos 30°} - \frac{b \cdot \tan \alpha}{6} = \frac{b}{6}\left(\sqrt{3} - \tan \alpha\right) \qquad (1.10)$$

The actual coordination is no longer 12 but 6 (Fig. 1.6b). The structure of the
brucite-like sheets is discussed in a further paragraph (Sect. 1.1.1.2)

At this point, it is possible to find out how the tetrahedral, octahedral and
interlayer sheets form different patterns of layer structure. The calculation of
the "unit cell" dimensions and volume is accordingly facilitated. The "unit
cell" concept will be used throughout this book; here, it is considered as
the smallest volume repeated in the two-dimensional space of the layer. The
calculation of the cell unit dimensions and volume will help in understanding
why some dimensional parameters are useful criteria for the X-ray diffraction
identification of clays, on the one hand, and how mineral density is determined,
on the other hand.

1.1.1.2
The Different Patterns of Layer Structure

1:1 Structure (no Interlayer Sheet): Kaolinite and Lizardite

In the unit cell of a kaolinite, four sites of the dioctahedral sheet are occupied
by Al^{3+} cations and two are vacant (Fig. 1.6a). The unit formula of kaolinite
is: $Si_4 O_{10} Al_4 (OH)_8$. In the trioctahedral sheet of a lizardite all six sites are
occupied by Mg^{2+} cations; the unit formula is: $Si_4 O_{10} Mg_6 (OH)_8$. The negative
charge of the oxygen anion framework is balanced by the positive charge of the
tetrahedral and octahedral cations. The crystal structure of 1:1 phyllosilicates
consists of five ionic planes. The actual a and b unit cell dimensions are,
respectively: $a = 5.15$; $b = 8.95$ Å for kaolinite and $a = 5.31$; $b = 9.20$ Å for
lizardite.

The distance between two neighbouring 1:1 layers corresponds to the thickness of the combined tetrahedral sheet+octahedral sheet (theoretically: 2.11 +
2.15 = 4.26 Å) to which is added the thickness of the interlayer spacing. The
latter depends on the length of the hydrogen bonds connecting the tetrahedral
sheet in one layer to the octahedral sheet in the neighbouring layer (about
3.0 Å according to Bailey 1980). The interlayer spacing of kaolinite is 7.15 Å,
and that of lizardite 7.25 Å for the pure magnesian end member; it increases
with the substitution rate of Mg^{2+} for Fe^{2+}.

Table 1.2. Calculation of the unit cellmass of kaolinite (M_{vk}) and lizardite (M_{vA})

Element	Mass	Stoich.coef	Kaolinite	Stoech.coef.	Lizardite
Si	28.09	4	112.36	4	112.36
Al	26.98	4	107.92		
Mg	24.30			6	145.8
O	16	18	28.8	18	28.8
H	1	8	8	8	8
Total			516.28		554.16

To make it simpler, let the unit cell be considered for a single layer, independently of polytypes (annex 2) as indicated in Fig. 1.1a. The a and b dimensions correspond to the actual ones and the c dimension is equal to the distance between two consecutive layers (actual value: $c.\sin\beta$); the calculation of its volume is given by: $a \times b \times c$. For kaolinite it is 329.83×10^{-30} m^3; for antigorite: 361.35×10^{-30} m^3. The density of these two minerals (ϱ_{vk} and ϱ_{vA} for kaolinite and antigorite respectively) may be calculated considering a number of unit cells equal to the Avogadro number ($N_A = 6.02252 \times 10^{23}$ mol^{-1}) as shown in Table 1.2

The densities ϱ_k and ϱ_A expressed in grams per cubic meters are calculated as follows:

$$\varrho_k = \frac{\text{unit cell mass}}{\text{unit cell volume} \times N_A} \tag{1.11}$$

and so

$$\varrho_k = 259.91 \times 10^4\, \text{g} \cdot \text{m}^{-3} \tag{1.12}$$

$$\varrho_A = 254.64 \times 10^4\, \text{g} \cdot \text{m}^{-3} \tag{1.13}$$

2:1 Structure (Without Interlayer Sheet): Pyrophyllite and Talc

The structure of 2:1 layers consists of seven ionic planes (Fig. 1.7a). The octahedral sheet in 2:1 layers is formed by two kinds of octahedra: cis-octahedra (2 M2 sites) in which (OH$^-$) groups form one side of a triangular face on the right or on the left, trans-octahedra (1 M1 site) in which (OH)$^-$ groups are located on the opposite vertices (Fig. 1.7b). Planes defined by the (OH)$^-$ groups when they are in the trans-position become planes of symmetry of the octahedral sheet (Fig. 1.7c). This is not true for the cis-position.

Pyrophyllite [$Si_4 O_{10} Al_2 (OH)_2$] is characterised by the presence of a vacancy in the trans-position. The actual unit cell dimensions are: $a = 5.160$; $b = 8.966$ Å (angles α and γ are close to 90°: $\alpha = 91.03°$ and $\gamma = 89.75°$). Bonding between neighbouring layers (from tetrahedral sheet to tetrahedral sheet) depends on van der Waals bonds. These bonds have a stable configuration when two neighbouring layers show a shift of about a:3 spacing along one of the ditrigonal symmetry directions. The thickness of the 2:1 layer and interlayer spacing yields the following value of $c.\sin\beta = 9.20$ Å ($c = 9.33$ Å; $\beta = 99.8°$).

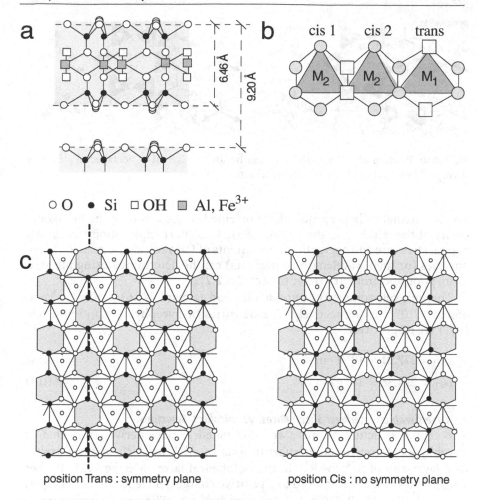

Fig. 1.7a–c. Structures of 2:1 dioctahedral phyllosilicates. **a)** Projection of the unit cell in the b–c plane (pyrophyllite). **b)** The three types of octahedra. **c)** Only the trans-position introduces a plane of symmetry in the octahedral sheet

The talc half-unit formula $[Si_4O_{10}Mg_3(OH)_2]$ points out the absence of octahedral vacancy. Deformations of the octahedral and tetrahedral sheets are limited and unit cell dimensions are close to the theoretical values of the 6-fold symmetry: $a = 5.29$; $b = 9.173$ Å. The c dimension is 9.460 Å.

The structure of the octahedral sheet in pyrophyllite and talc implies different energetic states for OH radicals. In the dioctahedral structure, their negative charge is compensated for by two neighbouring cations, each of them providing one-half of positive charge. As the vacancy breaks the balance of the repulsive forces, the O–H bond is inclined to its direction (Fig. 1.8a). In the tetrahedral structure, each OH is balanced by three bivalent cations, each of them providing one-third of the positive charge. In the latter case, the H^+ pro-

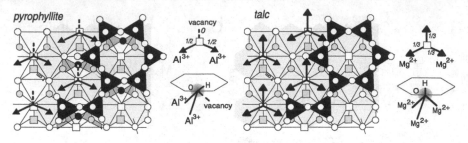

Fig. 1.8a,b. Position of cation – OH – cation bonds. **a)** Case of a dioctahedral structure (pyrophyllite). **b)** Case of a trioctahedral structure (talc)

ton is systematically perpendicularly oriented in the centre of the hexagonal cavity of the tetrahedral sheet (Fig. 1.8b). This short explanation shows that crystal structure and energetic environments of OH radicals are related. These energies (or their wavelength equivalents) can be determined using infrared absorption spectrometry (refer to Sect. 2.1.2.2).

The unit cell volumes of pyrophyllite and talc are $425.63 \times 10^{-30}\,\text{m}^3$ and $459.05 \times 10^{-30}\,\text{m}^3$ respectively. The densities of pyrophyllite (ϱ_{vP}) and talc (ϱ_{vT}) are respectively:

$$\varrho_{vP} = 277.21 \times 10^4\,\text{g}\cdot\text{m}^{-3} \tag{1.14}$$

$$\varrho_{vT} = 274.36 \times 10^4\,\text{g}\cdot\text{m}^{-3} \tag{1.15}$$

2:1 Layers with an Interlayer Sheet: Micas, Vermiculites, Smectites

The crystal structure of micas and dioctahedral clays derives from that of pyrophyllite through cation substitutions of Al^{3+} for Si^{4+} in the tetrahedral layer and of R^{2+} for R^{3+} in the octahedral layer (Méring 1975; Walker 1975). These substitutions lead to a positive charge deficiency in the 2:1 layer: $[(Si_{4-x}Al_x)O_{10}(R_{2-y}^{3+}R_y^{2+})(OH)_2]^{(x+y)-}$ per half-cell. The crystal structure of micas and trioctahedral clays derives from that of talc through tetrahedral and octahedral substitutions. In the latter, bivalent cations may be replaced by trivalent cations and vacancies ($\square$). The general unit formula becomes: $[(Si_{4-x}Al_x)O_{10}(R_{3-y-z}^{2+}R_y^{3+}\square_z)(OH)]^{(x-y+2z)-}$.

The charge deficiency of the 2:1 unit is balanced by the addition of a cation interlayer sheet in the crystal structure. The number of interlayer cations depends on their valency and on the interlayer charge value:

– dioctahedral minerals: brittle micas (margarite), $x + y = -2$, balanced by 1 Ca^{2+}; micas (muscovite, phengite, celadonite), $x + y = -1$, balanced by 1 K^+; vermiculites or dioctahedral smectites whose respective charge $x + y = -(0.7 - 0.6)$ or $x + y = -(0.6 - 0.3)$ is balanced by K^+, Ca^{2+}, Mg^{2+} or Na^+,

– trioctahedral minerals: micas (phlogopite, biotite), $x - y + 2z = -1$, balanced by 1 K^+; vermiculites and trioctahedral smectites whose respective charge

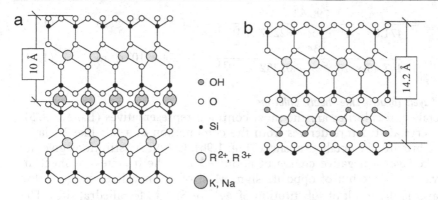

Fig. 1.9a,b. Crystal structures derived from the 2:1 layer. a) Presence of an interlayer sheet (micas, vermiculites, smectites). b) Presence of a brucite-like sheet (chlorites)

$x - y + 2z = -(0.7 - 0.6)$ or $x - y + 2z = -(0.6 - 0.3)$ is balanced by K^+, Ca^{2+}, Mg^{2+} or Na^+.

The interlayer cations are situated in the ditrigonal cavities outlined by two opposite tetrahedral sheets (Fig. 1.9a). If the negative charge of the 2:1 layer is high, they serve as "locks" strongly bonding these layers together. No expansion of the interlayer sheet is possible.

The a and b dimensions as well as angle β of the unit cell (for a single 2:1 layer whatever the polytype is; annex 2) depend on the substitution rate in tetrahedral and octahedral sheets (Table 1.3). The c dimension of vermiculites and smectites varies as a function of the number of sheets of polar molecules such as water, glycol, glycerol, and alkylammonium (refer to Sect. 1.1.1.3).

The unit cell molar volumes of muscovite $[(Si_6Al_2)O_{20}(Al_4)(OH)_4K_2]$ and phlogopite $[(Si_6Al_2)O_{20}(Mg_6)(OH)_4K_2]$ are respectively:

$$V_{mus} = 5.19 \times 9.04 \times 10.04 = 471.05 \times 10^{-30} \, m^3 \tag{1.16}$$

$$V_{phlo} = 5.314 \times 9.204 \times 10.171 = 497.46 \times 10^{-30} \, m^3 \tag{1.17}$$

These values have to be multiplied by 2, 3 or 6 for 2M and 2Or, 3T or 6H polytypes respectively). Their density is given by:

Table 1.3. Unit cell parameters of 2:1 phyllosilicates with an interlayer sheet

Parameter	Dioctahedral				Trioctahedral	
a (Å)	5.19	5.23	5.18	5.17	5.33	5.34
b (Å)	9.00	9.06	8.99	9.08	9.23	9.25
c (Å)	20.00	10.13	9.6; 14.4; 16.8	9.6; 15.4; 17.1	20.1	9.6; 14.9; 16.4
β	95.7°	100.92°	–	–	95.1°	97°

$$\varrho_{mus} = \frac{796.62}{471.05 \times 10^{-30} \times 6.02252 \times 10^{23}} = 280.81 \times 10^4 \, g \cdot m^{-3} \qquad (1.18)$$

$$\varrho_{phlo} = \frac{834.5}{497.46 \times 10^{-30} \times 8.02252 \times 10^{23}} = 278.54 \times 10^4 \, g \cdot m^{-3} \qquad (1.19)$$

2:1:1 Layers (Brucite-Like Sheet): Chlorites

Trioctahedral chlorites are the most common representatives (Bailey 1975). Their crystal structure derives from the combination of a talc-like 2:1 layer with a brucite-like octahedral sheet (Fig. 1.9b). Cation substitutions give the talc-like layer a negative charge of about −1 and the brucite-like sheet an equivalent charge but of opposite sign. Most of the negative charge in the 2:1 layer is the result of substitution of Al^{3+} for Si^{4+} in tetrahedral sites. The octahedral sheet usually has a low charge because the positive charge excess due to the replacement of bivalent cations ($R^{2+}=Mg^{2+}, Fe^{2+}, Mn^{2+}$) by trivalent cations ($R^{3+}=Al^{3+}, Fe^{3+}$) is balanced by the positive charge deficiency related to the presence of vacancies (unoccupied sites): $x - y + 2z \cong 0$. The Coulomb attraction between the 2:1 unit and the brucite-like sheet is strong. Therefore, the interlayer spacing remains at 14.2 Å; no expansion by adsorption of polar molecules is possible.

The composition of the brucite-like sheet is poorly known because it escapes the usual investigation means. Nevertheless, it most probably has no vacancy and has a positive charge excess due to the replacement of bivalent cations ($R^{2+} = Mg^{2+}+Fe^{2+}+Mn^{2+}$) by trivalent cations ($R^{3+} = Al^{3+}+Fe^{3+}$). According to Deer et al. (1962), the unit cell parameters of single chlorite layers (whatever the polytypes) are: $a = 5.3$; $b = 9.2$ and $c = 14.3$ Å, $\beta = 97°$. The unit cell volume is then 697.27×10^{-30} m^3. The density of this "theoretical" chlorite is $\varrho_{chl} = 264.68 \times 10^4 \, g \cdot m^{-3}$.

The two other varieties of chlorite are:

– donbassite, dioctahedral variety whose structure is derived from a pyro-phyllite-like layer with the addition of a gibbsite-like octahedral sheet;

– sudoite, di-trioctahedral variety, whose structure is derived from a pyro-phyllite-like layer with the addition of a brucite-like octahedral sheet.

Structure with Channelways: Palygorskite and Sepiolite

The crystal structure of these two minerals differs from that of 1:1, 2:1 or 2:1:1 phyllosilicates. The $[SiO_4]^{4-}$ tetrahedra periodically point outward and inward in groups of four (palygorskite) or six (sepiolite), thus making the octahedral sheet discontinuous. Therefore, tetrahedra form chains that are similar to those of amphiboles (Fig. 1.10). These chains extend along the x-direction (i.e. parallel to the a unit cell dimension), which gives the crystals a fibrous appearance. The octahedral sheet is close to the dioctahedral type. Its constituent cations Mg, Al, Fe^{2+} and Fe^{3+} are so ordered that the vacant site is at the centre of the chain. The overall negative charge results from substitutions of Al for Si in tetrahedra and of R^{3+} for R^{2+} in octahedra and is generally weak.

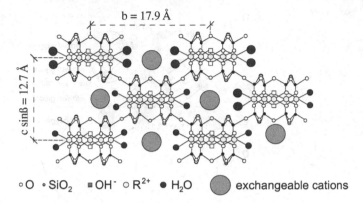

°O •SiO₂   ▪OH⁻ ○R²⁺ ●H₂O   (⬤) exchangeable cations

Fig. 1.10. Crystal structure of palygorskite. Channelways of rectangular section contain molecules of water and exchangeable cations that compensate for the positive charge deficiency of the tetrahedral and octahedral sheets

It is balanced by exchangeable cations that are located in channelways with water molecules.

1.1.1.3
A Strange Clay Mineral Property: the "Swelling" of the Interlayer Sheet

The "swelling" property is determined by the ability of cations to retain their polar molecule "shell" (water, glycol, glycerol) within the interlayer environment (Douglas et al. 1980). This property does not exist if the charge of the layer is too high (micas, chlorites) or zero (pyrophyllite, talc). More simply, this property is characteristic of di- and trioctahedral smectites and vermiculites (Table 1.6). Polar molecules are organised into layers whose number varies inversely with the interlayer charge:

1. charge contained between 0.8 and 0.6: 1 layer of polar molecules (di- or trioctahedral vermiculites and high-charge beidellites). High-charge saponites absorb 1 to 2 layers of polar molecules.

2. charge contained between 0.6 and 0.3: 2 to 3 layers of polar molecules (beidellites, montmorillonites and saponites, stevensites);

In the interlayer zone, cations are framed by ethylene glycol molecules, which are weakly bonded to the surface of tetrahedral sheets (hydrogen bonds). Like the water molecules, they are organised into more or less continuous layers.

Adsorption of polar molecules (water or organic molecules) alters the c dimension either progressively and regularly or in stages (Fig. 1.11). The total expansion is equal to the sum of the individual expansion of each layer able to absorb a varying number of water (0 to 3) or ethylene glycol layers (0 to 2). For clay with layers having a given charge, the number of absorbed water layers depends on two factors: the nature of the interlayer cation (Table 1.4) and the

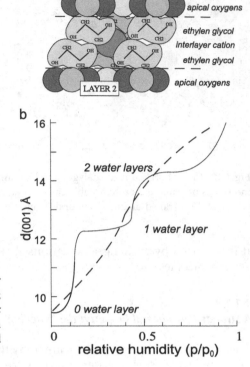

Fig. 1.11a,b. Adsorption of polar molecules. **a)** Two ethylene glycol layers; **b)** Relations between the crystal *c* dimension and the water saturation state of the interlayer zone. p/p_0: partial pressure of water

Table 1.4. Relationship between the number of water or ethylene glycol layers and basal spacing (c dimension) in 2:1 clay minerals (revised from Suquet et al. 1975). M: montmorillonite; B: beidellite; S: saponite; V: vermiculite

Hydration state	c dimension (Å)	Exchangeable cations				
		Li$^+$	Na$^+$	K$^+$	Ca^{2+}	Mg^{2+}
Infinite	–		MBSV	MB	M	
3 water layers	16.5–17				MB (S)	MBS
2 water layers	14–15	V	BSV	M	(B) SV	(S) V
1 water layer	12.3		BSV			
0 water layer	10			V		

Etylene glycol saturation state	c dimension (Å)	Exchangeable cations				
		Li$^+$	Na$^+$	K$^+$	Ca^{2+}	Mg^{2+}
2 glycol layers	16.9–17.1	M	MB	M	MB	MB
	16.4–16.6	SB	S	S	S	S
	16.1	V	V		V	
1 glycol layer	14.3–15.2			B		V
0 glycol layer	10			V		

partial pressure (p/p_0) of water or ethylene glycol. A stage variation of the total expansion implies that all the layers behave homogeneously and fix 1 or 2 water or ethylene glycol layers for a given range of p/p_0. On the contrary, a progressive variation implies that clay behaves as a mixed-layer mineral composed of layers having 0, 1 or 2 polar molecule interlayer sheets.

Now that the main types of phyllosilicate layers have been determined, the investigation scale can move from one angström to one nanometer, and the way they are stacked in the phyllosilicate crystals can be addressed. At this scale, we will discover the huge diversity of clay species. This will lead us to discriminate between crystals, particles and aggregates. The X-ray diffraction identification of clay minerals remains the guiding principle.

1.1.2
Crystal – Particle – Aggregate

A crystal, even as tiny as commonly found in clay species, is a three-dimensional object. It is composed of several layers that can be stacked in different ways. Theoretically, the number of layers should determine the crystal thickness. However, as simple as it may sound, the thickness of clay mineral crystals is difficult to measure routinely because of the presence of crystal defects. One must keep in mind that XRD only determines the size of the scattering coherent domains. Initially, the stacking sequences will be considered perfect and the effects of the presence of crystal defects will be ignored. Accordingly, emphasis will be put on the two typical structures that are commonly found in clay species depending on layer composition.

1.1.2.1
Stacking Sequence of Layers of Identical Composition: Polytypes

Phyllosilicates form crystals that are limited externally by crystalline faces whose shape and dimensions depend on growth processes (see Sect. 1.2.3). The thickness depends on the number of stacked layers. Simple-species crystals are composed of layers of identical chemical composition. For a given composition, different stacking modes are possible; each of them corresponds to a polytype. A polytype is considered here as a one-dimensional polymorph: the density stays constant, because the unit cell dimensions of each layer remain unchanged. On the contrary, the "structural formula" depicts the composition of the "over-unit cell" and changes with the polytype (annex 2).

For a given layer, octahedra show two different positions referred to as I and II in Fig. 1.12a. Position II can be superposed on position I after a rotation of 60°. Tetrahedral sheets, represented by hexagons formed by apical oxygens, exhibit a $\frac{a}{3}$ spacing shift whose direction depends on the position of the octahedral sheet (I or II). The angle between both directions of this shift is 60°. The nomenclature of polytypes is standardised: the number on the left indicates the number of layers per over-unit cell in the stacking sequence, the letter indicates the crystal system, the indexed number on the right gives the number

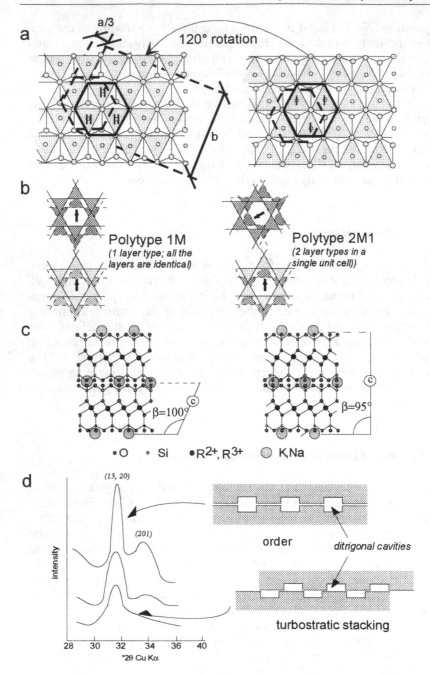

Fig. 1.12a–d. 1M and 2M1 mica polytypes. **a)** The two groups of octahedral sites. **b)** Formation of the 1M polytype by stacking without rotation of the same type of layer and of the 2M₁ polytype by alternation of type I and II layers and rotation of 120°. **c)** Projection in the a–c plane of the crystal structure of both polytypes showing the significant change in the value of angle β. **d)** Order and turbostratism in smectites from Mamy and Gaultier (1976)

of the stacking solution for the same symmetry and the same number of layers contained in the over-unit cell.

Considering the hexagonal symmetry of octahedral sheets, configurations I and II both present three rotational possibilities. Therefore, two consecutive layers can be superposed with rotations of 60°, theoretically yielding six possibilities: 0, 60, 120, 180, 240, and 300°. Owing to the ditrigonal symmetry, rotations of 60 and 300° on the one hand and of 120 and 240° on the other hand yield equivalent positions. Smith and Yoder (1956) have shown that there are theoretically 6 polytypes whose number of layers goes from 1 to 6. Of these, the most common are the 1 M and 2 M_1 polytypes, which are composed of configuration I layers stacked without rotation (0°) and with a rotation of ±120° respectively (Fig. 1.12b, annex 2). Angle β varies according to the stacking mode of layers (Fig. 1.12c): the 1M polytype is strongly monoclinic ($\beta = 100°$), the 2M_1 polytype is almost hexagonal ($\beta = 95°$). Smectites and more particularly montmorillonites form crystallites characterised by the turbostratic stacking sequence of a small number of layers (non-rational rotation between layers within the stacking sequence).

Attention should be paid to smectites and particularly to the montmorillonite group whose layers are stacked with random rotations. This stacking type, called "turbostratic", is detected by X-ray diffraction by a wide band (13, 20) asymmetric towards wide angles, and by arcs instead of points on electron diffraction patterns. The turbostratism degree decreases when layers are re-directed by drying-wetting cycles (Mamy and Gautier 1976). Indeed, this energy supply allows a reorganisation of the layer stacking leading the ditrigonal cavities of two consecutive layers to fit facing each other (Fig. 1.12d).

The higher the order degree of the polytype, the greater is the number of (h/d) planes. Nevertheless, these peaks, corresponding to weak atomic density planes, are of low intensity. They are obtained by randomly oriented powder diffraction. Determination of polytypes of the different species of phyllosilicates is easy (Brindley and Brown 1980; Moore and Reynolds 1989): e.g. micas,

Table 1.5. (*hkl*) peaks, characteristic of the main mica polytypes, from Moore and Reynolds (1989). For the 1M_d polytype, peaks of the 1M polytype grow weaker or disappear with the exception of those asterisked ($k \neq 3n$)

	1M			2M_1	
hkl peaks	d (Å)	Intensity	(*hkl*)	d (Å)	Intensity
11$\bar{1}$	4.35	15	111	4.29	10
021	4.12	10	022	4.09	10
11$\bar{2}$	3.66	50	11$\bar{3}$	3.88	30
112	3.07	50	023	3.72	30
11$\bar{3}$	2.92	10	11$\bar{4}$	3.49	30
023	2.69	20	114	3.20	30
131*	2.450	11	025	2.98	35
13$\bar{2}$*	2.405	4	115	2.86	30
13$\bar{3}$*	2.156	20	11$\bar{6}$	2.79	25

kaolin-group minerals, chlorites, serpentines, etc. As an example, the peaks of the 1M and $2M_1$ mica polytypes are given in Table 1.5. When layers are stacked both by random translations and rotations ($n \times 60°$), the 1M polytype is considered to become disordered ($1M_d$ polytype). The $k \neq 3n$ peaks grow wider and weaker. Some of them completely disappear.

The characteristics of the polytypes described by X-ray diffraction provide an average picture of the studied crystals. The relations of this picture with the arrangement of layers at the level of the individual crystal are revealed by high-resolution transmission electron microscopy (HRTEM) examination. In this regard, Alain Baronnet's work represents a major source, which any phyllosilicate (hence clay) mineralogist should definitely refer to. A summary of these data can be found in Baronnet (1997).

1.1.2.2
Stacking Sequence of Layers of Different Compositions: Mixed-Layer Minerals

Interstratification is very common in clay minerals, whether naturally occurring or obtained by experimental synthesis. The mixed-layer minerals (MLM) are easily identified using XRD. They exhibit specific rational or non-rational series of diffraction bands, depending on their crystal structure being regular or not, respectively. In both cases, XRD patterns are significantly different from those of pure species.

Conditions of Interstratification

The most commonly described two-component mixed-layer minerals are illite and dioctahedral smectite, kaolinite and dioctahedral smectite, chlorite and saponite. The condition that apparently best explains their frequency is the slight difference between the a and b dimensions of the two types of layers. Mixed-layer minerals formed by the stacking of trioctahedral and dioctahedral layers are unquestionably rare. Recent studies show that, even though rarely described in the literature, naturally occurring three-component mixed-layer minerals may be more abundant than commonly thought (Drits et al. 1997).

A mixed-layer mineral is identified when its components, their proportions, and the degree of order of their stacking sequence have been determined. Let's consider a mixed-layer mineral composed of two components A-B occurring in varying relative proportions W_A and W_B. It will be fully described if succession probabilities of A and B layers ("nearest-neighbour") are known: P_{AA}, P_{AB}, P_{BA} and P_{BB}. Generally, these six parameters are linked by four independent relationships:

$$W_A + W_B = 1 \tag{1.20}$$

$$P_{AA} + P_{AB} = 1 \tag{1.21}$$

$$P_{BA} + P_{BB} = 1 \tag{1.22}$$

$$W_A P_{AB} = W_B P_{BA} \tag{1.23}$$

So there are six variables and four non-redundant equations which permit their calculation if two are fixed. Usually, the composition as well as one of the junction probabilities is fixed (W_A = 0.4 and P_{BB} = 0.8, for instance). The development of probability calculations can be found in classical books (Brindley and Brown 1980 for instance).

Random Stacking Sequence (R0)

In a random stacking sequence, an A layer may be followed by an A or a B layer without any forbidden sequence (Fig. 1.13a). The succession probability of A and B layers depends only on the relative proportions W_A and $W_B = 1 - W_A$. Therefore, the probability for A to follow B is given by $P_{AB} = W_B$; we know that $P_{AB} + P_{AA} = 1$, so $P_{AA} = W_A$. The variation of P_{AA} as a function of W_A of the random stacking sequence is represented by a straight line whose slope is equal to 1 (Fig. 1.13b).

Maximum Degree of Order (R1)

If $W_B \leq 0.5$, the maximum degree of order is reached if the probability of finding a B-B pair is zero (Fig. 1.13a). Let's consider a 10-layer stacking sequence containing 40% of B; a sequence such as B-A-B-A-A-B-A-B-A-A may exist, but a sequence such as B-A-A-A-B-B-A-B-A-A is forbidden. As W_B decreases, the probability of formation of B-A-B-type sequences becomes zero. A long-distance order may be established with at least two then three consecutive

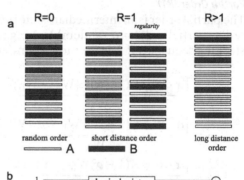

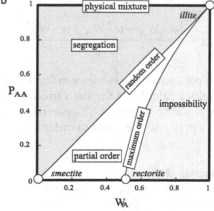

Fig. 1.13a,b. Interstratification.
a) Schematic diagram of randomly ordered, ordered and regularly ordered interstratifications of two types of A and B layers such as illite (10 Å) and smectite (17 Å). b) Relationship between succession probability of two illite layers (P_{AA}) and illite content (W_A) establishing mixing, segregation and interstratification domains (domain in *white*)

A after each B. Although the actual influence of the B layer on the subsequent layers is not established in terms of chemical or physical interactions, A-rich crystal structures can still be described using these long-distance ordering modes: order for a triplet R=2; order for a quadruplet R=3 etc.

In R1 stacking sequences, if A is the most abundant type of layer ($W_A \geq 0.5$), the maximum order implies that no BB layer pair can exist ($P_{BB} = 0$). Therefore, the following relations may be inferred:

$$P_{BB} = 0 \tag{1.24}$$

$$P_{BA} = 1 \tag{1.25}$$

$$P_{AB} = \frac{W_B}{W_A} \tag{1.26}$$

$$P_{AA} = 1 - P_{AB} = \frac{(W_A - W_B)}{W_A} \tag{1.27}$$

The variation of P_{AA} as a function of W_A for the maximum order is given by the curve originating in the composition of rectorite in the case of illite/smectite mixed layers (Fig. 1.13b).

Partial Order (R1)

The partial order is an intermediate state between random order and maximum order: partial order = α random stacking sequence + $(1 - \alpha)$ maximum order stacking sequence. If A is the most abundant type of layer, then:

$$P_{AA} = \frac{\left[\alpha \cdot W_A^2 + (1 - \alpha) \cdot (W_A - W_B)\right]}{W_A} \tag{1.28}$$

$$P_{AB} = \frac{\left[\alpha \cdot W_A + (1 - \alpha) \cdot W_B - \alpha \cdot W_A^2\right]}{W_A} \tag{1.29}$$

$$P_{BA} = \frac{\left[\alpha \cdot W_A + (1 - \alpha)W_B - \alpha \cdot W_A^2\right]}{W_B} \tag{1.30}$$

$$P_{BB} = \frac{\left[\alpha \cdot W_A^2 - \alpha \cdot W_A + \alpha \cdot W_B\right]}{W_B} \tag{1.31}$$

The α parameter varies between 0 and 1, which requires that P_{AA} vary between the values calculated for the random stacking sequence and the maximum order stacking sequence. The partial order domain is limited by the maximum order curve and the random order straight line (Fig. 1.13b).

Segregation (R1)

This type of stacking sequence is intermediate between random state and physical mixing (Fig. 1.13b). Consequently P_{AA} must be fixed as a function of

W_A in order to be located within the domain of segregation:

$$P_{AA} = \alpha + (1 - \alpha) \cdot W_A \tag{1.32}$$
$$P_{AB} = (1 - \alpha) \cdot W_B \tag{1.33}$$
$$P_{BA} = (1 - \alpha) \cdot W_A \tag{1.34}$$
$$P_{BB} = \alpha + (1 - \alpha) \cdot W_B \tag{1.35}$$

Probabilities do not describe the heterogeneity of natural clay minerals in which at least two parameters may vary: the number of layers in a stacking sequence, and the proportion of each type of layer within this stacking sequence. Variations of the second parameter are described by Markovian probabilities applied to quasi-homogeneous structures (same number of layers in the stacking sequence for all crystals). In this manner, proportions of the various possible types of stacking sequence can be calculated in a population of crystals exhibiting the same number of layers and the same degree of order: randomly ordered (Fig. 1.14a), and ordered (Fig. 1.14b). This probabilistic theory is detailed in a book by Drits and Tchoubar (1990). It is used as a basis by modelling software to calculate diffraction patterns, for instance NEWMOD (Reynolds 1985) or MLM2C (Plançon and Drits 2000).

Méring (1949) has proposed an elegant method for readily identifying randomly ordered or ordered mixed-layer minerals by X-ray diffraction. Indeed, he has shown that when the components of the mixed-layer mineral show neighbouring peaks, the latter interfere forming a wider diffraction band with an intermediate angular position. This position varies with the respective amounts of the two components. Therefore, except in the case of perfect regularity, mixed-layer minerals can be identified by a non-rational series of peaks. Only regularly ordered mixed-layer minerals show rational series, as do pure minerals. Let's consider illite – smectite mixed layers. The rational series characterise the following components:

- ethylene glycol-saturated smectite: 17 Å, 8.5 Å, 5.67 Å, 3.4 Å . . .
- illite: 10 Å, 5 Å, 3.33 Å . . . .
- rectorite (50% illite, 50% smectite): 27 Å, 13.5 Å, 9.00 Å, 6.75 Å, 5.40 Å, 4.50 Å . . .

Randomly ordered mixed-layer minerals are characterised by a non-rational series of peaks at 17, 10 to 8.50, 5.67 to 5.00, and 3.40 to 3.33 Å. The greater the illite content, the more the 2nd-, 3rd- and 4th order diffraction bands shift towards the typical positions of illite at 10, 5 and 3.33 Å respectively (Fig. 1.14c). Irregularly ordered mixed-layer minerals with more than 50% illite may be considered as randomly ordered rectorite and illite mixed layers (Drits et al. 1994). Application of Méring's law permits determination of the non-rational series of their characteristic peaks (Fig. 1.14d). Based on this principle, several methods for determining the illite content of illite-smectite mixed layers have been proposed (Srodon: 1980, 1981, 1984; Watanabe 1988). The same reasoning applies to other types of mixed layers (chlorite-smectite,

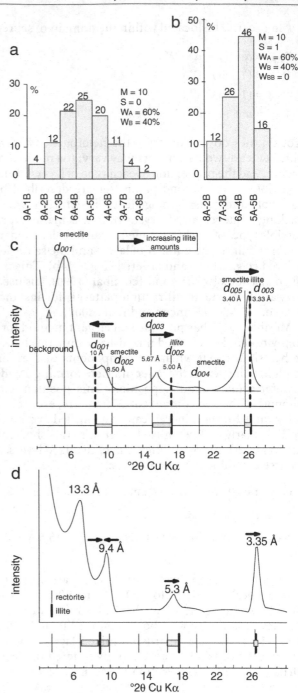

Fig. 1.14a–d. Probabilities of occurrence of crystals exhibiting differing compositions in a structure with two types of layers (a and b) whose composition is fixed: 60% of a and 40% of b from Drits and Tchoubar (1990). The calculation of percentages of the various possible stacking sequences in 10-layer crystals is given by the Markovian probability: **a)** Randomly ordered mixed-layer minerals (R0 or S0). **b)** Ordered mixed-layer minerals (R1 or S1). **c)** Application of Méring's law (1949) for a diffraction pattern of randomly ordered illite-smectite mixed layer. **d)** Diffraction pattern of an ordered mixed-layer mineral

kaolinite-smectite …) by adopting the rational series characteristic of each component.

Three-Component Mixed-Layer Minerals and Non-Markovian Probabilities

Recent works dedicated to illitisation in diagenetic series (Drits et al. 1997) show that mixed-layer clays cannot be properly described in a 2-component structure such as illite + smectite (2 ethylene glycol layers). The position and width of X-ray diffraction peaks are reproduced well only if a third component is introduced: vermiculite (1 ethylene glycol layer). Layer stacking sequences are governed by the same probabilistic laws (Plançon and Drits 2000). Although far beyond the scope of this book, these works can by no means be ignored, even if identification of 3-component mixed-layer minerals is for the time being relatively complex. An expert system and MLM3C calculation software for three-component diffraction patterns have been developed by these authors (Plançon and Drits 2000).

The calculation of the theoretical diffraction pattern that best fits the experimental diffraction pattern modifies its interpretation to a significant extent. Accordingly, Claret (2001) has shown that the diagenetic series of the eastern part of the Paris Basin and of the Gulf Coast (USA), considered as a classical transition from randomly ordered (I/S R=0) to ordered (I/S R=1) illite-smectite mixed layers, may be viewed as a mixture of smectite, I/S R=0 with a high illite content (65–70%) and illite. Furthermore, he has shown that the superstructures visible in samples collected from the basis of the sedimentary series of the Gulf Coast cannot be reproduced by calculations using Markovian probabilities. These works could change our way of thinking of clay diagenesis.

1.1.2.3
Crystals

Despite their very small size, the crystals of clay mineral species exhibit a particular morphology. The crystal shape and size depend on the physico-chemical conditions prevailing during their growth. The statistical analysis of some shape or size parameters is a useful tool for determining these conditions. However, before any conclusion is drawn, one must ensure that the individual particles observed with a Transmission Electron Microscope (TEM) are not polycrystalline.

Crystal Shape

Clays exhibit very different shapes according to the mineral species. Furthermore, for a given species, shapes may change depending on conditions of crystallisation: temperature, chemical composition and pH of solutions, crystallisation duration. Most of the time, size and shape are not independent. Consequently a reliable identification of the species by the shape of their crystals is very difficult even if some general features can be drawn.

Generally, smectite crystals are small (crystallites below 1 µm) and very thin (a few nm, i.e. a few layers). Morphology varies with the cation that

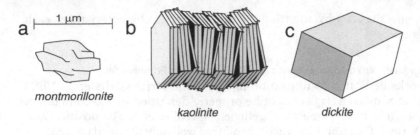

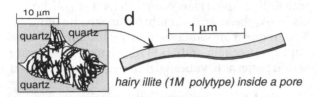

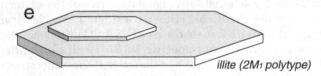

Fig. 1.15a–e. Examples of morphology of crystals of a few clay species in diagenetic series:
a) Veil: montmorillonite within a bentonite. **b)** Booklets: kaolinite within a diagenetic
sandstone. **c)** Rhomb: dickite within a sandstone. **d)** "Hairy" illite within a sandstone pore.
e) Hexagonal plates of $2\,M_1$ illite (intense diagenesis)

saturates the interlayer sheet. Montmorillonite and trioctahedral smectites
(saponites, stevensites) look like flakes whose rims are wound around themselves (Fig. 1.15a). Beidellite often exhibits a morphology of hexagonal laths.
These crystal habitus are to be related to the degree of turbostratism in the
layer stacking.

Kaolinite crystals are bigger than smectite crystals and very often exhibit
the shape of hexagonal prisms. In soils, particularly in tropical climates, these
prisms are flattened and exhibit defects due to substitution of Al^{3+} for Fe^{3+},
or even corrosion pits (dissolution). They are bigger and more regular in
diagenetic formations. They grow thicker until they take the shape of "books"
several tens of microns thick (Fig. 1.15b). They progressively transform into
dickite whose rhombohedral morphology is very characteristic (Fig. 1.15c).

Illite or illite/smectite mixed layers in diagenetic series exhibit hairy crystals
covering pore walls in sandstones (Fig. 1.15d). Each "hair" corresponds to
a very elongated and thin crystal that ends in a rectangle or triangle. This
particular morphology disappears to the benefit of thicker hexagonal shapes
in zones where diagenesis is more intense (Fig. 1.15e).

Experimental works on kaolinite (Fialips et al. 2000) and illite (Bauer et al. 2000) show that crystal morphology depends on supersaturation and pH conditions. Metastable shapes appear first. They are generally elongated, showing that growth is faster along some crystallographic directions. This research subject is very promising because it allows the message contained in the morphology of clay crystals to be deciphered.

Crystal Size

For a given species of clay observed in a rock, the crystals do not all exhibit the same size. The statistical distribution of their dimensions is a major indicator of their crystallisation process. The shape parameters (length/width ratio) and the size (area of the (001) faces) are measured on images obtained by transmission electron microscopy (TEM). Sample preparation comes up against considerable difficulties relating to interactions between crystals: they often coalesce during grid drying. Complex particles instead of isolated crystals are often observed if care is not taken. The true size dimension would be volume rather than area. However, measuring thickness for individual clay particles is difficult and cannot be applied routinely to a great number of crystals (Blum 1994). Fortunately, most often, the area of the (001) faces varies with volume reducing the experimental error.

Variations in size bring about changes in shape. Thus, under diagenetic conditions, the morphological modifications of crystals of illite/smectite mixed layers are different according to whether they were formed in sandstones or in shales (Lanson and Champion 1991; Varajao and Meunier 1995). The overall evolution (Fig. 1.16) tends towards isometric shapes as the size increases but both series are perfectly distinct (Lanson and Meunier 1995).

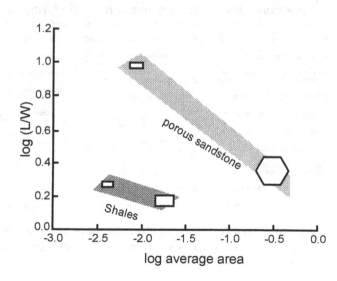

Fig. 1.16. Variation in the morphology of crystals of illite/smectite mixed layers as a function of their size in porous sandstones and shales. L/W: length/width ratio for each crystal or particle. For a perfect hexagon, L/W=1; mean surface area: length × width

Crystal Defects

Among the defects that may disturb the tridimensional regularity of the stacking sequences along the *c* direction, two are seemingly very frequent: lateral compositional changes of layers and intercalation of exotic layers. The intercalation of chlorite layers (14 Å) in illite layer stacking sequences can be observed. Depending on the abundance and distribution of these "exotic" layers, three stages can be distinguished: (1) a simple defect: a chlorite layer grows within an illite crystal by migration of the stacking defect (Fig. 1.17a); (2) an intercalation forming domains of illite/chlorite interlayering (Fig. 1.17b); (3) the segregation of illitic and chloritic domains forming a polycrystalline particle (Fig. 1.17c and d). Such intergrowths are frequently observed under diagenetic or very low-grade metamorphic conditions (Giorgetti et al. 1997). The presence of these exotic layers, whose X-ray identification is difficult when there are few of them, brings changes in the particle or crystal chemical composition. That is the reason why electron microprobe analyses (a few μm^3) should be carefully interpreted. The use of suitable chemical diagrams permits identification of the phase mixtures by the position of projections of the chemical compositions between pure end members.

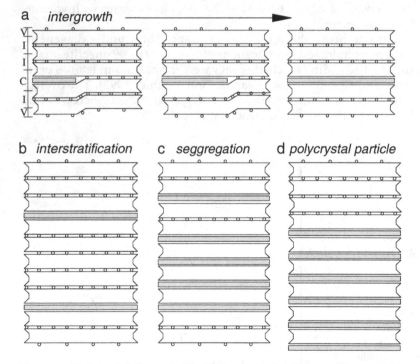

Fig. 1.17a–d. Crystal defects. **a)** Chlorite intergrowth within an illite crystal by migration of the stacking defect. **b)** Illite-chlorite interlayering. **c)** Segregation of chloritic domains. **d)** Illite-chlorite polycrystalline particle

Technically the high-resolution electron microscopy observation of defects in the layer stacking sequence of an I/S crystal is difficult because smectite dehydrates in the ultra-high vacuum. The thickness of its layers then falls from 15 then 12.5 down to 10 Å. They become almost impossible to distinguish from illite layers. Nevertheless, some techniques enable smectite layers to retain a thickness greater than 10 Å (Murakami 1993).

X-ray diffraction may also be used to detect the presence of crystal defects when applied to randomly oriented powders. Indeed, crystal defects reduce the size of the coherent scattering domains in all directions. In particular, this changes the profile (intensity and width) of some *hkl* peaks and may even make them disappear.

1.1.2.4
Particles and Aggregates

Observation of isolated crystals of clay species is sometimes difficult, and artefacts cannot always be avoided. The particles deposited on grids for TEM observation consist of several crystals sedimented on top of each other. These are artefacts that must be separated from the real particles formed in rocks by epitaxial crystallisation on the (001) faces (Fig. 1.18a) either by crystal aggre-

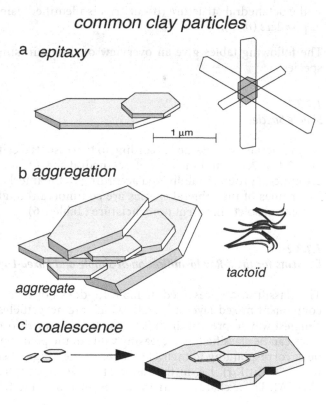

common clay particles

a *epitaxy*

1 µm

b *aggregation*

aggregate

tactoïd

c *coalescence*

Fig. 1.18a–c. The main types of particles and aggregates of clay minerals. a) Epitaxy, i.e. growth on a crystalline support. b) Aggregation of rigid or flexible particles or crystallites (tactoid or quasi-crystal networks). c) Coalescence: neighbouring crystals are joined by the growth of common layers

gation (Fig. 1.18b) or by coalescence (Fig. 1.18c). Kitagawa (1993) describes the coalescence process of sericite crystals in the course of their growth. Crystal defects can interrupt the periodicity along the three directions of space (Meunier et al. 2000). Particles, like twinned crystals, often exhibit reentrant angles allowing the boundaries of coalescing crystals to be recognised. These natural particles, contrary to artefacts, cannot be dispersed by chemical or ultrasonic treatment.

Aggregates are units of higher organisation in which crystals and/or particles are bonded together by Coulomb forces or by hydroxide or organic matter deposits. Smectite (particularly montmorillonite) crystallites are very thin and are bound together into a tactoid (Na-saturated state) or quasi-crystal (Ca-saturated state) network. This type of aggregate has a great swelling capacity when wet (see Figs. 1.5–1.11).

1.1.3
Identification Keys for Simple and Mixed-Layer Species of Clay Minerals

Knowing the basic crystal structures of clay minerals, some identification keys can be given on the basis of their XRD properties:

- the layer spacings (d_{001}) are measured on oriented samples;

- the octahedral structure (di- or tri-) is identified using randomly oriented powders (d_{060}).

The following tables give an overview of the main simple or interstratified species.

1.1.3.1
Simple Species

Clay minerals are classified according to two essential criteria: the type of layer (1:1, 2:1 or 2:1:1) and the type of octahedral sheet (di- or trioctahedral). The 2:1 mineral group is subdivided according to the interlayer charge (IC) value. The names of the mineral species are common although some of them have become obsolete in recent nomenclature (Table 1.6).

1.1.3.2
Elements for the X-Ray Identification of Simple and Mixed-Layer Species

The classification presented in Table 1.6 does not take into account the two-component mixed-layer minerals, which are nevertheless very common. The simplest way to present them is to use their X-ray diffraction characteristics. They can be classified in decreasing value of the position of their most intense peak corresponding to their 00l planes in the range of small angles (from 2.5 to 13 °2θ Cu Kα), i.e. in the range of the greatest interlayer spacing values (35–7 Å). These characteristics are presented in Table 1.7.

Table 1.6. The principle simple (non-interstratified) clay species

Crystalline features	Dioctahedral minerals	Trioctahedral minerals
	1:1 Minerals	
	Kaolinite	**Serpentine**
Electric charge of layer #0 $1T+1O+Int.Sp.=7\,\text{Å}$	Kaolinite, dickite, nacrite halloysite (7 or 10 Å)	Amesite, berthierine, chrysotile, antigorite, lizardite, cronstedtite, greenalite, ...
	2:1 Minerals	
Electric charge of layer=0 $1T+1O+1T+Int.Sp. = 9\,\text{Å}$	**Pyrophyllite**	**Talc**
Electric charge of layer: -0.2 to -0.6 $1T+1O+1T+Int.Sp.= 10 \rightarrow 18\,\text{Å}$ Int.Sp: cations±hydrated (Ca, Na) (Ch:10 Å; $2H_2O$:14 Å; EG: 17 Å)	**Smectites** Al: montmorillonite, beidellite Fe: nontronite	**Smectites** Mg: saponite, stevensite, hectorite
Electric charge of layer: -0.6 to -0.9 $1T+1O+1T+Int.Sp.=10 \rightarrow 15\,\text{Å}$ Int.Sp: cations±hydrated (Ca, Na) (Ch:10 Å; 2H2O:14 Å; EG: 14 Å)	**Vermiculites**	**Vermiculites**
Electric charge of layer=-0.9 to -0.75 $1T+1O+1T+Int.Sp.=10\,\text{Å}$ Int.Sp: not hydrated cations (K)	**Illite, glauconite**	
Electric charge of layer=-1 $1T+1O+1T+ Int.Sp.=10\,\text{Å}$ Int.Sp: not hydrated cations (K, Na)	**Micas** Al: muscovite, phengite, paragonite, Fe: celadonite	**Micas** Mg-Fe: phlogopite, biotite, lepidolite
Electric charge of layer=-2 $1T+1O+1T+ Int.Sp.=10\,\text{Å}$ Int.Sp: not hydrated cations (Ca)	**Brittle micas** Al: margarite, clintonite	
	2:1:1 Minerals	
$1T+1O+1T+1O(Int.Sp)=14\,\text{Å}$ Variable layer electric charge Int.Sp: octahedral layers (brucite- or gibbsite- like)	**Dioctahedral chlorites** Donbassite	**Trioctahedral chlorites** Diabantite, penninite, chamosite, brunsvigite, clinochlore, thuringite, ripidolite, sheridanite
	Di-trioctahedral chlorites Cookeite, Sudoite	
	2:1 Minerals **(fibrous structure)**	
		Palygorskite Sepiolite

Table 1.7. Position of peaks of mixed-layer clay minerals classified in decreasing interlayer spacing values of 00l planes

Position (Å)	Mineral	Position (Å)	Mineral
30–35	C/S R1 (E.G)	9.9–10.7	I/S R>=1>90% illite (Nat)
26–30.5	C/S R1 (Nat)	9.9–10.3	I/S R>=1>90% illite (E.G)
24–30	I/S or T/S (R1, E.G)	9.63–9.70	Paragonite
22–28	M/V or M/C (R1)	9.55–9.60	Margarite
20–27	I/S or T/S (R1, Nat)	~ 9.34	Talc
		~ 9.20	Pyrophyllite
18.5–16	Smectite-rich R0 mixed-layer minerals (E.G)	8.9–9.9	I/S R1 (E.G)
		8.60–8.90	T/S R1 (E.G)
17.5–16.5	Smectite (E.G)	8.50–9.25	I/S or T/S (R0, E.G)
14.5–16	C/S R1 (E.G)	8.40–8.60	Smectite (E.G)
13.7–15	C/S R1 (Nat)	8.10–9.00	T/S R1 (E.G)
14–15	Smectite or smectite-rich R0 mixed-layer mineral (Nat)	7.50–8.20	C/S R1 (E.G)
		7.20–8.50	K/S R0 (E.G)
		7.10–8.50	C/S R0 (E.G)
14–14.5	Vermiculite	7.5–8.2	C/S R1 (E.G)
14–14.35	Chlorite		
12.9–13.0	Smectite with 1 water layer	7.47–7.70	Sepiolite
12.8–13.8	T/S R1 (E.G)	7.20–7.50	Halloysite at 7 Å
12.0–12.7	T/S R0 (Nat)	7.00–9.00	M/C (>30% chlorite)
12–12.45	Sepiolite	7.20–7.50	Vermiculite (2nd order)
12–12.45	Smectite with 1 water layer (Nat)	7.20–7.36	Serpentine
10–14.5	M/V or M/C (R1 or R0)	7.20–7.22	Greenalite
10.2–14.35	I/S R1 (E.G)	7.20–8.50	K/S R0 (E.G)
10.2–14.35	I/S R1 (Nat)	7.10–8.50	C/S R0 (E.G)
		7.13–7.20	Kaolinite
10.3–10.5	Palygorskite (= attapulgite)	7.00–7.14	Chlorite
		7.00–7.13	Mineral at 7 Å
10.2–14.35	I/S R1 (E.G)		
10.2–14.35	I/ S R1 (Nat)		
10.1	Glauconite, halloysite at 10 Å		
10–10.1	Illite		
9.9–10.1	Dioctahedral (muscovite, phengite, celadonite) or trioctahedral mica		

Nat, "natural" sample;
E.G., ethylene-glycol saturated sample;
M/C, mica/chlorite mixed layer;
I/S, illite/smectite mixed layer;
C/S, chlorite/smectite mixed layer;
K/S, kaolinite/smectite mixed layer;
T/S, talc/smectite mixed layer;
R0, randomly ordered mixed-layer mineral;
R1, R>1, ordered mixed-layer mineral

1.2
Nucleation and Crystal Growth: Principles

Introduction

One cannot understand why, in a rock sample, a mineral exhibits a given composition, size and shape without knowing its formation history, from its initiation (nucleation) to its final appearance resulting from growth and dissolution episodes. This approach is classically used when dealing with metamorphism (Kretz 1994), but not with surface phenomena (supergene or hydrothermal alterations, diagenesis, sedimentation). This is because the size of clays, most often lower than 1 micron, is an obstacle to direct observation. Nevertheless, technical improvements and recent advances in our understanding have allowed the nucleation and growth processes of these minerals to be addressed.

These processes require ion or molecule exchanges between three partners: growing crystals, dissolving crystals and the solutions in which they are soaking. Insight into how these exchanges occur may come from two types of investigations: those conducted when the system is viewed statically (based on energy and matter balances) or dynamically (based on dissolution rates of matter-providing minerals, growth rate of developing minerals, and velocity of transfers in solutions). From the static standpoint alone, the nature and quantity of formed minerals are defined by the phase rule (equilibrium thermodynamics). The phase rule proves useful but insufficient in the clay mineral field. Energy and matter balances are a baseline essential to the reconstruction of processes, but still they do not yield all of the clues. The clay mineral field is *par excellence* one of reaction paths punctuated with instability or metastability states (kinetics). The crystallochemical nature and the size of crystals on the one hand, and the formation of assemblages of various species on the other hand are governed by kinetic laws.

1.2.1
Basics

The following fundamental statements must be familiar to any clay scientist. They are summarized here but can be found in greater details in many books. Among them, those by Putnis and McConnell (1980) and Stumm (1992) are particularly recommended.

1.2.1.1
Equilibrium – Supersaturation – Undersaturation

Dissolution and precipitation are the mineral reaction processes that take place in rocks submitted to alteration or diagenesis. Dissolution amounts to a cations-protons exchange between the solid and the solution. Precipitation (nucleation then growth) corresponds to the opposite exchange (Fig. 1.19a). One can see the significance of the undersaturation and supersaturation concepts, and it

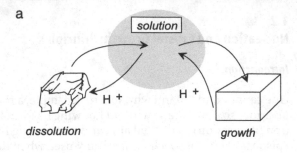

a

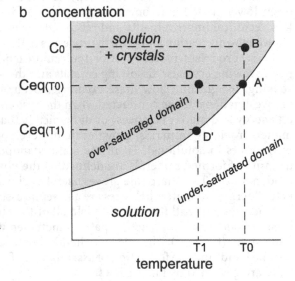

Fig. 1.19a,b. Dissolution –
precipitation. a) The three
"partners" in matter transfers.
b) Dissolution and precipitation
domains in a concentration-
temperature space (Baronnet
1988)

is easy to understand that pH conditions are among those factors controlling the course of these processes. Baronnet (1988) explains in a very straightforward way how two bivariant domains (supersaturation and undersaturation) and one univariant domain (curve establishing the boundary between both domains) are contained in the concentration-temperature space. For a fixed temperature T_0, equilibrium is reached at A' (Fig. 1.19b). Supersaturation is obtained only by evaporation of the solvent, which has the effect of increasing the concentration of elements in solution. Under natural conditions, this can be observed in the boiling zones of geothermal fields or in evaporitic basins, for instance. Undersaturation is obtained by increasing the solvent mass: the mixing process of warm and saline solutions with deeply percolating meteoric waters is frequently observed in geothermal fields or in diagenetic basins. For a fixed concentration C_0, the supersaturation and undersaturation domains correspond to low and high temperatures, respectively. The temperature of fluids in natural environments varies during time and space in the Earth's crust: generally, it decreases near the surface.

When, at a fixed temperature, supersaturation is reached, crystals tend to grow so that concentration is lowered, getting closer to the value of the equilibrium concentration at this temperature (C_{eq}). Conversely, in the undersaturation domain, crystals dissolve, thus increasing the concentration of elements in solution until equilibrium has been reached. At equilibrium, the amounts of dissolved solid and precipitated solids are equivalent. For a temperature T_0, supersaturation σ reached at point B (Fig. 1.19b) is then expressed as follows:

$$\sigma_{T_0} = \frac{C_0 - C_{eq}}{C_{eq}} \tag{1.36}$$

1.2.1.2
Elements of Nucleation Theory

Like all the other mineral species formed from aqueous solutions, clay minerals precipitate when a supersaturation state is reached either by direct precipitation (homogeneous nucleation) or precipitation on a previously existing solid phase (heterogeneous nucleation). In both cases, the stages are identical: (1) nucleation starting with the formation of embryos of which only "survive" those reaching the critical size (nuclei) and (2) crystal growth. Whatever the type of nucleation, the energy balance (ΔG) consists of two terms: (1) the total energy of inner bonds, which increases proportionately to the nucleus or crystal volume (ΔG_v) and (2) the energy of the solid-solution interface, which increases with the nucleus or crystal surface area (ΔG_s):

$$\Delta G = \Delta G_v + \Delta G_s \tag{1.37}$$

Homogeneous Nucleation
Calculation of ΔG_v   For a given ion or molecule, the chemical potential is different between the supersaturated solution (μ_l) and the solid (μ_s). By definition, μ_s is equal to the chemical potential of the saturated solution (crystal – solution equilibrium). This difference of energy is multiplied by the number n of ions or molecules that leave the solution to enter the solid:

$$\Delta G_v = -n \left(\mu_l - \mu_s \right) = -nkTLn \left(\frac{a_l}{a_s} \right) \tag{1.38}$$

k: Boltzmann's constant

a_l and a_s: ion or molecule activities in the solution and the solid respectively

If one considers that the ratio of activities is not very different from the ratio of concentrations and that the system is not far from equilibrium ($a_l = a_s + \Delta a$), one can write:

$$\frac{a_l}{a_s} \approx \frac{C_0}{C_{eq}} \quad \text{hence} \quad \frac{a_l}{a_s} \approx 1 + \sigma \tag{1.39}$$

For simplification, let's assume that the nucleus is a sphere with radius r. Its volume $v = (4/3)\pi r^3$. The number n of ions or molecules is given by the division of $(4/3)\pi r^3$ by the molecule or ion volume v in the crystal: $n = v/v$.

Calculation of ΔG_s   The surface area of a spherical nucleus is given by: $S = 4\pi r^2$. If $\bar{\sigma}$ represents the free energy of the crystal – solution interface per surface unit, the term ΔG_s is $S\bar{\sigma}$ hence $4\pi r^2 \bar{\sigma}$. It is positive (energetical cost of the outer surfaces).

Calculation of ΔG   The energy of formation of the nucleus ΔG is then expressed as follows:

$$\Delta G = -\frac{4}{3}\pi r^3 kTLn(1 + \sigma) + 4\pi r^2 \bar{\sigma} \tag{1.40}$$

It is negative according to thermodynamics conventions.

The $\Delta G = \Delta G_v + \Delta G_s$ function is represented in Fig. 1.20: it increases when the radius varies from 0 to the critical value r^* (embryos tend to resorb) then decreases for upper values (embryos grow). Knowing that the evolution of any system occurs with free energy minimisation, one can understand the reason why some embryos dissolve whereas others grow. Simply calculate the derivative of the function above: $\frac{d(\Delta G)}{dr}$. If it is positive, ΔG increases with the radius, so embryos are unstable and dissolve; if the derivative is negative, ΔG decreases as the radius increases, so embryos grow. The value of the critical

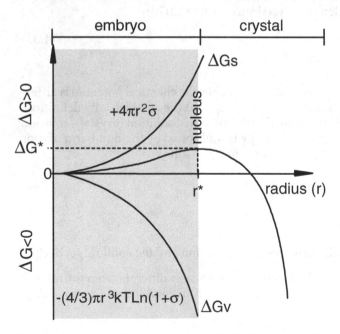

Fig. 1.20. Representation of the variation in the free energy of nuclei as a function of their radius r

radius r^* and the corresponding energy ΔG_{r*} are given when the derivative $\frac{d(\Delta G)}{dr} = 0$:

$$r^* = \frac{2v\bar{\sigma}}{kTLn(1 + \sigma)} \qquad (1.41)$$

$$\Delta G_{r*} = \frac{16\pi v^2 \bar{\sigma}^3}{3\left[kTLn(1 + \sigma)\right]^2} \qquad (1.42)$$

σ belonging to the denominator, the embryo critical size (r^*) and the nucleation barrier energy (ΔG^*) are as small as it is high.

Heterogeneous Nucleation

Embryos form on a solid support that acts as a "catalyst" of the nucleation by reducing the energy barrier. Then, the critical radius is more easily reached (Fig. 1.21a). If the structure of the surface of the solid support is close to that of some atomic planes of the clay mineral crystal lattice, then the interfacial energy between the two solids is lower than that between the clay mineral and the solution. Nucleation then takes place on the support at a lower supersaturation rate.

When the embryo reaches the critical radius r^*, it becomes a nucleus able to grow to form a crystal. Growth is governed by the interfacial energy between

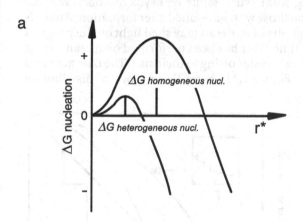

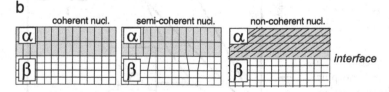

Fig. 1.21a,b. Heterogeneous nucleation. a) Highlighting of the solid substrate as a catalyst. b) Schematic representation of the main types of interfacial relationship between the embryo and its support

the two phases. This energy depends on the type of boundary separating both phases: to coherent, semi-coherent or non-coherent boundaries (Fig. 1.21b) correspond energy values of 10–50, 100–300 and 500–1,500 erg cm^{-2} respectively (1 erg = 10^{-7} J). In all cases, heterogeneous nucleation begins when the interfacial energy between the two solids is lower than that between the precipitated crystal and the solution. Nucleation is triggered by a supersaturation rate lower than that required for a direct precipitation from a solution. This has been applied for fibrous illite formation in oilfield sandstones (Wilkinson and Haszeldine 2002).

Compared with homogeneous nucleation in which the surface energy ΔG_s is $S\bar{\sigma}$, heterogeneous nucleation involves the energy of three types of interfaces: nucleus-solution (n-sl), nucleus-solid substrate (n-sb) and substrate-solution (sb-sl):

$$\Delta G_s = S_{\text{n-sl}}\bar{\sigma}_{\text{n-sl}} + \left(\bar{\sigma}_{\text{n-sb}} - \bar{\sigma}_{\text{sb-sl}}\right) S_{\text{n-sb}} \tag{1.43}$$

The catalytic effect of the substrate requires that $\bar{\sigma}_{\text{n-sb}} \langle \bar{\sigma}_{\text{sb-sl}}$. In this case, the surface energy ΔG_s is reduced. Consequently, the activation energy of nucleation is reduced too.

Crystal Number – Time Relationship

The direct study of nucleation in natural rocks is obviously impossible. Indeed, only those crystals stemming from "successful" embryos are observed. No information can be obtained on those who have failed after formation. Analysis through frequency-size histograms of minerals may shed light on the progress of nucleation as a function of time. This has been performed on metamorphic rocks for which crystal counting is easier owing to their size. The data acquired certainly apply to clay rocks. Figure 1.22 shows three types of distribution

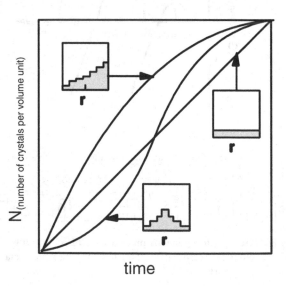

Fig. 1.22. Reconstruction of nucleation rates in metamorphic rocks through frequency-size histograms of minerals (r: radius of the circumscribed circle)

corresponding to three different velocity ranges: increasing with time, constant or passing through a maximum (Kretz 1994).

1.2.1.3
Fundamentals of Growth Theory

General Mechanisms

Once again, many lessons about the growth of clays in rocks submitted to alteration or diagenesis can be drawn from the studies of metamorphic rocks. Several processes can take part in the growth of minerals. If working "in parallel", the fastest one controls the growth rate, but, if working "in series", it is the slowest one. Kretz (1994) showed that interface-controlled and diffusion-controlled processes can take over from one another during the growth history of porphyroblasts in metamorphic environments (Fig. 1.23). This far-reaching work has found many applications in the studies about growth processes of illites (Nagy 1994).

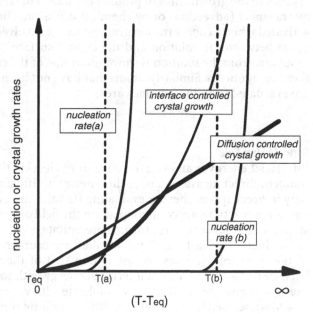

Fig. 1.23. Processes controlling the growth of porphyroblasts in metamorphic rocks (Kretz 1994)

Interface-Controlled Growth   When the nucleus of a phase β starts growing on a phase α, the size of the $\alpha - \beta$ interface is increased. The growth rate is controlled by the transfer of atoms from α to β through the interface. This transfer comes up against an energy barrier, which represents the resistance to atom incorporation in the growth sites.

Diffusion-Controlled Growth   In most nucleation-growth processes, the composition of product crystals differs from that of dissolving crystals. Consequently the process necessitates a transfer of atoms from the dissolution sites of reactant

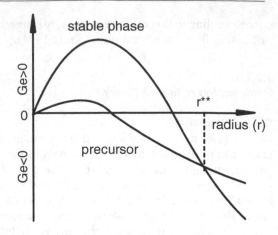

Fig. 1.24. Schematic representation of the variation in the free energy of formation of crystal nuclei as a function of size. From 0 to r^{**}, the formation of the precursor is faster than that of the stable phase, and vice versa for $r > r^{**}$

crystals to the growth sites of product crystals. This transfer takes place either by transport (advection) or by chemical diffusion. Diffusion in a 3D-space is activated when a concentration gradient of the involved chemical components exists between the solution and the crystal surface. This happens when the input rate from the solution is slower than that of the crystal incorporation for these components. Similarly, atoms that have not been "consumed" by growth are evacuated from the interface area.

Ostwald Step Rule

Interfacial energy is an essential term in nucleation thermodynamics. Unfortunately, direct measurement of this energy is difficult and for the time being only indirect approaches for evaluating its value are available. Successive precipitation reactions are commonplace in the field of clay minerals. The Ostwald step rule states that the first-formed precipitate is the one with the highest solubility hence the least stability. It is called a precursor. The nucleation kinetics of the most soluble phase is faster than that of the next less soluble phase because it shares the weakest interfacial energy with solution (Fig. 1.24). A fine example is given by the reaction of kaolinite with very strong KOH solutions (1 to 4 M) described by Bauer et al. (1998). Kaolinite transforms into K-feldspar going through intermediate compounds: mica, KI zeolite then phillipsite.

1.2.2
Nucleation and the First Growth Stages of Clay Minerals

The nucleation and crystal growth fundamentals established above permit presentation of some examples showing how these phenomena have been investigated in clays formed under experimental and natural conditions.

1.2.2.1
Experimental Syntheses

Experimental Conditions
Most of the time, syntheses are performed in closed systems from solid materials selected for their compositional similarity to the minerals being synthesised. These solids must have a large specific surface to achieve an efficient reaction with the solutions (mineral or glass powders, gels). They must have a higher solubility than clay minerals. Gels are usually preferred to glasses. They are obtained by the reaction of sodium metasilicate with metallic salts in proportions that are different for 1:1 or 2:1 minerals (Decarreau 1980):

1:1 minerals $2 SiO_2 Na_2 O + 2 Al(NO_3)_3 \rightarrow Si_2 Al_2 O_7 + 6 NaNO_3 + H_2 O$

2:1 minerals $4 SiO_2 Na_2 O + 3 MgCl_2 + 2 HCl \rightarrow Si_4 Mg_3 O_{11} + H_2 O + 8 NaCl$

The resulting gels are strongly hydrated. Glasses can be made from those gels by heating them at 600 °C for several hours. Gels or glasses are subsequently brought into contact with distilled water on the basis of 5 to 10 g of solid for 1 litre of water. Trioctahedral Mg-bearing clay minerals are obtained very easily at ambient temperature by simple reaction in a beaker. This is the case of the magnesian minerals saponite and stevensite. The other species are synthesised at higher temperature and pressure. This technique has been used to synthesise beidellite, saponite and nontronite (Grauby et al. 1993, 1994).

Relations Between Solution pH and Crystal Growth
The synthesis mechanism can be summed up as follows (Decarreau and Petit 1996): gel or glass dissolution→supersaturation of the solution for the clay mineral → nucleation → crystal growth.

Therefore, the crystallisation rate is at first high because the supersaturation rate is at maximum. As the mass of the solid material increases, the supersaturation rate decreases and the crystallisation rate is reduced. This explains the logarithmic pattern of the curves representing the crystal mass as a function of time (Fig. 1.25)

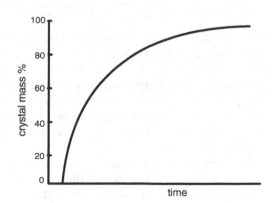

Fig. 1.25. Variation in the mass of crystallised clay minerals as a function of time for a synthesis experiment at fixed temperature

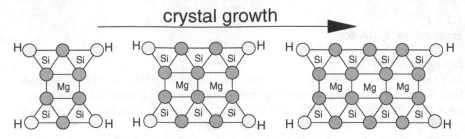

Fig. 1.26. Schematic representation of the growth of a magnesian phyllosilicate crystallite. The chemical balance shows that H^+ protons are released into solution: $[Si_4O_{10-\varepsilon}Mg_3 (OH)_{2+\varepsilon}]^{\varepsilon+} \rightarrow Si_4O_{10}Mg_3(OH)_2 + \varepsilon H^+$. *Full circles*: oxygens with two compensated valences; *empty circles*: oxygens likely to form a silanol group by fixation of a H^+ of the solution (Decarreau and Petit 1996)

Gels obtained by coprecipitation for magnesian minerals contain embryos of trioctahedral phyllosilicates whose mean diameter does not exceed 10 Å. Those are the embryos from which the clay mineral grows (Decarreau and Petit 1996). The O^{2-} ions that are not common to two tetrahedra (boundary ions) have an uncompensated valence balanced by the H^+ cations of the solution (refer to Sect. 5.1.1.2). Crystal boundaries are then formed by silanol groups: Si–OH. Consequently, the chemical balance of the synthesis reaction of trioctahedral phyllosilicates is written as follows:

$$4\,SiO_2Na_2O + 3\,MgCl_2 + 2\,HCl + \varepsilon H_2O$$

$$\rightarrow [Si_4Mg_3O_{10-\varepsilon}(OH)_{2+\varepsilon}]^{\varepsilon+} + \varepsilon(OH)^- + 8\,NaCl \qquad (1.44)$$

This reaction shows that the pH of solutions increases during the nucleation period of phyllosilicates. The pH values measured during the synthesis experiments range from 9 to 10.

The smaller the microcrystal, the greater the amount of silanols/oxygens (Fig. 1.26). Therefore any growth releases H^+ protons into solution (Decarreau and Petit 1996):

$$[Si_4O_{10-\varepsilon}Mg_3(OH)_{2+\varepsilon}]^{\varepsilon+} \rightarrow Si_4O_{10}Mg_3(OH)_2 + \varepsilon H^+ \qquad (1.45)$$

Subsequently, growth takes place following two processes: (1) increase in the size of the layer (increase in the coherent scattering domain size along the a or b direction), and (2) increase in the number of layers in the stacking sequences. The first process brings about the decrease in the number of silanol groups Si – OH, which is verified with infrared absorption spectrometry by measuring the ratio of band surfaces at 900 and 1,010 cm^{-1} (Fig. 1.27a). This results in the release of H^+ cations into the solution and hence pH decreases (Fig. 1.27b).

The composition of the solutions in contact with forming clay minerals can also be used as an indicator of their growth. Indeed, comparing their ionic activity product (I.A.P) with their equilibrium constant K_{eq} is enough. For

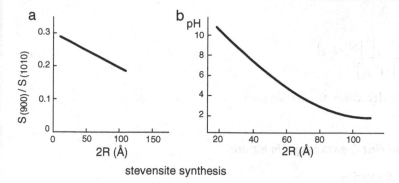

stevensite synthesis

Fig. 1.27a,b. Release of H⁺ cations during the growth of magnesian phyllosilicates. a) Decrease in the number of Si–OH groups estimated from the ratios of band surfaces at 900 and $1{,}010\,\mathrm{cm}^{-1}$ as a function of the coherent scattering domain size in the layer plane. b) Variation in pH as a function of the coherent scattering domain size in the layer plane (Decarreau and Petit 1996)

a magnesian phyllosilicate, K_{eq} is calculated as follows (refer to Sect. 3.1.2.1):

$$\mathrm{Si_4O_{10}Mg_3(OH)_{2\,(c)} + 6\,H^+{}_{(aq)} \rightarrow 3\,Mg^{2+}{}_{(aq)} + 4\,SiO_{2\,(aq)} + 4\,H_2O_{(l)}} \quad (1.46)$$

hence

$$K_{eq} = \frac{[\mathrm{Mg^{2+}_{aq}}]^3\,[\mathrm{SiO_{2aq}}]^4}{[\mathrm{H^+_{aq}}]^6} \quad (1.47)$$

The chemical analyses of the solutions show that the content of Si and Mg (or other bivalent cations such as Ni^{2+}) does not vary much in comparison with pH. Therefore, the difference between the values of IAP and K, which has not yet reached the value of K_{eq}, can be considered as related to the progress of the growth reaction (Fig. 1.28):

$$[\mathrm{Si_4Mg_3O_{10-\varepsilon}(OH)_{2+\varepsilon}}]^{\varepsilon+}_{(c)} + 6 - \varepsilon\,H^+{}_{(aq)}$$
$$\rightarrow 3\mathrm{Mg^{2+}}_{(aq)} + 4\mathrm{SiO_{2\,(aq)}} + 4\mathrm{H_2O_{(l)}} \quad (1.48)$$

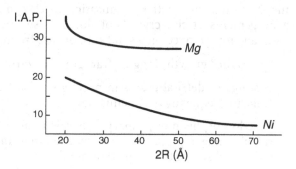

Fig. 1.28. Variation in the ionic activity product (I.A.P) as a function of the coherent scattering domain size in the layer plane for synthetic magnesian (Mg) and nickel-bearing (Ni) stevensites

hence

$$K = \frac{\left[Mg^{2+}{}_{aq}\right]^3 \left[SiO_{2aq}\right]^4}{\left[H^+{}_{aq}\right]^{6-\varepsilon}} \tag{1.49}$$

The value of ε decreases as crystals grow.

1.2.2.2
Nucleation and First Growth Stages in Nature

Homogeneous Nucleation
Examples of homogeneous nucleation in the natural environment essentially refer to the direct precipitation of clay minerals from solutions having reached the required supersaturation by evaporation or by temperature decrease (see Fig. 1.19). Two cases may occur:

1. supersaturated solution → magnesian clay minerals + salts or other silicates

2. supersaturated solution + aluminous silicates → magnesian clay minerals

The first process can particularly be observed in active geothermal fields in zones where physical conditions permit boiling of the solution. This is the case of the Milos site in Greece where seawater starts boiling at −900 m thus causing the coprecipitation of saponite, actinolite and talc (Beaufort et al. 1995). This phenomenon is responsible for the clogging of the pipes of geothermal power plants. Indeed, steam must be depressurised in order to be usefully recovered. Pressure drops bring about supersaturation and cause the formation of zoned deposits within pipes: aragonite alternating with a magnesian clay (Beaufort, personal communication). The second process can be observed in environments where an intense evaporation of seawater or of hydrothermal solutions occurs. Stevensite deposits form in marine evaporitic environments (Trauth 1977). In continental environments, saline soils and vertisols of arid and semi-arid areas constitute favourable environments for the formation of sepiolite-palygorskite or montmorillonite.

Heterogeneous Nucleation
Although traces of a nucleation process are very discrete, they can be recognised through the spatial organisation of clay minerals in the rock. Indeed, arrangements of clay crystals obviously resulting from an oriented growth from an embryo fixed on a solid support are frequently observed by:

- "palisadic" growth clogging fractures in hydrothermal systems;

- coatings of detrital grains in porous sandstones submitted to diagenesis. Chlorite frequently occurs this way.

- pseudomorphism of primary minerals in altered rocks. For example, vermiculite may replace biotites of weathered granites and retain the external shape and size of the original mineral.

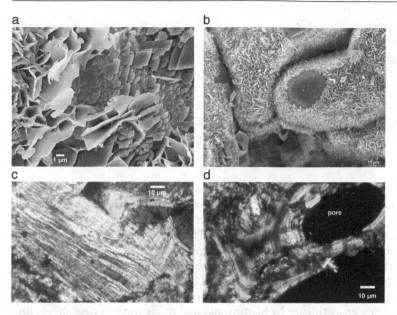

Fig. 1.29a–d. Commonly occurring microstructures of clay minerals in rocks. **a)** "Palisadic" clogging of fractures (hydrothermal alteration). **b)** Coatings of crystals oriented perpendicular to the surface of the supporting grain (diagenesis). **c)** Pseudomorphic replacement of a biotite by a trioctahedral vermiculite (supergene alteration). The *clear patches* are kaolinite. **d)** Argillan (cutan) in a soil

Palisadic growth and coatings are very easy to distinguish from deposits that line fracture or pore walls (cutans). They result from the sedimentation of particles transported by fluids. They commonly occur in illuviation soils where clays form stable suspensions in solutions (Fig. 1.29).

1.2.3
Growth of Clay Minerals

Now that chemical exchanges during the growth process are understood, the question remains: how do clay minerals grow physically? In other words, how are chemical components incorporated on a crystal face? These questions are still under study. However, some aspects have been partially answered.

1.2.3.1
The Role of Screw Dislocations

Growth Under Experimental Conditions
Generally a crystal is a polyhedron whose faces, edges or vertices are the elements in contact with the ambient solutions. The rate of growth and the resulting general shape of the crystal will be notably different depending on

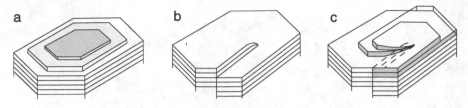

Fig. 1.30a–c. Growth principle. **a)** Phyllosilicate growing sheet after sheet. **b)** Emerging of a screw dislocation. **c)** Spiral growth from the screw dislocation

the location of growth. So are determined faces of type F (flat), S (stepped) or K (kinked), containing 2, 1 or 0 strong Si–O bond chains inside the face respectively. The greater the growth rate, the smaller the resulting face surface. The F faces grow slowly and are therefore larger. This is the case of phyllosilicates that exhibit a large development of (001) faces.

The experimental syntheses of micas (Baronnet 1976) show that two processes take place in the development of the (001) faces. First, plate-shaped embryos occur by heterogeneous or homogeneous tri-dimensional nucleation (Fig. 1.30a). They grow thicker by bi-dimensional nucleation of new sheets on the (001) faces. Then, beginning with a certain stage of development, a screw dislocation may emerge on (001) and (00$\bar{1}$) faces (Fig. 1.30b). The spiral growth takes place from this dislocation (Fig. 1.30c).

Recent experimental synthesis work improves our understanding of some of the causes controlling crystal morphology: chemical parameters and time. Fialips et al. (2000) have shown that the shape of kaolinite crystals depends on the pH of solutions: isometric hexagonal crystals form under acidic conditions, and laths form under basic conditions. The inhibition of the growth of some faces causing the elongation of crystals into laths is due to the adsorption of foreign cations on their external surfaces. Bauer et al. (2000) have shown that lath-shaped illite appears first as a metastable state. Then, for longer crystal growth periods, hexagonal particles are formed.

Growth in Natural Environments

The phyllosilicates formed under natural conditions systematically exhibit traces of spiral growth. Several methods are used to show the growth steps on the (001) faces. One of the most commonly used, before the advent of the atomic force microscopes (AFM), was gold deposition on particle growth steps for transmission electron microscopy observation. The principle was discovered by Gutsaenko and Samotoyin (1966). It is summed up in Fig. 1.31 from Kitagawa's procedure (1997).

Observations on kaolin-group minerals (Sunagawa and Koshino 1975), pyrophyllites, illites (Kitagawa 1992; 1995), and illite/smectite mixed layers (Kitagawa and Matsuda 1992) show the presence of spiral-shaped steps on the (001) faces of crystals. The crystal growth takes place from emerging screw dislocations on these faces. That illite-smectite mixed layers, whether 1M or 2M

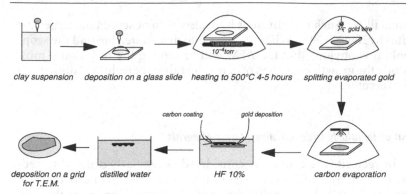

clay suspension   deposition on a glass slide   heating to 500°C 4-5 hours   splitting evaporated gold

carbon coating   gold deposition

deposition on a grid   distilled water   HF 10%   carbon evaporation
for T.E.M.

Fig. 1.31. Crystal surface decorating method by nucleation of gold grains on the edges of the growth steps (Kitagawa 1997)

polytype, exhibit identical growth patterns is noteworthy (Fig. 1.32a–b). Coalescence can be explained by the formation of steps covering crystals in contact with each other (Fig. 1.32c). This has been elegantly demonstrated for micas by Nespolo (2001).

The widely used AFM now provides very accurate images of the surface of clay minerals. Blum (1994) has revealed spiral-shaped growth steps originating from an emerging screw dislocation on the (001) faces of more or less isometric

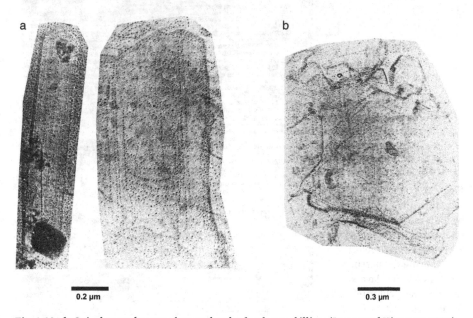

a   b

0.2 µm   0.3 µm

Fig. 1.32a,b. Spiral growth steps observed on hydrothermal illites (Inoue and Kitagawa 1994). a) 1M illites. b) 2M illite

hydrothermal illites. Such growth steps have also been observed on diagenetic fibrous illites by Nagy (1994). Without going into details well beyond the scope of this book, it is obvious that the AFM will be among the techniques most suited to study the growth processes of clay minerals once technical difficulties have been overcome.

1.2.3.2
The Problem of Regularly Ordered Mixed-Layer Minerals

Corrensite or rectorite are classically considered as regularly ordered mixed-layer minerals composed of 50% chlorite or mica and 50% tri- or dioctahedral smectite respectively. Following this assumption, they are no more than ordered mixed-layer minerals with a specific composition. As such, they have

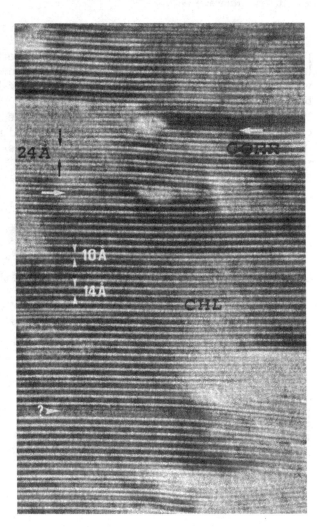

Fig. 1.33. Corrensite crystal structure. Image of layers by high-resolution electron microscopy (Beaufort et al. 1997). Crystal defects (*white arrows*) interrupt these 24 Å thick layers showing that "chlorite" (14 Å) and "smectite" (10 Å) sub-layers are not independent

been viewed as the midway stage in the transformation of smectites into chlorite or mica. Recent studies on corrensite have modified this viewpoint. It has been shown indeed that the crystal structure is not explained by two independent layers stacking one atop the other (one of chlorite type, the other of smectite type), but rather by a single structure whose thickness (24 Å in the dehydrated state) is equal to the added thicknesses of a chlorite layer and a smectite layer (Beaufort et al. 1997). Corrensite is a true mineral whose structure rests on the stacking sequence of identical layers composed themselves of two 2:1 sub-layers. This has been inferred from two observations: (1) high-resolution electron microscopy observations show that there is no such thing as a "smectite" layer independent of a "chlorite" layer, and (2) each stacking defect interrupts both types of layers simultaneously (Fig. 1.33). Corrensite crystal structure is developed in Sect. 2.2.3.2 (Fig. 2.29). Logically, its growth should form steps at 24 Å.

Microtopography of mica-smectite mixed layers (rectorite) shows that crystal growth takes place along spiral-step pairs originating in emerging screw dislocations on the (001) faces. Unfortunately, the height of these steps has not been reliably measured yet: 10 Å or 20 Å (Kitagawa and Matsuda 1992). Their shape is clearly less regular than that of spirals observed on micas, illites or even I/S mixed layers. They are rounded with some straight parts but never exhibit hexagonal shapes.

The formation of regularly ordered mixed-layer minerals by natural or experimental alteration of pre-existing phyllosilicates (biotite, muscovite, chlorite) usually leads to pseudomorphism (change in the crystal lattice with conservation of the original crystal habit). HRTEM observations show that inheritance is not limited to the outward form, and that it is expressed by the persistence of a polytype that is uncommon in the new crystal species (Baronnet 1997). The author gives simple criteria to discriminate between solid-state transformation (SST) and dissolution-recrystallisation (DR) mechanisms. They are summed up in Table 1.8.

Table 1.8. Criteria for discriminating between the two main intermineral transformation mechanisms from Baronnet (1997). SST: solid state transformation; DR: dissolution-recrystallisation

Criteria	SST	DR
Crystal habit	Retained	Lost
Topotaxy	Yes	Yes (epitaxy) or no
Polytype inheritance	Yes	No
Chemical and structural homogeneity	No	Yes
Chemical inheritance	Yes	No
Inheritance of crystal defects	Yes	No

1.2.4
Ripening Process

How can chemical exchanges between the solution and crystals be related to the physical process controlling crystal growth? For a given species, clay particle populations are characterized by the distribution of some size or shape parameters. The statistical study of these distributions gives some genetic information, and more specifically, it helps to understand the effects of ripening.

1.2.4.1
Ostwald Ripening Rule

Interfacial Tension

Let's consider a sphere of liquid (β) in contact with its vapour (α) surrounded by a bulk mass of β. If one assumes that the molar energy of β in the environment is lower than it is within the sphere by a quantity ΔG, the transfer of n moles of β from the environment into the sphere will bring about an increase in the mass hence in the total Gibbs energy of the sphere: $dG = \Delta G dn$, where n is the number of moles of β in the sphere and G is the Gibbs free energy of the sphere. Let r be the radius of the sphere, A the surface of the sphere, V^β the molar volume of β and $\sigma = \frac{dG}{dA}$ interfacial tension, it yields:

$$\Delta G = \frac{\sigma dA}{dn} \tag{1.50}$$

$$\Delta G = \sigma \frac{\frac{dA}{dr}}{\frac{dn}{dr}} \tag{1.51}$$

The volume of the sphere is

$$nV^\beta = \frac{4}{3}\pi r^3 \tag{1.52}$$

$$n = \frac{1}{V^\beta}\frac{4}{3}\pi r^3 \tag{1.53}$$

using the differential, it yields:

$$\frac{dn}{dr} = \frac{1}{V^\beta}4\pi r^2 \tag{1.54}$$

The surface of the sphere is: $A = \pi r^2$. Using the differential, it yields:

$$\frac{dA}{dn} = 8\pi r \tag{1.55}$$

Thus ΔG can be expressed as follows:

$$\Delta G = \frac{2V^\beta \sigma}{r} \tag{1.56}$$

or, for a spontaneous displacement of matter from the sphere to its environment:

$$\Delta G = -\frac{2V^\beta \sigma}{r} \tag{1.57}$$

The Gibbs-Thompson equation shows that excess Gibbs free energy of a small sphere is a consequence of its interfacial tension. The lower the radius, the higher the interfacial tension. This means that the "energetical cost" of small particles is higher than that of big ones. Thus, for a given mineral species crystallizing in natural or experimental systems, small particles are unstable with respect to biggest ones.

The Ripening Process

In an isothermal closed system containing a crystalline species represented by a small-sized precipitate, one observes, for a constant mass, a decrease in the number of crystals and an increase in their size over time. This phenomenon was first described by Ostwald (1900). This implies that some small-sized particles dissolve while others grow. In total, the interfacial free energy of the system decreases. Dissolution and reprecipitation (but not nucleation) take place simultaneously because of the solubility-size dependence of particles as described by the Gibbs-Thompson-Freundlich equation:

$$\ln \frac{s_1}{s_2} = \frac{2\bar{\sigma}v}{kT_a}\left(\frac{1}{r_1} - \frac{1}{r_2}\right) \tag{1.58}$$

s_1 solubility of the particle with radius r_1 (m)

s_2 solubility of the particle with radius r_2 (m)

$\bar{\sigma}$ solid-solution interfacial specific free energy ($\mathrm{J\,m^{-2}}$)

v molecule volume for the solid ($\mathrm{m^3}$)

k Boltzmanns' constant ($\mathrm{J\,K^{-1}}$)

T_a temperature (K)

The term $\frac{2\bar{\sigma}v}{kT_a}$ is greater than 0 so a particle with radius r_1 is less soluble than another one with radius r_2 if $r_1 > r_2$. If, at a given time t, the relative supersaturation is $\sigma = \frac{c-c_\infty}{c_\infty}$ (c_∞ being the solubility of a particle with infinite radius), the critical radius r^* of a particle with solubility c at equilibrium with the environment can be calculated as follows:

$$Ln\frac{c}{c_\infty} = Ln(1+\sigma) = \frac{2\bar{\sigma}v}{kT_a r^*} \tag{1.59}$$

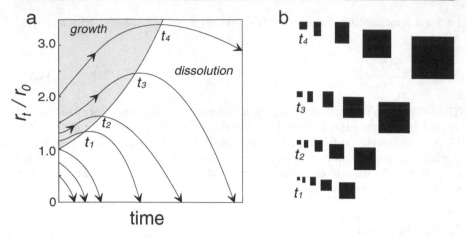

Fig. 1.34a,b. Ostwald ripening process. a) Variation in the particle size as a function of time (four stages from t_1 to t_4). The size is normalised to the particle radius at the given time t_0. b) Conservation of the size distribution

hence

$$r^* = \frac{2\bar{\sigma}v}{kT_a Ln\left(1 + \sigma/\right)} \tag{1.60}$$

Thus, coming back to the Gibbs-Thompson-Freundlich equation in which r_1 and r_2 are replaced by r and r^* respectively, it appears that particles with radius r lower than r^* will dissolve whereas those with radius r greater than $r*$ will grow: this is the "ripening" process.

Nevertheless, as the mass transfer is a continuous process, it leads to the increase of the mean radius $\bar{r}$ of the particle population. Consequently, the mass of the precipitate increases and the supersaturation rate decreases progressively. This results in the dissolution of those particles whose radius was controlled by an early (hence greater) given supersaturation rate (Fig. 1.34a). Therefore, a particle with a radius greater than the critical radius at a given time t_1 begins to grow, but then its growth rate decreases until it has become zero (derivative equal to zero), and finally it dissolves (whereas particles with greater radius grow) (Fig. 1.34b).

During the Ostwald ripening process, the particle size continuously increases. The statistical distribution theoretically tends to a steady state, which is emphasised in a normalised frequency vs. size diagram: $\left(\frac{f_r}{f_{\max r}}\right)$ as a function of $f(\omega) = \frac{1}{\varpi\beta\sqrt{2\pi}} \exp\left[-\left(\frac{1}{2\beta^2}\right)\left(\ln(\varpi) - \alpha\right)^2\right]$. The shape of the representative curve of this function depends on the mechanisms controlling the chemical element transfer between dissolving and growing particles. In the stability fields of clay minerals (porous rocks, presence of fluids), two mechanisms are to be considered: (1) the volumic diffusion of molecules (or ions) within the solution; (2) the spiral growth (or dissolution) from screw dislocations of particles.

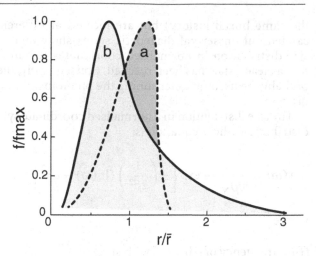

Fig. 1.35. Theoretical particle size distribution expressed in normalised coordinates during the ripening process controlled either by the volumic diffusion of molecules or ions in the solution (*curve a*) or by the spiral growth or dissolution from screw dislocations (*curve b*)

Transfer mechanisms are controlled by the slowest process. The theoretical shape of the size distribution expressed in normalised coordinates for each of both processes has been given by Chai (1975) for calcites and by Baronnet (1976) for phlogopites (Fig. 1.35).

1.2.4.2
Analysis of Crystal Habits and Sizes: Application to I/S

The Ostwald ripening process imposes very restrictive conditions, which are usually not met in natural environments (Baronnet 1994). However, the size distribution analysis of particles of illite/smectite mixed layers (I/S) in hydrothermal and diagenetic environments (Inoue et al. 1988; Eberl and Srodon 1988; Eberl et al. 1990; Lanson and Champion 1991; Varajao and Meunier 1994) shows that a ripening phenomenon does exist. Indeed, small-sized lath-shaped smectite-rich I/S particles transform progressively into large-sized hexagon-shaped illite particles. Decomposition of X-ray diffraction patterns (Lanson and Champion 1991) permit classification of particles into three families: (1) smectite-rich mixed-layer minerals, small size, elongated laths (I/S); (2) small-sized and thin illites (small-sized coherent scattering domain) + illite-rich I/S, large size, elongated laths (PCI: "poorly crystallised illite"); (3) illites showing two different habits: thick lath-shaped or hexagonal particles (WCI: "well crystallised illite"). Transmission electron microscope observations permit identification of five classes of particles: (1) small-sized elongated laths; (2) large-sized elongated laths; (3) large and thick laths; (4) small isometric particles; and (5) large isometric or even hexagonal particles. The size increases progressively following the sequence described in Figs. 4.26 and 8.16.

The factors controlling this ripening process are determined by the size distribution analysis of I/S particles extracted from rock samples that have undergone diagenesis at various degrees. In other words, these rock samples belong to a similar type (shales or sandstones) and to a similar series (with the same burial history) but are located at different depths. Observations carried out on several diagenetic series show that, whatever the depth, the size distribution in normalised coordinates remains constant. This implies that a steady state has been reached. Consequently, the shape of the curve can probably be used for determining the phenomenon that controls the ripening process.

The size distribution in a normalised coordinate system follows a lognormal distribution whose equation is:

$$f(\omega) = \frac{1}{\varpi\beta\sqrt{2\pi}} \exp\left[-\left(\frac{1}{2\beta^2}\right)(\ln(\varpi) - \alpha)^2\right] \tag{1.61}$$

where

$f(\omega)$ frequency of particles with size ω

β^2 $\sum[\ln(\omega) - \alpha]f(\omega)^2$ variance of frequency logarithms

α $\sum[\ln(\omega)f(\omega)]$ mean of frequency logarithms

The lognormal shape is the closest to the curve representing the spiral growth or decrement in Fig. 1.35. Inoue and Kitagawa (1994) have shown the presence of such spirals not only in illite isometric crystals but also in lath-shaped I/S crystals (Fig. 1.32).

Spiral growth is not the only process that leads to the increase in the particle size. Coalescence phenomena have been frequently observed (Whitney and Velde 1993; Kitagawa 1995). Additional layers permit adjacent particles to be joined as was shown by Nespolo (2001) for micas (Fig. 1.36).

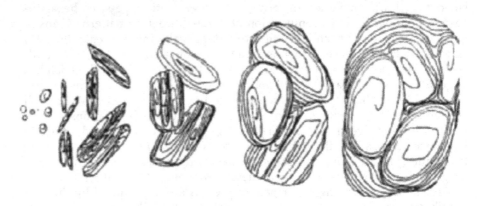

Fig. 1.36. Schematic representation of the growth and coalescence mode of "sericite" crystals ("illite-phengite" originating in geothermal systems in Japan) (Kitagawa 1993)

1.2.5
Growth of Mixed Crystals and Particles

Crystal growth has hitherto been described for chemically and structurally homogeneous clays (composed of a single layer type). Of course, this is an ideal but not exclusive situation. Indeed, under experimental and natural conditions, more complicated clay particles are formed. Some of them are mixed layer minerals, yet others must be considered as polycrystalline particles. This distinction enhances the importance of crystal defects.

1.2.5.1
Growth of Not-Regularly Ordered Mixed-Layer Minerals

Growth of One of the Components
Mixed-layer minerals cannot have their true crystal structure defined by the description of stacking modes based on statistics as presented above. They are mostly identified by their diffraction patterns on oriented samples, in which the crystal structure and interparticle effects are superimposed. One result of the work of Nadeau et al. (1984) was a question of whether mixed-layer minerals are true crystalline entities or just preparation artefacts. Any answer to this question is acceptable only if the crystal structure considered for the mixed-layer minerals is consistent not only with the characteristics of diffraction patterns and the chemical compositions, but also with the growth process of crystals.

From the standpoint of crystal growth, the structure of illite/smectite mixed layers (I/S) is hardly conceivable with two types of compositionally different layers (due to problems of supersaturation). It can be represented more easily by a model in which high-charge interfaces have the composition of a half-illite (Fig. 8.18). This model is consistent with the growth of a single phase during diagenesis: illite. It is also consistent with the chemical compositions of the fluids collected in pores that are supersaturated with respect to that particular phase. This model rests on an important concept: the 2:1 layer polarity (Cuadros and Linares 1995). The tetrahedral charges are not equivalent in both sheets of the layer. Polarity then imposes a growth model for layers that are arranged symmetrically about the interlayer sheet (Fig. 1.37a).

Solid state Transformation
In the series of mixed-layer minerals, notably I/S of diagenetic formations, the process most commonly thought of to explain the decrease in the smectite content is the solid state transformation (SST). Theoretically, neither the habit nor the number of layers of mixed-layer mineral crystals can be changed by such a process. A smectite layer transforms into an illite layer through ion diffusion (Altaner and Ylagan 1997). Even if the size and thickness of mixed-layer mineral crystals obviously change with increasing physico-chemical conditions (temperature, Al-K chemical potentials) in diagenetic or hydrothermal alteration environments, this process cannot be totally ruled out (Drits et al.

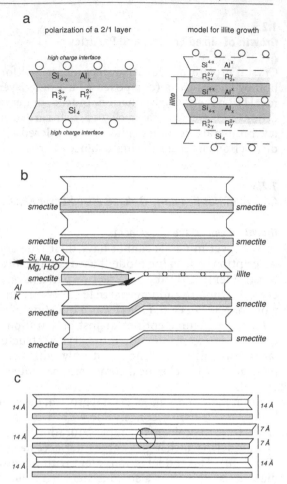

Fig. 1.37a–c. Growth of not-regularly ordered mixed-layer minerals: example of il-lite/smectite mixed-layers. a) Po-larity (Cuadros and Linares 1995) leads to a growth model based on 2:1 layers arranged sym-metrically about the interlayer sheet. b) Solid state transfor-mation (modified according to Altaner and Ylagan 1997). c) Berthierine-chamosite transi-tion by inversion of tetrahedra (Xu and Veblen 1996)

1998). The replacement of a smectite layer by an illite layer leads to stacking defects and reduces the coherent scattering domain size that determine the intensity of X-ray diffraction peaks (Fig. 1.37b).

On the other hand, the solid-state transformation process has been thor-oughly documented in the 14 Å berthierine-chlorite transition (chamosite) by high-resolution electron microscope observations (Xu and Veblen 1996). Every other octahedral sheet of the 7 Å layers becomes the brucite-like sheet of the 14 Å layers by inversion of tetrahedra and diffusion of H^+ protons. This observed process imposes on the one hand that the lengths of ionic- and Coulomb-type bonds permit conservation of the geometry of the octahedral sheets and on the other hand that the chemical composition of the octahedral and brucite-like sheets are consistent (distribution of the charges related to ionic substitutions).

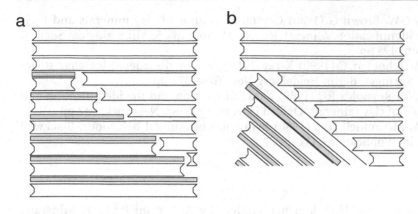

Fig. 1.38a,b. Polymineral particles. **a)** The layers of both species are intercalated or in the continuation of each other. **b)** The two species are separated by an intercrystalline joint

1.2.5.2
Polymineral Particles

Particles result from the coalescence of compositionally different crystals. In the case of phyllosilicates, this phenomenon is frequent and greatly disturbs the crystal size analyses as well as the calculation of unit formulae from chemical compositions. Thus, a number of solid solutions determined by classical analyses or even by electron microprobe are only in fact the result of the mixing of two phases associated in particles. The most common example is observed in deep diagenetic or metamorphic formations: chlorite-illite particles cannot be separated by classical techniques.

The particle layers may be arranged according to three different patterns during the growth process: (1) perfect mixed layering, (2) perfect continuity through a single layer (Fig. 1.38a) and (3) separated by a crystal defect (intercrystalline joint) (Fig. 1.38b). Undetectable by X-ray diffraction, which only identifies the mixture of species (illite + chlorite), these particles are revealed by transmission electron microscopy. All patterns intermediate between intercalation of a few exogene layers (Fig. 1.17), the genuine mixed layering and the segregation of compositionally different domains are possible.

Suggestions for Additional Reading

Crystal Structures of Clays

Bailey SW (1988) Hydrous phyllosilicates (exclusive of micas). Rev Mineral 19:725

Bouchet A, Meunier A, Sardini P (2001) Minéraux argileux – Structures cristallines – Identification par diffraction de rayons X. Avec CD-ROM. Editions TotalFinaElf, 136 pp

Brindley GW, Brown G (1980) Crystal structures of clay minerals and their X-ray identification. Mineral Society Monograph 5, Mineralogical Society, London, 495 pp

Drits VA, Tchoubar C (1990) X-ray diffraction by disordered lamellar structures. Springer, Berlin Heidelberg New York, 371 pp

Moore DM, Reynolds RC (1989) X-ray diffraction and the identification and analysis of clay minerals. Oxford University Press, New York, 332 pp

Putnis A, McConnell JDC (1980) Principles of mineral behaviour. Blackwell Scientific Publications, Oxford, 257 pp

Softwares

ERIX, ENRIX, ARGIRIX Bouchet A, Meunier A, Sardini P (2001) Minéraux argileux. CD-ROM. Editions TotalFinaElf, 136 pp

MLM2C, MLM3C Plançon A, Drits VA (2000) Phase analysis of clays using an expert system and calculation programs for X-ray diffraction by two- and three-component mixed layer minerals. Clays Clay Miner 48:57–62

NEWMOD Reynolds RC Jr (1985) NEWMOD a computer program for the calculation of one-dimensional diffraction patterns of mixed-layered clays. Reynolds RC, 8 Brook Rd, Hanover, NH, USA

Crystal Growth

Baronnet A (1988) Minéralogie. Dunod, 184 pp

Putnis A (2003) Introduction to mineral sciences. Cambridge University Press, 457 pp

Electron Microscopy

Nagy KL, Blum AE (1994) Scanning Probe Microscopy of Clay Minerals. CMS workshop lectures. The Clay Mineralogical Society, Boulder, 239 pp

CHAPTER 2

Crystal Chemistry of Clay Minerals

Following the "guiding line" for the first part of this book (Chaps. 1 to 5), the phyllosilicate framework being in mind, it is now possible to study how the chemical composition of clay minerals is related to their crystal structure. Some species have a typical composition while others are much more variable. Why? How is it, components whose ionic radii are very different are incorporated inside the crystal lattice? Is it possible to distinguish the clay species including mixed layer minerals using their chemical composition? Answers to these questions, which are so important for mineralogists, chemists or geologists implies the use of specific techniques particularly, electron microprobe, analytical electron microscope (AEM), spectroscopies.

Because of their so particular crystal structure, phyllosilicates are able to interstratify different layer types in their stacking sequence (mixed layer minerals). This is why this chapter is divided into two parts devoted to the description of solid solutions and mixed layering respectively. Some practical techniques for useful chemiographic projections are proposed.

2.1
Solid Solutions

2.1.1
Introduction

Prior to studying the chemical composition of solid solutions and the mixed layering of phyllosilicates, it is necessary to properly define the various organisation levels of these minerals. Summarising Chap. 1, five levels can be distinguished:

- *Sheets*: The tetrahedral and octahedral sheets are each formed by three ionic planes. Each cation is framed by O^{2-} or $(OH)^-$ anions. An exception arises in the interlayer sheet that is considered as being formed only by those cations that compensate for the layer charge.

- *Layers*: Layers result from the association of several ionic sheets, according to a limited number of combinations (Fig. 2.1). Palygorskite and sepiolite are fibrous minerals that correspond to another type of organisation, which will not be addressed here.

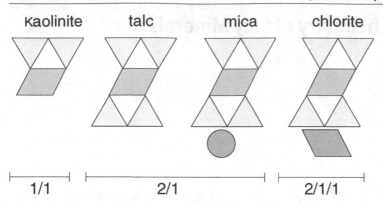

Fig. 2.1. Schematic representation of the four layer types according to the McEwan 1/1, 2/1 and 2/1/1 patterns

- *Crystals*: Crystals consist of a varying number of layers (three or four for smectites up to several tens for chlorites or corrensites). If these layers have the same chemical composition, stacking modes determine polytypes. If they have different compositions, then they are mixed-layer minerals. Clay minerals exhibit small crystals (< 2 μm) whose *hk0* faces are rarely formed. Many of them contain numerous stacking defects.

- *Particles*: Several crystals can coalesce by epitaxy on the (001) faces, thus forming a particle recognisable by its reentrant angles. Particles can be mono-or polyphased.

- *Aggregates*: Crystals or particles are often bonded together randomly by weak bonds, or by iron oxides or organic matter. The shape and size of aggregates vary significantly. Some molecular water can be stored in the mesoporosity.

It is considered here that the study of solid solutions relates to the following organisation levels: sheets, layers and crystals, exclusive of particles (and hence of aggregates). This requires great caution in the use of wet method chemical analyses on sample weighing a few milligrams, and of microprobe analyses on targets measuring a few μm³. Similarly, great care is necessary when relating microprobe analyses to the crystal structure of mixed-layer minerals by X-ray diffraction (XRD), which is performed on a few milligrams of matter, that is to say on a greater number of particles.

One of the main questions raised by the chemical composition of clay minerals is how to distinguish between solid solutions and mixed layering. Any variation up to the layer organisation level can only be described in terms of a solid solution, mixed layering being by definition a stacking sequence of several layers (Fig. 2.2a). Therefore, ambiguity arises at the crystal level. The solid solution is considered here to represent a random distribution of the substitution of one ion for another. Conversely, mixed layering represents the maximum degree of segregation leading to compositionally different layers. Between these two end members, minerals exhibit intermediate situations more difficult to classify such as, for instance, K–Na substitutions in the muscovite-paragonite series.

2.1.2
The concept of Solid Solution Applied to Phyllosilicates

2.1.2.1
Crystallographic Definition of Solid Solutions

Isotypes and Isomorphs
Three criteria permit identification of isotypic mineral species:

1. analogous structural formulae, even if the chemical compositions are different;

2. identical space symmetry groups;

3. crystal structures formed by comparable coordination polyhedra.

Isotypism does not imply that cell parameters are identical between both species. On the other hand, the a/b, b/c and c/a ratios are constant, and α, β and γ angles are identical.

Isomorphous minerals first meet the requirements of isotypism. Besides, they are composed of cations and anions whose ionic radii are sufficiently close to allow both mineral species to form mixed crystals. Mixed crystals then correspond to a solid solution whose crystal symmetry does not change. Only cell parameters vary continuously as a function of the replacement of one ion by another (Vegard 1928). A typical solid solution is that of the olivine series contained between the stoichiometric end members of forsterite (SiO_4Mg_2) and fayalite (SiO_4Fe_2). The replacement of one ion by another, or isomorphous substitution, requires that their valencies be identical and that the difference in their ionic radii be less than 15% (Goldschmidt's rules).

Ionic Radii and Coordination
Pauling's empiric rules, stated in 1929, permit determination of coordination domains from simple geometric relationships based on the ratio of the cation and anion ionic radii (R_c) and (R_a), respectively. Thus, tetrahedra (4-fold coordination) exist if $0.225 \leq R_c/R_a \leq 0.414$ whereas octahedra exist if $0.414 \leq R_c/R_a \leq 0.732$ (Fig. 2.2b). Since the ionic radii of the main ions taking part in the crystal structure of clays are known (Table 1.1), determination of those ions capable of substituting for each other in the three main coordinations involved in phyllosilicates (4, 6 and 12) is easy.

This explains why–under natural conditions of low temperatures ($T \leq 350\,^{\circ}C$)-replacement of Si^{4+} cations in tetrahedra is ensured by Al^{3+} or Ge^{4+}, which have the same ionic radius (0.53 Å), and much more rarely by Fe^{3+}, which is 20% "bigger" ($r_i = 0.63$ Å). The substitution possibilities are more extensive in octahedra whose inner cavity is larger. Divalent Mg^{2+}, Fe^{2+}, Co^{2+},

a

b

Ra: anion ionic radius
Rc: cation ionic radius

12-fold coordination

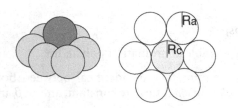

$$\frac{Rc}{Ra} = 1$$

6-fold coordination

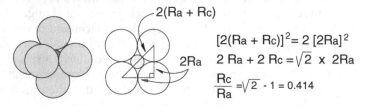

$$[2(Ra + Rc)]^2 = 2\,[2Ra]^2$$

$$2\,Ra + 2\,Rc = \sqrt{2} \times 2Ra$$

$$\frac{Rc}{Ra} = \sqrt{2} - 1 = 0.414$$

4-fold coordination

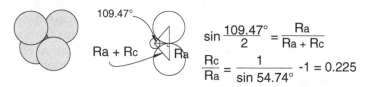

$$\sin \frac{109.47°}{2} = \frac{Ra}{Ra + Rc}$$

$$\frac{Rc}{Ra} = \frac{1}{\sin 54.74°} - 1 = 0.225$$

Fig. 2.2a,b. Solid solutions. **a)** Schematic representation of (1) a random solid solution, (2) a random solution with segregation of compositionally different domains in the layer and (3) mixed layering. **b)** Relationship between coordination and ionic substitution according to Pauling's Rules

Ni^{2+}, Zn^{2+}, Cu^{2+} readily substitute for each other (homovalent substitutions) because their ionic radii are contained between 0.83 Å (Ni^{2+}) and 0.92 Å (Fe^{2+}). This is also valid for trivalent Al^{3+} and Fe^{3+}. Trivalent R^{3+} can be replaced by divalent R^{2+} (heterovalent substitutions).

The Three Types of Solid Solution

Substitution Solid Solutions These are formed by the isomorphous replacement of ions or atoms in equivalent sites of the crystal structure. In the case of ions, valency must be identical. This is the case of Fe^{2+} replacing Mg^{2+} in the octahedral sheet of phyllosilicates (the difference in their ionic radii is only 7.5%). Geometrical distortions are slight and the local charge balance is maintained.

Addition Solid Solutions In the case of heterovalent substitutions, the electrical neutrality may impose the presence of additional ions that are located in particular sites. This is the case of the ions in the interlayer sheet of phyllosilicates.

Omission Solid Solutions As heterovalent substitutions change the charge balance, the latter is restored by the incomplete occupation of a crystallographic site (vacancies). This is the case of trioctahedral chlorites in which the substitution rate of R^{3+} for R^{2+} in the octahedral sheet imposes vacancies (symbolised by $\square$).

Ionic Planes in the Crystal Structure of Clays The crystal structures of phyllosilicates belong to three different patterns (Fig. 2.1): 1:1 (kaolinite, serpentine), 2:1 (pyrophyllite, talc, muscovite, biotite), and 2:1:1 (chlorite). The number of anionic (P_a) and cationic (P_c) planes depends on the type of structure (Fig. 2.3):

- 1:1 $\Rightarrow 3P_a + 2P_c$;

- 2:1 (pyrophyllite, talc)$\Rightarrow 4P_a + 3P_c$;

- 2:1 (muscovite, smectite, biotite)$\Rightarrow 4P_a + 4P_c$;

- 2:1:1 (chlorite)$\Rightarrow 6P_a + 5P_c$.

Each of these planes can display cation substitutions. Anion substitutions are relatively rare at low temperature. They generally occur by replacement of hydroxyl radicals $(OH)^-$ by F^- or Cl^-. Consequently, these substitutions are ignored here so that emphasis is put on cation substitutions, which are much more important in clays.

Octahedral Substitutions

In octahedral sheets, the three possible types of substitutions are:

- homovalent substitutions: $Al^{3+} \Leftrightarrow Fe^{3+}$ or $Mg^{2+} \Leftrightarrow Fe^{2+}$,

- heterovalent substitutions: $Fe^{3+} \Leftrightarrow Fe^{2+}$,

- dioctahedral-to-trioctahedral transition: $\square \Leftrightarrow Fe^{2+}$.

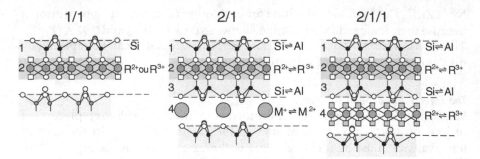

Fig. 2.3. The 1:1 minerals have two cationic planes (one with tetrahedral coordination, one with octahedral coordination). 2:1 minerals have three cationic planes (two with tetrahedral coordination, one with octahedral coordination). An additional cationic plane exists when the 2:1 structure is not electrically neutral (interlayer sheet). The 2:1:1 minerals have four cationic planes (two with tetrahedral coordination, two with octahedral coordination)

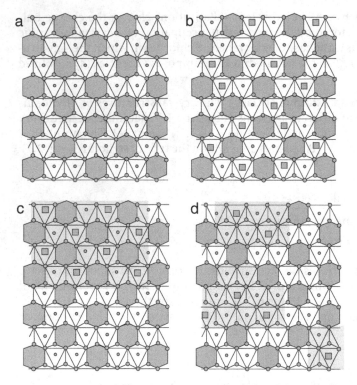

Fig. 2.4a–d. The octahedral sheet of a dioctahedral phyllosilicate (here, vacancies are in the *trans* position) can exhibit several types of organisation: **a)** no substitution; **b)** randomly distributed substitutions; **c)** substitutions segregated into zones (composition clusters); **d)** additional cations occupying the *trans* vacancies and forming trioctahedral zones (structural clusters)

The main distribution modes of cation substitutions in octahedral sheets are summarised in Fig. 2.4. The easiest way to study these distributions is the use of infrared spectroscopy that permits short-distance order-disorder relationships to be shown (see Sect. 2.1.2.2).

The octahedral sheet of phyllosilicates can exhibit several types of organisation:

a) no heterovalent substitutions. This is the case of 1:1 minerals (kaolinite, serpentine) and 2:1 minerals without interlayer sheet (pyrophyllite, talc). The rate of homovalent substitutions ($Al^{3+} \Leftrightarrow Fe^{3+}$ or $Mg^{2+} \Leftrightarrow Fe^{2+}$) remains relatively low;

b) randomly distributed substitutions. This is the structure of the true solid solutions typical of micas (phengites) or smectites;

c) substitutions distributed into zones (composition clusters);

d) additional cations occupying vacancies: the rate of octahedral occupancy continuously varies between 2 and 3 (structural clusters).

Now "maps" of the composition of the octahedral sheet can be drawn by combining data from different spectroscopy techniques and distribution calculations based on Monte Carlo simulations (Cuadros et al. 1999). The order-disorder concept is fundamental for characterising the crystal structure (Plançon 2001). It facilitates the modelling of the statistical distribution of the octahedral and a better understanding of how charges are divided within the layers.

Tetrahedral Substitutions

In low-temperature natural minerals, the ionic substitution most frequently observed in the tetrahedral sheet is Si^{4+} replaced by Al^{3+}. The difference of valency between both ions produces a negative charge (positive charge deficiency) and changes the symmetry of the tetrahedral sheet (Veder 1964;

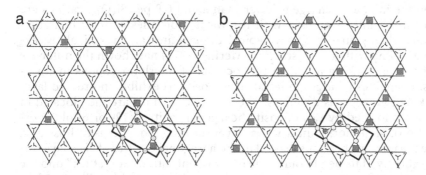

Fig. 2.5a,b. The tetrahedral sheet of phyllosilicates can exhibit substitutions **a)** distributed randomly; and **b)** in an ordered fashion. Two Al tetrahedra cannot be direct neighbours (Lowenstein's rule)

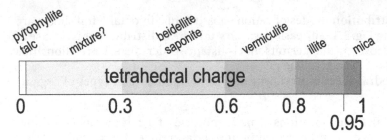

Fig. 2.6. Variation range of the tetrahedral charge expressed per Si_4O_{10}. The origin of the charge is the substitution of Al^{3+} for Si^{4+}

Hazen and Wones 1972). According to Lowenstein's rule, two Al tetrahedra cannot be direct neighbours (Herrero et al. 1985, 1987).

The distribution of substitutions is studied by NMR spectroscopy (see Sect. 2.1.2.2). The distribution can be random or ordered (Fig. 2.5). The substitution rate calculated on the Si_4O_{10} basis can theoretically vary from zero (pyrophyllite, talc) to 1 (micas). It can reach 2 in some phyllosilicates (margarite). The fixation mechanisms of interlayer cations are controlled by the location and intensity of these negative charges. The most common clay minerals are contained in the variation range 0 to 1 (Fig. 2.6): smectites, vermiculites and illites.

Substitutions in the Interlayer Sheet
Cation substitutions in the interlayer sheet are much more varied than those in tetrahedral and octahedral sheets for three reasons:

1. the 12-fold coordination of interlayer sites, even if reduced to a 6-fold co-ordination because of the loss of hexagonal symmetry to the benefit of a ditrigonal one, produces large-sized sites able to fix cations such as Sr^{2+}, Ba^{2+}, K^+, Rb^+ or Cs^+, whose ionic radius is between 1.75 and 2.02 Å (Fig. 1.5).

2. when the interlayer charge is lower than about 0.8 per Si_4O_{10}, cations are fixed in the interlayer zone with the water molecules of their hydration sphere. They gain a property typical of smectites and vermiculites: exchangeability with ions in solution. Méring (1975) has shown the influence of the layer-stacking mode on the location of interlayer cations in smectites. Beidellites (charges in tetrahedral position) exhibit an ordered layer stacking sequence, as do micas. The interlayer cations are then located in the space formed by two opposite hexagonal cavities (12- or 6-fold coordination). Montmorillonites, whose layer stacking sequence is turbostratic, have the interlayer cations located in the hexagonal cavities of each layer independently of its neighbours. The monovalent cations are "stuck" into the hexagonal cavity even when montmorillonite is hydrated (see Fig. 1.12d).

3. The water molecules fixed about the bivalent interlayer cations, whatever the type of smectite considered, are arranged in a configuration close to

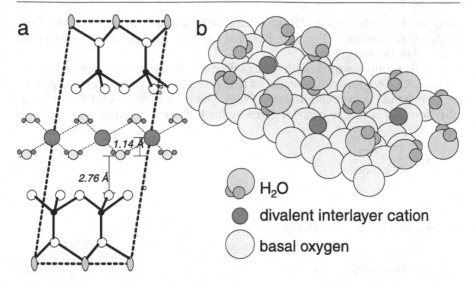

Fig. 2.7a,b. Schematic representation of water molecule organization in the interlayer space. **a)** Two water layers in a vermiculite (from Walker 1961). **b)** The "ice-like" arrangement of water molecules surrounding the bivalent interlayer cations in the basal oxygen plane of 2:1 layers

that of ice (Sposito 1989; Mercury et al. 2001). They form a non-planar hexagonal lattice whose vertices are alternately connected to the cavities of the tetrahedral sheets of the opposite layers (Fig. 2.7).

Let's consider here that only those cations which are irreversibly fixed, i.e., non-exchangeable, take part in the definition of solid solutions. Practically, that amounts to limiting solid solutions to illites and micas and exclude smectites and vermiculites. The technique most suited to the study of interlayer sheets is the far-infrared absorption spectrometry.

2.1.2.2
Spectrometries: a Tool for Exploring Solid Solutions

Absorption Spectrometry in the Near Infrared (10,000–4,000 cm⁻¹) and the Middle Infrared (4,000–400 cm⁻¹)

For any temperature higher than the absolute zero, ions vibrate within the crystal structures. These vibrations are characterised each by an energy or its wavelength equivalent; they cause the chemical bonds to be stretched and bent. The wavelength ranges are those of the near infrared. Consequently, any resonance between the vibration of ions and the infrared is expressed by the absorption of the energy in a given wavelength range. Spectrometry defines absorbed wavelengths and permits identification of the ionic bonds involved. That amounts to exploring the clay crystal structure through functional groups considered as independent. The structural OH are mostly used: Al-OH-Fe^{3+} or Al-OH-Al for instance (Fig. 2.8). Detailed explanations for clay minerals can be found in Wilson (1994).

The OH group can be considered as a dipolar harmonic oscillator with force constant K. An approximate value of K is given by the valency exchanged in the O–H bond (Robert and Kodama 1988). The wave number of the stretching vibration ν_{OH} of this hydroxyl is given by the following formula:

$$\nu_{OH} = \frac{1}{2\pi}\sqrt{\frac{K_{OH}}{\mu_{OH}}} \tag{2.1}$$

where

μ_{OH} reduced mass of the OH group

$$\mu_{OH} = M_O \cdot M_H / (M_O + M_H) \tag{2.2}$$

In the crystal structure of talc [$Si_4O_{10}Mg_3(OH)_2$], each hydroxyl is bonded to three Mg^{2+} cations and points to the centre of the hexagonal cavities that are determined by the basal oxygens of the 6SiO_4 rings (Fig. 2.8). In such a structure,

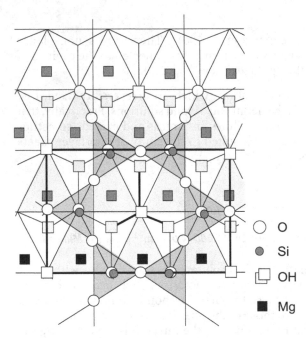

O   O

Si   Si

OH   OH

Mg   Mg

Fig. 2.8. Hydroxyls in a talc-like structure are bonded to three bivalent octahedral cations (Mg^{2+}). The proton is thus equally repelled and is positioned at the centre of the hexagonal cavities of the tetrahedral sheets

the two valencies of a Mg^{2+} cation are shared between the 6 oxygens of the octahedron (4 apical oxygens and 2 OH), or $2/6^+$ for each bond. In this manner, one oxygen is bonded to three Mg^{2+} ions and receives $3 \times 2/6^+$, or one valency unit (v.u). The other valency of this oxygen is used in bonds with the hydrogen (hydroxyl group) or with the Si^{4+} ion in the tetrahedra.

The crystal structures of pyrophyllite and talc (2:1 minerals) do not have either heterogeneities in the tetrahedral sheet (no substitutions of Al^{3+} for Si^{4+}) nor interlayer ions to exert a repulsion on the H^+ proton from hydroxyls. Under such conditions, the interactions undergone by the proton are simply those with the oxygen from the hydroxyl. The corresponding wave number is $v_{OH} = 3,677\ cm^{-1}$. The presence of cations in the interlayer sheet with the proton from the hydroxyl causes an increase in the force constant of the O–H bond. The higher the valency of an interlayer ion, the higher the wave number. For example, phlogopite has two characteristic absorption bands:

$$(Si_3Al)\,O_{10}Mg_3\,(OH)_2K \rightarrow v_{OH} = 3,724\ cm^{-1} \tag{2.3}$$

$$(Si_3Al)\,O_{10}Mg_3\,(OH)_2Ba_{0.5} \rightarrow v_{OH} = 3,745\ cm^{-1} \tag{2.4}$$

The absorption bands corresponding to all types of bonds existing in the crystal structure have thus been gradually determined (Besson et al. 1987). The most frequently occurring are indicated in Table 2.1. This method is used to define the distribution modes of the Al, Fe and Mg octahedral cations in smectites (Madejova et al. 1994; Vantelon et al. 2001) or in illite-smectite mixed layers (Cuadros and Altaner 1998).

Table 2.1. Absorption band values for the main chemical bonds in the crystal structure of phyllosilicates

Chemical bond	Benching vibrations		Stretching vibrations	
	Wavenumber v (cm^{-1})	References	Wavenumber v (cm^{-1})	References
MgAlOH	3,687	Farmer (1974); Besson et al. (1987)		
Al2OH	3,620–3,640	Farmer (1974); Besson et al. (1987)	930	Farmer (1974); Russel et al. (1970)
	3,634–3,636	Grauby et al. (1975); Besson et al. (1987)		
$Fe^{3+}2OH$	3,565	Goodman et al. (1976); Besson et al. (1987)	816	Farmer (1974); Russel et al. (1970)
	3,555–3,558	Grauby et al. (1975); Besson et al. (1987)		
AlOHFe^{3+}			873	Farmer (1974); Russel et al. (1970)

Absorption Spectroscopy in the Far-Infrared (400–10 cm⁻¹) and Raman Spectroscopy

The cations that compensate for the negative charge of the 2:1 layer form the interlayer sheet. The coordination for each one of the cations is theoretically equal to 12: six A–O bonds (A: cation; O: basal oxygen) with the upper layer, and 6 with the lower layer. In true crystal structures, the coordination is mostly reduced to six because of the distortion of the hexagonal cavities, which lose their hexagonal symmetry (Fig. 2.7). All cations vibrate within their coordination polyhedron according to two modes: vibration in the (001) plane and perpendicular to the (001) plane.

The absorption frequencies due to A–O vibrations vary with the charge (Z) and mass (M) of the compensating cation (as a function of $\sqrt{\frac{Z}{M}}$). The mass considered is that of the cation with the water molecules of its hydration sphere (e. g.: Na^+, $2H_2O$). The laws of variation are extracted from Laperche's work (1991) and shown in Fig. 2.9.

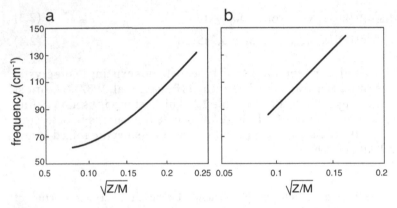

Fig. 2.9a,b. Variation in the wave number of the A-O vibration in vermiculites as a function of $\sqrt{\frac{Z}{M}}$, for Na^+, NH^{4+}, K^+, Rb^+, Cs^+, Ca^{2+}, Sr^{2+} and Ba^{2+} (after Laperche 1991). a) A–O vibration in the (001) plane. b) A–O vibration perpendicular to the (001) plane

The wave number of the stretching vibrations of A–O bonds vary for a given interlayer cation as a function of the composition of the 2:1 layer. This can be particularly observed in K-micas in which the wavenumber of the K–O bond varies as a function of the length of this bond, namely as a function of the rotation angle of tetrahedra (Fig. 1.3). If the angle is zero, all K–O bonds are short (3.15 Å). If this angle is increased, the difference between short and long bonds becomes more pronounced, with the short one possibly decreasing down to 2.85 Å for an angle of 12°.

One of the major difficulties involved in the interpretation of infrared spectra is the superimposition of a very wide band due to water absorption upon the absorption bands of hydroxyls. To improve analyses, dehydration of the sample – with the risk of disturbing its structure – or replacement of hydrogen by deuterium is then necessary. One of the interests of Raman spectroscopy (inelastic light-scattering spectrometry) is the observation of the vibrations of hydroxyls without particular preparation, even in hydrated samples. The work carried out on the kaolin-group minerals is a good example of the contribution of this spectroscopy technique to the knowledge of their crystal structures (Frost 1995; Frost and van der Gaast 1997). This technique also permits characterisation of the organisation of the molecules absorbed between the layers of these minerals, such as urea (Frost et al. 1997). The infrared absorption spectroscopy coupled with the Raman diffusion spectroscopy is a very fruitful technique that should be more widely exploited in the future.

High-Resolution NMR Spectroscopy

High-resolution nuclear magnetic resonance (NMR) provides information on the distribution of the Si^{4+} and Al^{3+} cations in the tetrahedral sheets of 2:1 phyllosilicates, insofar as their iron content does not exceed a few percent. Samples are placed in a very high magnetic field on the order of about 10 Teslas. The energy levels associated with the different orientations of the nuclear magnetic moments are separated. Samples are then irradiated by a radio frequency field of several tens of MHz (frequency varies with the nucleus involved and with the intensity of the external magnetic field applied). Irradiation causes transitions between these different energy levels, resulting in a resonant absorption whose spectral characteristics (frequency, intensity) yield information about the environment of the studied ion. Thus, the coordination, nature and number of the adjacent cations, as well as the order-disorder relations and the deformation of polyhedra, can be determined. The nuclei useful for studying the tetrahedral sheets of phyllosilicates are ^{29}Si and ^{27}Al. The position of each peak (chemical displacement) is given in ppm in relation to the spectra of standards: tetramethylsilane (TMS) for ^{29}Si and a solution of aluminium chloride for ^{27}Al.

The NMR signal for tetrasilicic phyllosilicates (talc, pyrophyllite) is unique at -97 ppm. This means that each tetrahedron is actually bonded to three other identical tetrahedra. Substitution of Al for Si yields additional bands whose positions depend on the number of Al adjacent to Si (Fig. 2.10). The intensity of these bands depends on the proportions of each cation configuration. Each additional Al adjacent to a Si (Si[3Si] Si[2Si, 1Al], Si[1Si, 2Al]) leads to a variation of about $+4$ ppm in the position of the corresponding band (Sanz and Robert 1992).

Aluminium can be found in two coordinations: tetrahedral ($+60$ to $+70$ ppm) and octahedral ($+7$ to $+8$ ppm). In tetrahedral sheets, aluminium is scattered so that it is bonded to three SiO_4 tetrahedra (Lowenstein's by-pass rule). Therefore, most of the time, the tetracoordinated aluminium yields one NMR band only. Nevertheless, the homogeneous charge scattering may be intermediate between the maximum scattering and that predicted by Lowenstein's rule (Herrero et al. 1989).

EXAFS (Extended X-ray Absorption Fine Structure) Spectroscopy

EXAFS spectroscopy is based on the absorption of X-rays by atoms present in the crystal structure. The absorption coefficient shows oscillations about

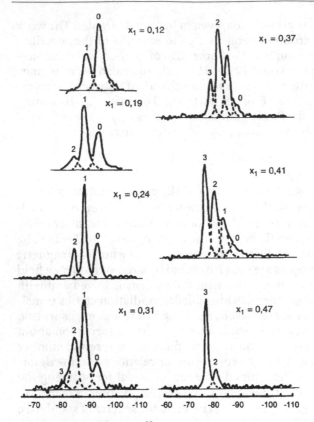

Fig. 2.10. NMR spectra of ^{29}Si (79.4 MHz) for 2:1 phyllosilicates (micas and saponites) in which the substitution rate x of Al for Si varies from 0.12 to 0.47. The number of Al bonded to Si is indicated by 0, 1, 2 or 3 (after Sanz and Robert 1992)

an ionisation threshold when the excitation is produced by photons whose energy ($h\nu$) is greater than the energy of an electron of a given level (E_0). The electron is then ejected with a kinetic energy $E_c = h\nu - E_0$. The pattern of the oscillations about the threshold depends on the electron structure and on the symmetry of the site of the atom that has absorbed the incident X rays. Therefore EXAFS spectroscopy provides information on the short-distance structure of the crystal considered. This is an efficient technique for studying the environment of the transition elements (Cr, Fe, Co and Ni) present in the octahedral sheet of phyllosilicates (Manceau and Calas 1985–1986; Manceau et al. 1985).

Mössbauer Spectroscopy

Mössbauer spectroscopy is based on the resonance absorption of a γ-emission stemming from a radioactive source and made polychromatic by the Doppler effect. As regards clay minerals, this technique is mostly used to determine two

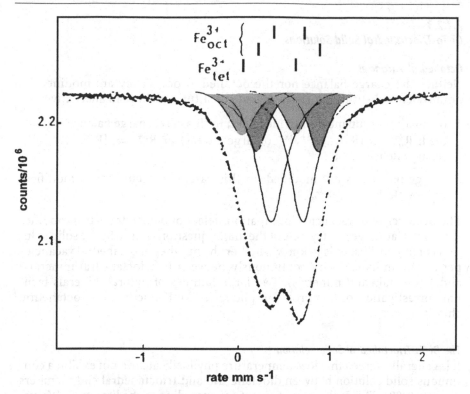

Fig. 2.11. Mössbauer spectrum of a nontronite performed at 77 °K showing the single band for tetrahedral Fe^{3+} and two bands for octahedral Fe^{3+}. The doublet whose quadrupole splitting is the slightest is attributed to $^{VI}Fe^{3+}$ in the M_2 site (*cis*), the greatest being attributed to $^{VI}Fe^{3+}$ in the M_1 site (*trans*)

parameters: (1) the relative Fe^{2+}–Fe^{3+} contents in the crystal structure, and (2) the occupation site and the environment of Fe^{3+} or Fe^{2+} cations. The stable Mössbauer nucleus is ^{57}Fe (natural abundance: 2.245%) and the radioactive source is ^{57}Co, whose half-life is 270 days.

In phyllosilicates, ferrous iron occupies octahedral sites; only ferric iron can be distributed between tetrahedra and octahedra. In the absence of bands–or even distinct shoulders–the presence of tetrahedral ferric iron is inferred only from a better fit of the experimental spectrum by a sum of quadrupole doublets, of which one corresponds to tetrahedral Fe^{3+}. In the octahedral sheet, Fe^{3+} in M_2 sites is identified by the doublet with the slightest quadrupole splitting, the greatest being attributed to Fe^{3+} in the M_1 site (definition of M_1 and M_2 sites in Fig. 1.7). As regards Fe^{2+}, the internal doublet (slight quadrupole splitting) is attributed to the M_1 site, the external doublet to the M_2 site (Fig. 2.11). Note that the octahedral sheet and the brucite-like sheet cannot be distinguished in chlorites.

2.1.2.3
Di-To-Trioctahedral Solid Solutions

Octahedral Vacancies
Neither the charge balance nor the octahedral occupancy are modified by homovalent substitutions. Heterovalent substitutions show an alternative:

1. octahedral occupancy is not modified; in this case, charge balance is modified: $R_n^{2+} \Rightarrow [R^{2+}_{n-x}R^{3+}_x]^{x+}$ (charge excess) or $R_n^{3+} \Rightarrow [R^{3+}_{n-x}R^{2+}_x]^{x-}$ (charge deficiency);

2. charge balance is not modified; in this case, site occupancy is modified: $R_n^{2+} \Rightarrow [R^{2+}_{n-x-0.5x}R^{3+}_x\square_{0.5x}]$.

The occurrence of vacancies (occupation defect of octahedral sites) modifies the crystal lattice geometry. One of the major questions raised by the solid solutions of phyllosilicates is to know whether the number of octahedral vacancies per Si_4O_{10} units can vary continuously between 0 (trioctahedral minerals) and 1 (dioctahedral minerals). Two large families of natural minerals facilitate investigation of this problem: phengite-like K-micas and trioctahedral chlorites.

The Di-To-Trioctahedral Solid Solution
It is generally agreed that low-temperature phyllosilicates do not exhibit a continuous solid solution between dioctahedral and trioctahedral end members (Güven 1988). This is the case of smectites according to the literature. Weaver and Pollard (1970) showed that trivalent cation amounts cannot be less than 65% of octahedral sites in dioctahedral smectites. This represents 1.3 R^{3+} cations on a Si_4O_{10} calculation basis. The remaining 0.7 vacant sites are occupied by R^{2+} cations. Nevertheless, octahedral occupancy can reach a level of 2.3. Foster (1960) has shown that natural trioctahedral smectites exhibit a minimum substitution rate of R^{3+} for R^{2+} of 33%, or a minimum number of 1.83R^{2+}.

Considering the small size of these minerals and the difficulty of purifying them, the chemical analyses may correspond to mixtures between both species. Therefore, the limits indicated in Fig. 2.12 are only estimates. Moreover, the differences of ionic radius between cations bring about deformations of coordination polyhedra. The accumulated elastic energy can break some bonds when substitutions become too numerous, thus causing the formation of crystal defects (Radoslovich and Norrish 1962).

Wiewiora's Projection System (1990)
A rigorous representation of the chemical composition of 2:1 minerals exhibiting substitutions of $2R^{3+} + \square \Leftrightarrow 3R^{2+}$ ($\square$: vacancy) should be a four-end member system (tetrahedra): $Si-2R^{3+}-3R^{2+}-\square$ (Fig. 2.13). This system, which is difficult to handle, has been elegantly resolved by Wiewiora (1990a). This author proposes a vectorial projection in an orthonormal plot in which the

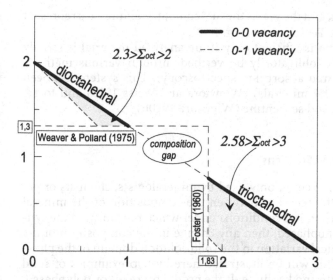

Fig. 2.12. Schematic diagram of octahedral occupancy calculated on the Si_4O_{10} basis of natural dioctahedral and trioctahedral smectites, based on data by Weaver and Pollard (1975) and Foster (1960). The *shaded area* represents the variation range of the number of vacant octahedral sites between 0 and 1. The composition gap is difficult to establish because of the possible mixtures between both mineral species

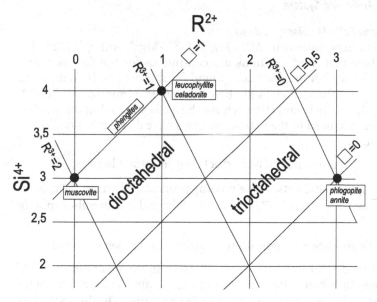

Fig. 2.13. Chemical projection system enabling substitutions of $2R^{3+}+\square \Leftrightarrow 3R^{2+}$ ($\square$: vacancy) to be represented. This system applies for unit ("structural") formulae calculated on the Si_4O_{10} basis for 2:1 minerals (Wiewiora 1990)

x-axis is the R^{2+} content, and the y-axis the Si^{4+} content. Oblique axes represent the vacancy rate and the Fe^{3+} content (Fig. 2.13).

This projection system is efficient only if the analysed material is strictly monophased. This must obligatorily be verified through various methods (X-ray diffraction, infrared absorption spectroscopy). This system has been adapted to chlorites (2:1:1 minerals) (Wiewiora and Weiss 1990) and to 1:1 minerals such as kaolin and serpentine (Wiewiora 1990b)

2.1.3
Experimental Study of Solid Solutions

Synthesis experiments are commonly used by mineralogists, chemists or geologists to investigate the relations between the composition of the mineral phases and the physicochemical conditions under which they form. If the synthesis products are monophased, then any change in the composition of the systems leads to a chemical variation in the solid solution domain of the phase under consideration. This will be illustrated here by two examples of solid solution studies: smectite and kaolinite. If the products are bi- or polyphased, any change in the composition of the system leads to partitioning of elements between the phases. This will be examined in Sect. 3.1.4 for kaolinite-Fe oxide equilibria.

2.1.3.1
Smectites in the Al–Fe–Mg System

Al–Fe^{3+}, Al–Mg, and Fe^{3+}–Mg binary systems
Mineral synthesis experiments in $Al^{3+}-Mg^{2+}$, $Fe^{3+}-Mg^{2+}$ and $Al^{3+}-Fe^{3+}$ binary systems show that solid solutions are continuous for the first two, and very extended with an intermediate compound for the last one (Grauby et al. 1993, 1994, 1995). Synthetised clays are beidellites (Al^{3+}), nontronites (Fe^{3+}), and saponites (Mg)-namely smectites whose charge is situated in tetrahedral sheets. Smectites produced in the various systems have distinct X-ray diffraction characteristics:

- Al–Mg: mixture of dioctahedral and trioctahedral crystal lattices;

- Fe^{3+}–Mg: a single crystal lattice type passing progressively from the dioctahedral end member (purely Fe^{3+}) to the trioctahedral end member (purely Mg);

- Al–Fe^{3+}: in the absence of magnesium, substitutions are very limited.

Apart from the solid solutions with poor extension in the neighbourhood of the pure end members beidellite and nontronite, an intermediate compound forms: $Al_{1.2}-Fe^{3+}_{0.8}$. The two emerging composition gaps might be due to the fact that the difference in ionic radius between Al^{3+} (0.53 Å) and Fe^{3+} (0.64 Å) is greater than 15%. All smectites are *trans* vacant, and Al and Fe are distributed randomly in the octahedral sheet.

Infrared spectra of both series Al–Mg and Fe^{3+}–Mg show $3Mg^{2+}OH$ vibrational bands corresponding to trioctahedral clusters, and $2R^{3+}OH$ vibrational bands corresponding to dioctahedral clusters. Also, ν_{OH} vibrational bands involving two neighbouring octahedra can be detected: one is occupied by R^{3+}, the other by Mg^{2+}. The existence of dioctahedral and trioctahedral domains of segregation whose size is too small to yield a coherent diffraction is inferred from all the X-ray diffraction and spectroscopic data.

Finally, experimental syntheses show that heterovalent solid solutions can be continuous between the dioctahedral and trioctahedral pure end members. The intermediate compositions between the end members are ensured by the mixture in variable proportions in the same layer of more or less extended di- or trioctahedral domains. Nevertheless, stability over time of these solid solutions remains poorly known. It is probably relatively transient, judging by the compositions of minerals formed in nature.

Al–Fe–Mg Ternary System

As soon as the environment contains magnesium, smectites will form, whatever their composition in the Al-Fe^{3+}-Mg system. The solid solution is complete. Only the beidellite-nontronite tie line in the Al-Fe^{3+} binary system exhibits incomplete solid solutions separated by composition gaps (Fig. 2.14). The dioctahedral ($R^{3+} = Al^{3+} + Fe^{3+}$)-to-trioctahedral ($Mg^{2+}$) transition takes place continuously by variation in the respective proportions of domains of each type. The size of these domains is always too small to produce a coherent X-ray diffraction. The crystallinity of smectites is improved when R^{3+} ions occur in small amount (Grauby et al. 1994).

The dioctahedral-to-trioctahedral transition imposes heterovalent substitutions of type $2R^{3+} + \square \Leftrightarrow 3R^{2+}$ ($\square$: vacancy). For local charge balance to be maintained, this substitution must take place in three neighbouring octahedral sites (Fig. 2.15). That is the reason why, even for low Mg contents, the structure is locally trioctahedral. One understands that this type of substitution allows for complete heterovalent solid solutions in clays with tetrahedral charge. Clays with octahedral charge allow for incomplete solid solutions only, because substitutions of R^{2+} for R^{3+} are those determining the value of their charge. Such substitutions in montmorillonites are known to vary between low-charge and high-charge values, namely between 0.3 and 0.6 R^{2+} per Si_4O_{10}.

2.1.3.2
Iron in Kaolinites

Kaolinite has long been considered not to display isomorphic substitutions. Nevertheless, the presence of Fe^{3+} ions has been systematically detected in those kaolinites resulting from weathering processes, and notably from lateritic soils (Muller et al. 1995; Balan et al. 1999). Thanks to the increasing use of spectroscopy techniques (infrared, EPR, Mössbauer in particular), numerous studies have shown that kaolinites formed in weathering conditions sometimes

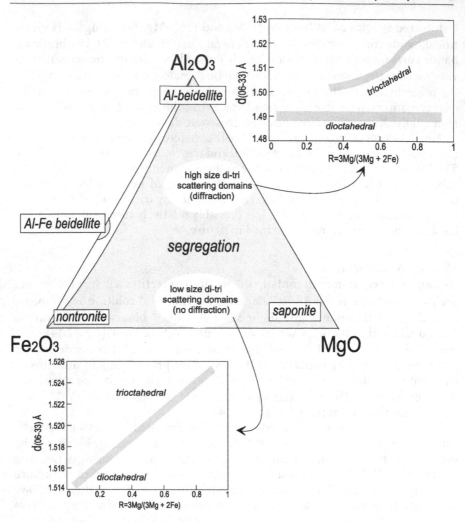

Fig. 2.14. Phase diagram schematically representing the results of experimental syntheses of smectites in the Al_2O_3–Fe_2O_3–MgO system (after Grauby et al. 1993, 1994; 1995 and Léger 1997)

contain a small amount of octahedral Fe^{3+}. Due to the difference in ionic radius between Al^{3+} and Fe^{3+} (0.61 and 0.73 Å, respectively, Table 2.2), this type of substitution brings about an increase in the b cell parameter (0.0066 Å per Fe^{3+}) and causes the occurrence of crystal defects (Brindley 1986; Petit and Decarreau 1990). The lattice coherence along the b axis decreases notably (Table 2.2). The growth rate and the size of such crystals decrease with the iron content. The Fe_2O_3 content of synthetic kaolinites may exceed 7% (Iriarte Lecumberri 2003).

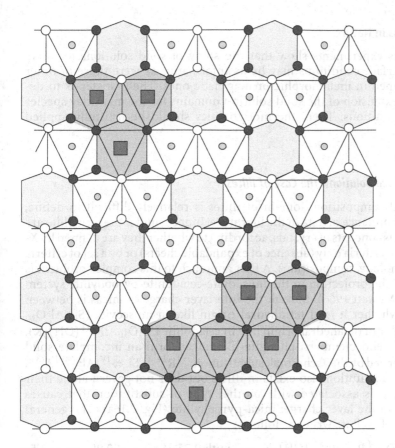

Fig. 2.15. The dioctahedral-to-trioctahedral transition imposes heterovalent substitutions of type $2R^{3+} + \square \Leftrightarrow 3R^{2+}$ ($\square$: vacancy). Therefore trioctahedral domains form locally

Table 2.2. Variation in the coherence of the crystal lattice of kaolinites as a function of the octahedral substitution rate of $Fe^{3+} \Leftrightarrow Al^{3+}$ along the b and c directions (Petit and Decarreau 1990)

Fe atoms (gel)	b axis (Å)	c axis (Å)
0.02	312	290
0.03	310	280
0.04	247	230
0.05	160	191
0.1	256	250
0.2	213	258
0.3	171	246
0.6	129	220

2.1.4
Solid Solutions in Nature

The synthesis experiments show that the study of solid solutions for clays formed in surface conditions may be as informative as were those of other minerals formed in metamorphic or magmatic ones. The first step is to determine the extension of the solid solution domains for the main clay species in natural conditions. Then, thermodynamics should be fruitfully applied (Chap. 3).

2.1.4.1
Extension of Solid Solutions: the Case of Illites

The chemical composition domain of illites is relatively difficult to define, despite the abundance of data in the literature. Indeed, under the term illite are hidden various concepts of crystals, according to whether they are defined by X-ray diffraction criteria only (absence of expandable sheets) or by a set of criteria (XRD, morphology, composition). A first approach to this complex problem is facilitated by the projection in the muscovite-celadonite-pyrophyllite system (Hower and Mowatt 1966). Indeed, the interlayer charge of micas is between 0.95 and 1, whether it is of tetrahedral origin like in muscovite ($Si_3 Al O_{10} Al_2 (OH)_2 K$) or of octahedral origin like in celadonite ($Si_4 O_{10} Al Fe^{2+}(OH)_2 K$). Between both end members, phengite-like micas form an incomplete solid solution governed by the Tschermak substitution: $^{VI}R^{2+}.^{IV}Si \Leftrightarrow ^{IV}Al^{3+}.^{VI}Al^{3+}$. This type of substitution also exists in illites but does not govern alone their composition. It is associated with another type of substitution that causes the decrease in the layer charge: mica-pyrophyllite (Fig. 2.16a). The general formula unit of an illite writes as follows:

$$[Si_{4-x}Al_x] O_{10} \left(R^{3+}_{2-y}R^{2+}_{y} \right) (OH)_2 K_{x+y} \quad \text{with } 0.75 \leq x+y \leq 0.95 \quad (2.5)$$

This formula can be distinguished from that of glauconites whose charge is predominantly octahedral. There is apparently no continuous solid solution between illite and glauconite. When the charge is low (≤ 0.75), the solid solution becomes seemingly continuous but the crystallochemical properties of these minerals are those of illite/smectite mixed layers.

Illites form in different natural environments: weathering environments (Meunier and Velde 1976), geothermal fields (Eberl et al. 1987), and diagenetic environment (Velde 1985). Represented in the $M^+-4Si-R^{2+}$ system, the range of solid solution of illites is divided into two parts according to their polytype $1M$ or $2M_1$ (Fig. 2.16b). The boundary between both parts corresponds to a layer charge close to 0.9 per Si_4O_{10} (Meunier and Velde 1989). For a given composition on the boundary line, the two polytypes coexist. One composition is noteworthy: that of illite, which grows under diagenetic conditions, whose charge is 0.9 and whose composition contains the maximum of substitutions of R^{2+} for R^{3+}:

$$[Si_{3.3}Al_{0.7}] O_{10} \left(Al_{1.8}Fe^{3+}_{0.05}Mg_{0.15} \right) (OH)_2 K_{0.9} \quad (2.6)$$

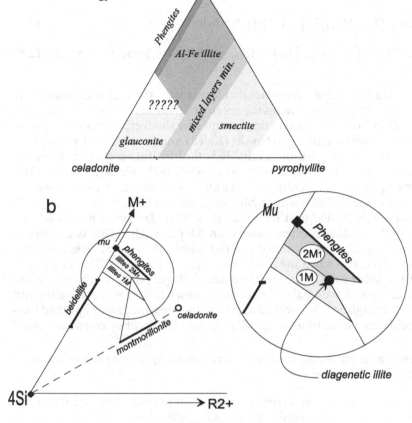

Fig. 2.16a,b. Chemical composition of illite. **a)** Solid solution in the muscovite – celadonite – pyrophyllite system (Hower and Mowatt 1966). **b)** Extension of solid solutions for 1M and 2M$_1$ illites in the M$^+$–4Si–R^{2+} system (Meunier and Velde 1989)

It must be noted that this value is noticeably different from that classically estimated by extrapolation from measured compositions of illite-smectite mixed layers (determined by X-ray diffraction). An interlayer charge of 0.75 corresponds to a 100% illite composition, i. e. non-measurable expandability (Hower and Mowatt 1966). The difference is related to the limitation of the method of measurement of the rate of expandable layer in illite-rich I/S by diffraction.

2.1.4.2
Glauconite – Celadonite – Ferric Illite

According to the international nomenclature (AIPEA committee), celadonite and glauconite are dioctahedral micas with the following theoretical composition (Odom 1987):

$$[Si_{4-x}Al_x] O_{10} \left(Mg_{1+y}Fe^{3+}_{1-y} \right) (OH)_2 K \quad \text{with } x (= y) < 0.2 \qquad (2.7)$$

$$[Si_{4-x}Al_x] O_{10} \left(R^{3+}_{2-y}R_{2+y} \right) (OH)_2 K_{x+y} \quad \text{with } x > 0.3 \text{ and } Fe^{3+} \gg Al \quad (2.8)$$

Glauconite is a low-temperature mica that forms at the sediment-seawater interface (the term glaucony designates the green pellets either rounded or retaining the form of bioclasts: sponge spicules, foraminifer shells etc). Celadonite-group minerals are widely occurring in basalts (Kaleda and Cherkes 1991 among others). The chemical compositions available in the literature seem to indicate the existence of a continuous solid solution between both micas (Andrews 1980; Newman and Brown 1987; Odin et al. 1988). Unfortunately, no definitive crystallographic evidence has been established until now (Fig. 2.17a).

Most of the published unit formulae show that glauconies have a K^+ ion deficiency: the content is often lower than 0.8 per Si_4O_{10}. This is due to the presence of smectite layers mixed layered within the crystal structure. The K^+ content decreases when the mixed layering rate increases (Thompson and Hower 1975; Velde and Odin 1975; Odin and Fullagar 1988). Noteworthy is the chemical composition domain of glauconies, which is situated between the end members mica (glauconite – celadonite) on the one hand, and high- and low-charge beidellite on the other hand (Fig. 2.17b). This has two consequences:

1. the smectite component, not the mica component, exhibits substitutions of Al for Si in tetrahedra;

2. smectite is not a nontronite; the Fe^{3+}-rich layer in a glaucony is that of mica. Therefore, the interstratified smectite is beidellite-like.

Ferric illites, sometimes designated under the term iron-bearing illites, glauconite micas (Kossovskaya and Drits 1970), and hyper-aluminous glauconites (Berg-Madsen 1983) do not form in the marine environment like glaucony but in the continental domain in salt lakes or lagoons (Porrenga 1968; Deconninck et al. 1988; Baker 1997) and in arid soils (Norrish and Pickering 1983; Jeans et al. 1994). The layer charge of ferric illites varies from 1 to 0.7 per Si_4O_{10}. The charge originates both in the tetrahedral sheet by substitutions of Al for Si and in the octahedral sheet by substitution of bivalent elements R^{2+} (Mg, Fe^{2+}) for trivalent elements R^{3+} (Al, Fe^{3+}). The iron content (Fe^{3+} + Fe^{2+}) varies between 0.5 and 1 ion per half-unit cell. The existence of a continuous solid solution domain between ferric illites and glauconites seems unlikely (Velde 1985), for crystallographic reasons. Indeed, the size of Fe^{3+} ions, and *a fortiori* Fe^{2+} ions, is very different from that of Al^{3+} ions; their replacement brings about significant distortions in the crystal lattice that produce defects. This explains the separation of chemical composition domains (Fig. 2.17c). This is the conclusion in the present state of knowledge.

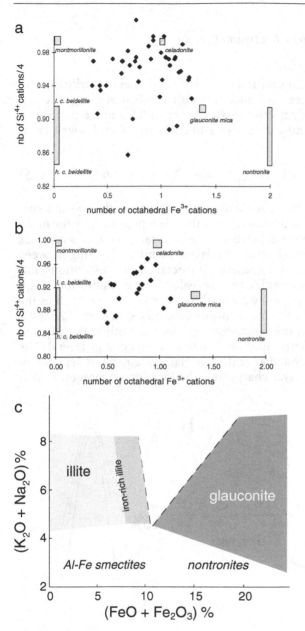

Fig. 2.17a–c. Solid solutions and mixed layering in the celadonite-glauconite group. **a)** Data from the literature (Andrews 1980; Newman and Brown 1987; Odin et al. 1988) show that a complete solid solution may exist between celadonite and glauconite micas. The solid solution extends towards the domain of beidellites as the K deficiency increases. **b)** The extension towards the domain of beidellites is not a solid solution; it is due to the mixed layering of beidellite and glauconite mica layers (Thompson and Hower 1975). **c)** Chemical composition domains of ferric illites and glauconies (after Velde 1985)

2.1.4.3
The Beidellite – Montmorillonite – Nontronite Domain

Beidellite-Montmorillonite

Theoretically, a complete solid solution exists between smectites with octahedral charge (montmorillonites) and smectites with tetrahedral charge (beidellites). The typical charge of smectites varies between 0.3 and 0.6 per Si_4O_{10} (Fig. 2.18a). The general unit formula of smectites can then be written as follows:

$$[Si_{4-x}Al_x]\,O_{10}\left(R^{3+}_{2-y}R^{2+}_y\right)(OH)_2M^+_{x+y}\quad \text{with } 0.3 \le x+y \le 0.6 \qquad (2.9)$$

This formula rests on a fundamental assumption: all layers of the crystal structure are identical and their individual composition is represented by the mean unit formula. In reality, the crystal structure of natural smectites is complex and exhibits either segregations inside the layers or mixed layering between compositionally different layers: tetrahedral or octahedral substitutions, low charge or high charge. This complexity is revealed by the application of appropriate tests for determining the value and location of charges–such as the Hofmann-Klemen test (beidellite/montmorillonite mixed layers: 17 Å/9.60 Å as shown in Fig. 2.18b) and the saturation by alkylammonium (Lagaly and Weiss 1969) or K ions (interstratification of low and high charge layers). The behaviour of smectites before and after the Hofmann-Klemen test depends on both the value of the mean layer charge and the location of charges. Sato et

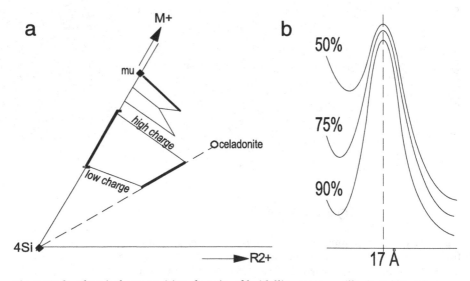

Fig. 2.18a,b. Chemical composition domain of beidellite-montmorillonite. **a)** Continuous solid solution between both end members whatever the charge. **b)** Identification of beidellite/montmorillonite mixed layers by the Hofmann-Klemen test. Percentages of beidellitelike layers are indicated

al. (1992) – who used the concept of expansion energy – show in a detailed study that for a given mean layer charge, the expandability decreases with an increasing proportion of tetrahedral charges.

Determination of the layer charge using the unit formula yields a mean value that does not take into account location. Indeed, the distribution of cation substitutions mostly produces heterogeneities at two organisation levels (Laird 1994; Mermut 1994):

- in smectites, some layers have a higher charge than their neighbour. Their expansion capacity is reduced and they irreversibly collapse at 10 Å when saturated by K^+ions;

- inside a layer, the distribution of substitutions of Si for Al in tetrahedra and R^{3+} for R^{2+} in octahedra leads to the formation of sites of different charges. This could be at the origin of the variation in the exchange energy from 5.7 to 10.9 kJ/eq measured by microcalorimetry (Talibudeen and Goulding 1983).

Combining tetrahedral and octahedral ionic substitutions, it is easy to show that the resulting charge emerging in the hexagonal cavities at the layer surface may have different values (annex 3).

Beidellite – Nontronite – Ferric Montmorillonite

The compositional range of solid solution for iron-bearing smectites, or non-tronites, is certainly relatively restricted about the theoretical end member $2Fe^{3+}$ (Güven 1988). Although some compositions intermediate between bei-dellites and nontronites can be found in the literature, the crystallographic data are too inaccurate for the existence of a complete solid solution between both species to be possibly admitted. In fact, the difference in ionic radius between Al^{3+} and Fe^{3+} ions is such that the replacement of the first by the second brings about an increase in the b cell dimension according to an empiric law of type: $b = 8.9270 + 0.1174 [Fe^{3+}]$ (Eggleton 1977; Brigatti 1983; Köster et al. 1999). When substitutions are too numerous, they produce significant distortions in the crystal lattice. The accumulated elastic energy is sufficient to break chemical bonds and create structure defects. The b cell dimension of beidellite (9.00 Å) and nontronite (9.12 Å) are too different to allow for a complete substitution. The large size of the nontronite cell explains the ability of these minerals to fix preferentially K^+ ions, whose ionic radius is significant: $r = 1.46$ Å (Velde 1985).

Köster et al. (1999) have given a convincing description of the montmoril-lonitic properties of an iron-rich smectite in peridotite nodules included in a basalt flow in Ölberg (Germany). The layer charge is essentially provided by substitution of Mg^{2+} ions for Fe^{3+} ions in the octahedral sheet. A solid solution does not seem to exist with nontronites, whose charge is located in tetrahedra.

2.1.4.4
Domain of Trioctahedral Smectites

Like their dioctahedral equivalents, trioctahedral smectites theoretically exhibit a range of solid solution bordered by the following compositional end members: tetrahedral high and low charges (saponite); octahedral high and low charges (stevensite – hectorite). The general unit formula of saponites is comparable to that of beidellites:

$$[Si_{4-x}Al_x]\, O_{10} \left(R_3^{2+}\right) (OH)_2 M_x^+ \quad \text{with } 0.3 \le x \le 0.6. \tag{2.10}$$

The layer charge is provided by the octahedral sheet in stevensite and hectorite, where Li^+ ions or vacancies replace divalent cations (Mg^{2+} or Fe^{2+}). The general unit formula may write as follows:

$$[Si_4]\, O_{10} \left(R_{3-y-z}^{2+}Li_y^+\square_z\right) (OH)_2 M_{y+2z}^+ \quad \text{with } 0.3 \le y + 2z \le 0.6 \tag{2.11}$$

In the M^+–$4Si$–$3R^{2+}$ system, the $R^{2+} \Leftrightarrow R^{3+}$ substitution in the octahedral sheet is not represented. Nevertheless, the compositional range of the theoretical solid solution as thus defined for trioctahedral smectites (Fig. 2.19a) is quite consistent with the chemical composition domain of natural minerals, in which iron is strictly in a ferrous state and aluminium is strictly located in tetrahedra.

Trioctahedral smectites, whose octahedral sheet exhibits $R^{2+} \Leftrightarrow R^{3+}$ substitutions, have a wide compositional range of solid solution. Figure 2.19b shows the extent of this range in the MR^3–$2R^3$–$3R^2$ system (Velde 1985). The general unit formula of these smectites writes as follows:

$$[Si_{4-x}Al_x]\, O_{10} \left(R_{3-y-z}^{2+}R^{3+}y\square z\right) (OH)_2 M_{x-y+2z}^+ \text{ with } 0.3 \le x - y + 2z \le 0.6 \tag{2.12}$$

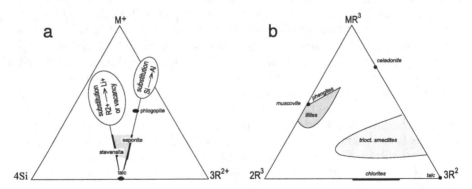

Fig. 2.19a,b. Representation of the compositional range of the theoretical solid solution of tetrahedral smectites. **a)** In the M^+–$4Si$–$3R^{2+}$ system, this range is limited by saponites (tetrahedral charge) and by the stevensite – hectorite series (octahedral charge provided by the $R^{2+} \Leftrightarrow Li^+$ substitution or by the presence of vacant sites). **b)** In the MR^3–$2R^3$–$3R^2$ system, this range has been empirically determined (Velde 1985)

2.1.4.5
Complex Solid Solutions: Chlorites (Al, Fe, Mg, □)

14 Å Chlorites

Among the most complex phyllosilicates, chlorites consist of a 2:1 layer associated with an additional octahedral sheet (brucite-like or gibbsite-like sheet) in the interlayer (see Sect. 1.1.1.2). These two structural units are bonded together by Coulomb attraction, the 2:1 layer being charged negatively, and the additional octahedral sheet having the same charge but of opposite sign. The negative charge of the 2:1 layer (about -1 per Si_4O_{10}) is provided essentially by $Si^{4+} \Leftrightarrow Al^{3+}$ substitution in the tetrahedral sheets. Whether the tetrahedral charge is fully compensated by the interlayer sheet ($+1$) or partially compensated by the octahedral sheet of the 2:1 layer is still not known.

Octahedral and brucite-like sheets can be dioctahedral or trioctahedral, theoretically exhibiting four possible structures: (1) tri-tri-, (2) di-tri-, (3) di-di-, and (4) tri-dioctahedral structures. In nature, the first type is by far the most abundant, the second and the third are rarer and the fourth is unknown. This structural specificity of chlorites allows for increased possibilities of ionic substitutions. Foster (1962) has shown that the Al content in octahedral sheets is mostly lower than the Al content in tetrahedral sheets. Fe^{3+} frequently occurs but can sometimes be replaced by Cr^{3+}. Finally, Bayley (1988) proposes a description of the chlorite composition with the following general unit formula:

$$\left[Si_{4-x}R_x^{3+}\right] O_{10} \left(R_{6-x-3y}^{2+}R_{x+2y\square y}^{3+}\right) (OH)_8 \tag{2.13}$$

The Wiewiora and Weiss projection system (1990) is perfectly suited to represent all substitutions of type $R^{2+} \Leftrightarrow R^{3+}$ and $R^{2+} \Leftrightarrow \square$ (Fig. 2.20).

The chemical composition domains of the main species of tri-tri-, di-tri- (sudoite) and di-dioctahedral (donbassite) chlorites are shown in Figs. 2.2–2.21a. In the present state of knowledge, nothing indicates that the theoretically defined compositional ranges of solid solutions (Fig. 2.20) are truly those of natural minerals. Indeed, those compositions controlled by accurate crystallographic analyses delimit discontinuous and much narrower compositional ranges (Fig. 2.21a).

7 Å Chlorites

Theoretically, the 7 Å chlorites form a family of trioctahedral minerals. They have a tetrahedral charge provided by substitution of Al^{3+} or Fe^{3+} for Si^{4+} ions. The so created deficiency is compensated for by an excess charge in the octahedral sheet provided by substitution of trivalent ions (Al^{3+}, Fe^{3+}) for bivalent ions (Fe^{2+}, Mg^{2+}). The maximum accepted value for these charges is close to 1. Therefore, their chemical compositions in the $Si-R^{3+}-R^{2+}$ systems are situated below the line going from kaolinite to serpentine (or ferrous equivalent: greenalite). Two series of 7 Å chlorites are formed according to the nature of the prevailing bivalent and trivalent cations:

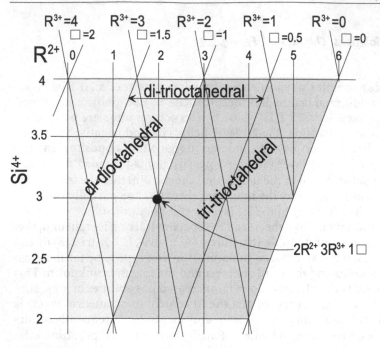

Fig. 2.20. Representation of the compositional ranges of solid solution of tri-trioctahedral, di-dioctahedral and di-trioctahedral chlorites in Si^{4+}–R^{2+} coordinates, according to the projection system of Wiewiora and Weiss (1990). $\square$: vacancy

- Mg–Al^{3+} series: aluminous serpentines. According to Velde (1985), the solid solution is continuous between the end members serpentine and Mg-berthierine (Fig. 2.21b)
- Fe^{2+}–Fe^{3+} series: Fe-berthierines form two distinct compositional groups according to Fritz and Toth (1997), one towards the end member kaolinite and the other one towards the end member greenalite (Fig. 2.21c).

Odinite is a 7 Å phyllosilicate whose structure is thought to be intermediate between di- and trioctahedral. A standard unit formula is given by Bailey (1988):

$$[Si_{1.788}Al_{0.212}]\,O_5\,\left(Al_{0.556}Fe^{3+}_{0.784}Fe^{2+}_{0.279}Mg_{0.772}Mn_{0.015}Ti_{0.016}\right)(OH)_4\,. \quad (2.14)$$

These minerals are still poorly known and are probably smectite-berthierine mixed layers (Velde, personal communication)

2.1.4.6
False Solid Solutions

The definition of the compositional ranges of solid solutions for natural or synthetic minerals using classical or microprobe chemical analyses is always

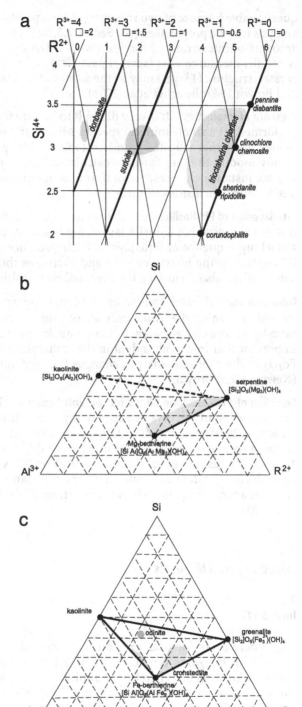

Fig. 2.21a–c. Solid solutions of chlorites. **a)** 14 Å chlorite compositions plotted in the diagram of Wiewiora and Weiss (1990). The *shaded zones* represent the chemical composition domains of tri-tri-, di-tri-and di-dioctahedral natural chlorites. **b)** 7 Å chlorites: the magnesian series seems to form a continuous solid solution between the end members berthierine and serpentine (after Velde 1985). **c** The iron-bearing series forms two distinct groups according to their Fe^{2+}–Fe^{3+} content (after Fritz and Toth 1997)

questionable because of two reasons: (1) even purified, the clay particles under analysis may be polyphased (see Sect. 1.2.5.2); (2) the presence of undetectable amounts of interstratified foreign layers is frequent. The scale difference between the measurements (several particles) and the interpretation in terms of crystal structure (1 layer) may be the source of several types of errors (Newman and Brown 1987; Bouchet et al. 2001).

Presence of Impurities   It is very difficult to be sure that the chemical analysis is performed on a single mineral species. Micro- and nano-inclusions of quartz, oxides, and other silicates – undetectable by X-ray diffraction – may significantly modify the unit formula calculated from a chemical analysis. That is why for instance, the presence of titanium in numerous phyllosilicates is often due to micro-inclusions of rutile.

Interlayering of Phyllosilicates   Thanks to the high-resolution electron microscopy, it is now known that chlorite layers, for instance, can be interlayered within a stacking sequence of illite layers. Their presence artificially diminishes the K^+ content in the interlayer zone and increases the proportion of Tchermak substitution, also distorting the assumed compositions of phengites.

Oxidation State of Iron   The major problem of microanalyses is the impossibility of determining the oxidation state of iron. Determination of the Fe^{3+}/Fe^{2+} ratio by simple charge balancing in a unit formula entails errors all the more important that iron is abundant in the mineral. Indeed, passing from FeO to Fe_2O_3 in the chemical analysis changes the contents of all the other elements (Newman and Brown 1987).

Selection of the Calculation Basis for the Unit Formula   This selection rests on the determination of the crystal structure by X-ray diffraction: 1:1 or 2:1 or 2:1:1 structure. It is unequivocal as long as minerals are not mixed layered. It becomes a source of errors when minerals of different type are mixed layered: 1:1 with 2:1 (example: kaolinite/smectite) or 2:1 with 2:1:1 (example: smectite/chlorite). Under such conditions, the calculation basis of the unit formula varies as a function of the relative proportions of the two components (Bouchet et al. 2001).

2.2
Mixed Layered Minerals

2.2.1
Introduction

The different clay species previously described are formed by the stacking of layers of the same type (1:1, 2:1 or 2:1:1) and having the same chemical composition domain. Therefore, if a solid solution does exist, the layer exhibits the same characteristics in the crystal. This is not the case for mixed-layer minerals, which are formed by the stacking of layers of different type or composition (illite/smectite I/S or chlorite/smectite C/S for the commonest).

In Sect. 1.1.2.2, the structure of mixed-layer minerals has been described only along the layer stacking direction (c direction). As for pure minerals, this data remains incomplete as long as the following is not known:

- How are cations, hence electric charges, distributed in these crystal structures?

- What is the distribution of the different chemical composition domains in the $a–b$ plane (Fig. 1.17)?

This section is dedicated to the chemical composition of mixed-layer minerals and is divided into three parts. The first part is a condensed presentation of the relations between their crystal structure and the distribution of cations. The second part gives a description of the chemical composition domains of the main groups of mixed-layer minerals found in nature. The third part gives an overview of the latest progresses, which view mixed-layer structures from the angle of crystal growth.

2.2.2
Crystallochemistry of Mixed-Layer Minerals

2.2.2.1
Definitions

Solid Solution and Mixed Layering
Solid solutions are considered here as the result of ionic substitutions governed by Goldschmidt's rule. Therefore, each layer in the stacking sequence, if compositionally different from its neighbours, stays in the same chemical composition domain. Only the distribution of these substitutions may explain the possible differences of composition from layer to layer. When compositions are not randomly distributed, then several types of layers exist in the stacking sequence. It is no longer a solid solution but a mixed-layer mineral even if the two layer types belong to the same species (low and high-charge smectites for instance).

In natural environments, mixed-layer minerals belong as much to the dioctahedral series (illite/smectite, kaolinite/smectite) as to the trioctahedral series (chlorite/saponite, talc/stevensite). Mixed-layer minerals comprising both trioctahedral and dioctahedral sheets are rarer. The reason for this is the incompatibility of $a–b$ dimensions. Indeed, an illite and a vermiculite whose (060) peaks are indexed at 1.499 and 1.542 Å, respectively, show a difference of b dimension equal to 9.252–8.994=0.387 Å. This means that a full-cell shift shall occur every 23.24 illite cells, or every 209 Å along the b direction.

McEwan Crystallites or Fundamental Particles
How should the crystal structure of a unit layer be considered? For 2:1 minerals (mica and smectite) for instance, two different conceptions are proposed:

1. "McEwan crystallites": the octahedral sheet is a symmetry plane in the 2:1 layer (Fig. 2.22a);

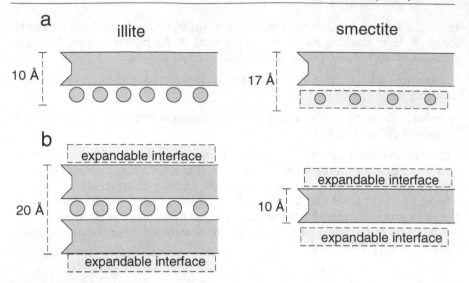

Fig. 2.22a,b. The two concepts of crystal structure of mixed-layer minerals. a) McEwan crystallite. b) Fundamental particle (Nadeau et al. 1984)

2. "fundamental particles" (Nadeau et al. 1984): the plane of symmetry is in the interlayer sheet (Fig. 2.22b). The model requires that the external 2:1 layers of fundamental particles have tetrahedral sheets with different charges. These layers are polarised.

Now it seems that, whatever the model selected, cations forming the 2:1 layer can be distributed either equally symmetrically to the octahedral sheet (non-polar layer), or unequally with occurrence of a polarity (see Sect. 1.2.5.2). Polarized layers should exist (Cuadros and Altaner 1998).

These two concepts allow for a perfect description of the layer stacking sequences along the c direction for illite/smectite mixed layers for instance:

– McEwan crystallite (Fig. 2.23a): stacking sequence of smectite (2:1 layer with expandable interlayer sheet at 17 Å) and illite (2:1 layer with non-expandable interlayer sheet at 10 Å);

– Fundamental particles (Fig. 2.23b): stacking sequence of illite fundamental particles separated by expandable interfaces (equivalent to smectite). The thickness of illite fundamental particles increases in mixed-layer minerals whose illite content is greater than 50%.

These two standpoints have been connected by showing that fundamental particles are sub-units of McEwan crystallites (Veblen et al. 1990). Whatever the model considered, the sequence of layers in the stacking arrangement is described in the same manner by the Markovian probability. However, both models remain speculative since they are not fully coherent with all the crystal characteristics and particularly their growth process.

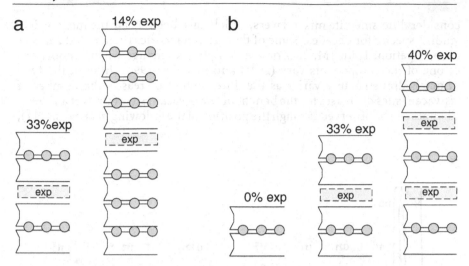

Fig. 2.23a,b. Correspondence between McEwan crystallites and fundamental particles (after Eberl and Blum 1993 revised). **a)** The rate of expandable layers in the McEwan crystallite depends on the thickness of fundamental particles. **b)** The rate of expandable layers depends on the number of fundamental particles

2.2.2.2
Chemical Composition and Structure of Mixed-Layer Minerals

Mixed-layer minerals are distinguished from pure clay species by the fact that the structure of their crystals cannot be considered simply as a repeating cell along the three space directions. The only exception could be given by regularly ordered mixed-layer minerals that exhibit typical superstructures. Nevertheless, the latter are increasingly viewed as pure mineral species and no longer as mixed-layer minerals. This is the case of corrensite for instance (Beaufort et al. 1997). At that point, we have to remember that the diffraction patterns of pure minerals show rational series of diffraction peaks whereas those of mixed-layer minerals show non-rational series.

For not-regularly ordered mixed-layer minerals, the number of compositionally different layers in the stacking sequence (montmorillonite and illite or saponite and chlorite) can be determined by the study of $00l$ peaks (cf. 1.1.2.2). Nevertheless, determination of the crystallochemical characteristics of these components – such as the accurate location of vacancies in dioctahedral layers (*cis* or *trans*) or the stacking modes of layers in the crystal (polytypism) – is not enough; hkl bands also have to be considered.

Relation Between a trans- or cis-Vacant Structure and Polytypism (Dioctahedral Species)

The crystal structure of dioctahedral mixed-layer minerals in the a–b plane can be described using two types of data. At the layer scale, the proportion of vacancies in *trans* or *cis* position, and the number and location of charges must be specified for the octahedral and tetrahedral sheets, respectively. Let's consider illite-smectite mixed layers, which have been by far the most widely studied species for decades. Some of their characteristics (octahedral *trans* or *cis* occupation; 1Md, 1M, 2M₁ or 2M₂ polytypes) change when proportions of one of the components vary (Drits and McCarty 1996). Indeed, the type of octahedral structure varies as the illite content increases: the number of *cis*-vacant sites decreases for the benefit of *trans*-vacant sites (Drits et al. 1996). Changes can be observed through the position of the following *hkl* peaks: (11$\bar{1}$),

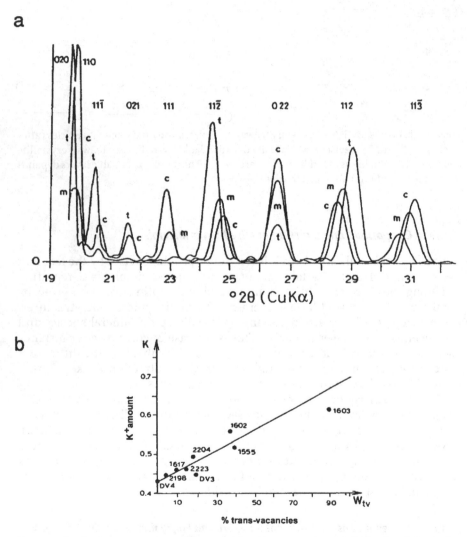

Fig. 2.24a,b. *cis*-Vacant to *trans*-vacant transition accompanying I/S illitisation (after Drits et al. 1996). **a)** Calculated X-ray diffraction patterns showing the characteristic peaks of *cis* (*c*), *trans* (*t*) or *cis-trans* structures in equal proportions (*m*). **b)** Relations between K content and rate of *trans*-vacant structure in I/S

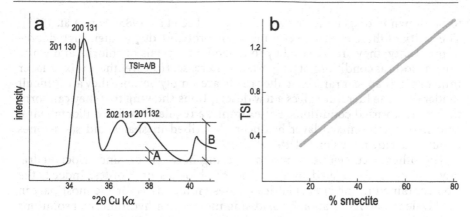

Fig. 2.25a,b. Turbostratism degree in illite /smectite mixed-layer minerals (after Reynolds 1992). a) Definition of the turbostratism index. b) Variation of the turbostratism index as a function of the smectite layer content

and the respective proportions of each site can be calculated (Fig. 2.24a). The linear relationship between the number of *trans*-vacant sites and the K content (Fig. 2.24b) shows that illitisation is actually the cause of this transformation. If so, it should be inferred that the octahedral structure changes from *cis* to *trans* during crystal growth. This will have to be investigated more deeply in the future.

In the same way, the stacking mode of layers, and consequently the crystal symmetry, change during I/S illitisation: first 1M, the polytype becomes $2M_1$. The transformation is practically complete in the diagenesis-anchizone transition zone, where illites are progressively replaced by phengites. How this transition is performed: dissolution of 1M – crystallization of $2M_1$, $2M_1$ overgrowth upon 1M or both? Again, this should be investigated in the future.

Shift associated with non-rational rotations may occur between two layers, thus introducing a defect in the regularity of the stacking sequence. This defect characterises the turbostratic stacking sequence. Reynolds (1992), defining turbostratism index, shows that this index regularly increases with the smectite content in I/S (Fig. 2.25). For pure smectites, stacking sequences are almost totally turbostratic in the natural state. Nevertheless, Mamy and Gautier (1976) have shown that the stacking order for a montmorillonite increases after saturation with K^+ ions and several wetting-drying cycles. The effect of these stacking defects on the X-ray diffraction patterns has been described by Drits et al. (1984).

Di-Trioctahedral Mixed-Layering

The difference between crystal cell sizes (mostly along the *b* dimension) has long been considered as an insuperable obstacle for the formation of mixed-layer minerals between dioctahedral and trioctahedral phyllosilicates. Nevertheless, the multiplication of high-electron microscope observations has allowed the complete series of mixed-layer minerals between illite and chlorite to be shown in diagenetic or volcanic rocks (Lee et al. 1985; Ahn et al. 1988). The rarity of these mixed-layer minerals is probably due to their high degree of instability; they are replaced by independent crystals of chlorite and white mica as soon as conditions of temperature increase. However these mixed-layer minerals may appear rarer than they really are simply because they are difficult to identify. The literature gives a few descriptions showing that they can form under most varied conditions. For example, a regularly ordered dioctahedral chlorite-smectite mixed layer has been described in shales and sandstones about a basaltic intrusion (Blatter et al. 1973).

The other occurrences of mixed layering between di- and trioctahedral phyllosilicates are related to alteration of chlorites or biotites. Indeed, the iron contained in the crystal lattice of these minerals is for the most part in the bivalent state. Alteration by oxidising meteoric or hydrothermal solutions causes a change in valency ($Fe^{2+} \rightarrow Fe^{3+}$), which locally produces positive charge excess.

Alteration in oxidising environment (supergene or hydrothermal) of biotites or chlorites leads, through the oxidation of Fe^{2+} ions, to the formation of dioctahedral vermiculites. The reaction $[Fe^{2+}OH]^+ \rightarrow [Fe^{3+}O]^+ + H^+ + e^-$ considerably changes the electric charge of the octahedral sheet and entails the expulsion of the most soluble cations: Mg and Fe^{2+}. The R^{3+} gain and the R^{2+} loss transform the trioctahedral lattice into a dioctahedral one (Herbillon and Makumbi 1975; Proust et al. 1986). This process leads to the formation of a series of di-tri mixed-layer minerals: chlorite or biotite-vermiculite. High-resolution microscope observations of altered biotites have shown that vermiculite layers "spread" within the mica structure (Banfield and Eggleton 1988). Regularly-ordered mixed-layer minerals are often identified in these series. This is a remarkable point whose signification is not fully understood till now: is it due to a periodical physical process or to asymmetries in the biotite or chlorite layer structures?

2.2.3
Composition of the Most Common Mixed-Layer Minerals

2.2.3.1
Illite/Smectite

Chemical Composition of Mixed-Layer Mineral Series
The chemical composition of illite/smectite mixed layers has been studied as a solid solution in the muscovite – celadonite – pyrophyllite ternary system (Hower and Mowatt 1966). This assumption has been used as a basis for calculating thermodynamic parameters of different types of mixed-layer minerals (Ransom and Helgeson 1993; Blanc et al. 1997). Although this kind of approach does not respect the definition of solid solutions as stated at the beginning of this chapter, the use of the muscovite – celadonite – pyrophyllite ternary system is absolutely relevant for the study of the variation in the rate of octahedral and tetrahedral substitutions as a function of the illite content. Despite the

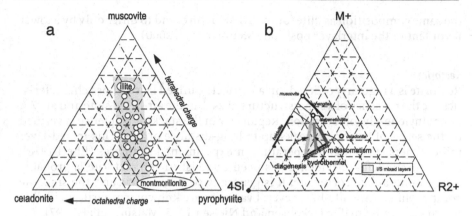

Fig. 2.26a,b. The chemical composition of illite/smectite mixed layers. **a)** Projection in the muscovite – celadonite – pyrophyllite system (Hower and Mowatt 1966). Points are distributed from the illite domain to the montmorillonite domain. **b)** Projection in the M^+– $4Si$–R^{2+} system. Three I/S series can be distinguished according to their origin: diagenesis in shales, high-temperature diagenesis in geothermal fields and metasomatic series in K-bentonites (Meunier and Velde 1989)

fact that I/S samples used by Hower and Mowatt (1966) were collected from different types of deposits, the muscovite – celadonite – pyrophyllite diagram shows that compositions range from the illite domain where the charge is mainly tetrahedral to the montmorillonite domain where the charge is essentially octahedral (Fig. 2.26a). The same tendency is observed in the series of samples collected from the same deposits (Meunier and Velde 1989). The study of I/S collected from shales in diagenetic formations, geothermal fields or metasomatised zones, has shown the following facts:

1. the mixed layered montmorillonite has a constant charge in the I/S series. The value of the charge depends on the type of deposit: low charge (0.3 per Si_4O_{10}) in diagenetic environments, medium charge (0.45) in geothermal fields, high charge (0.6) in metasomatised zones.

2. the mixed layered illite, whatever the I/S series, has a fixed composition (Fig. 2.26b).

Whether viewed as a solid solution or as a mixed layering, the chemical composition of I/S corresponds to a mixture of two components: an illite of fixed composition and a montmorillonite whose charge depends on the physicochemical conditions imposed in the deposit. Nevertheless, refined X-ray diffraction analyses show the existence of a third component: vermiculite (Shutov et al. 1969; Foscolos and Kodama 1974; Drits et al. 1997). It does not appear clearly in the chemical compositions either because it occurs in too small amount to be detected or, more subtly, because its composition is intermediate between those of smectites and illites forming the mixed-layer mineral. In reality, the vermiculite identified with X-rays is likely to have about the same composition as illite for the 2:1 structure, and to differ only by a lower K content in the interlayer position (Meunier et al. 2000).

Rectorite
Rectorite is a regularly ordered mica – smectite mixed layer whose characteristics are the presence of a superstructure at 24 Å in the air-dried state and at 27 Å after ethylene-glycol saturation. Regularity in the stacking sequence is verified by the series of harmonic peaks up to long-distance orders. This mixed-layer mineral persists at high temperature in experimental syntheses (Eberl 1978). Although no definite evidence has been established yet, rectorite is probably a true mineral like corrensite. It is generally sodic or calcic and the smectite component is beidellite-like. Several varieties are known: margarite-beidellite or paragonite-beidellite (Jakobsen and Nielsen 1995; Matsuda et al. 1997).

The presence of rectorite in the chloritoid zone (Paradis et al. 1983) argues for a mineral different from the regularly ordered illite/smectite mixed layers found in diagenetic environments. The latter are potassic and generally contain a little more illite than smectite (60%) and their smectite component is rather montmorillonite-like. They used to be designated under the term allevardite, which unfortunately has been dropped.

2.2.3.2
Saponite/Chlorite

Phase Mixtures or Mixed Layering?
Chlorite/saponite mixed layers have been classically considered as forming a continuous series between smectite and chlorite (Chang et al. 1986; Robinson and Bevins 1994), in which corrensite (regularly ordered mixed-layer mineral with 50% saponite and 50% chlorite) represents an intermediate stage in the conversion series. The trioctahedral series was considered as the symmetrical figure of the I/S dioctahedral series.

This assumption has not been confirmed by the observations carried out on the variations in the saponite contents of C/S in geothermal environments. The series of mixed-layer minerals are discontinuous (Inoue et al. 1987; Inoue et Utada 1991). The apparent continuity of the chemical compositions between the two end members saponite and chlorite is more the result of the mixture of smectite-rich phases with corrensite and of corrensite with chlorite rather than of the mixed layering in all proportions of chlorite and smectite. Projected in the $M^+-4Si-3R^{2+}$ system, these compositions show the influence of mixtures (Fig. 2.27).

Corrensite
Whether the series of mixed-layer minerals are continuous or discontinuous, the interpretation of the crystal structure of corrensite used to be based on the assumption of a regular stacking sequence of saponite and chlorite layers. During conversion, the ratio of chlorite layers increases progressively at the

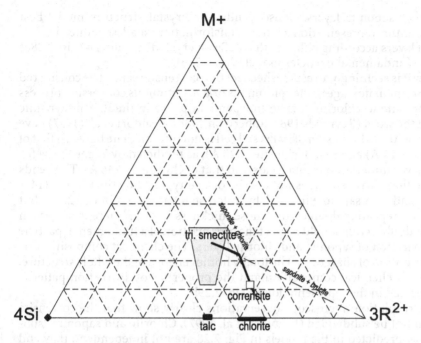

Fig. 2.27. Projection in the M$^+$–4Si–3R^{2+} system of the chemical compositions of a series of saponite/chlorite mixed layers (*thick lines*) resulting from the hydrothermal alteration of volcanic rocks of the Caldera Ohyu in Japan (Inoue et al. 1984). *Both lines* show the presence of mixtures of saponite+corrensite and corrensite+chlorite

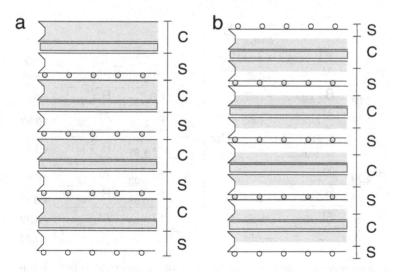

Fig. 2.28a,b. The two possible types of crystal structure for corrensite. **a**) Model based on the regular alternation of chlorite and saponite layers (McEwan's crystallite). **b**) Model based on the crystal structure of phyllosilicates

expense of saponite layers. Consequently, the crystal structure model best suited for the representation of this transformation shall separate the two types of layers according either to the McEwan crystallites pattern (Fig. 2.28a) or to the fundamental particles model (Fig. 2.28b).

Today it is seemingly an established fact that corrensite cannot be considered as an intermediate stage in the continuous or discontinuous conversion process from saponite to chlorite. It is an independent phase in the thermodynamic sense of the word (Reynolds 1988; Robertson 1988). Beaufort et al. (1997) have shown that the *b* dimension of its crystal cell is not intermediate between that of saponite (9.21 Å) and that of chlorite (9.25 Å), as predicted by Vegard's law for mixed-layer minerals. It is much closer to that of chlorite (9.246 Å). This tends to prove that corrensite does not consist of two types of distinct layers (14 Å chlorite and 10 Å saponite in the dehydrated state) as shown in Fig. 2.29a, but rather of a repeating double unit whose thickness is 24 Å. There is no reason why this double layer should respect the organisation of a fundamental particle with separation of saponite and chlorite layers. The chemical composition, as the distribution of charges, can be totally homogeneous in the whole structure. The overall chemical composition and the properties of diffraction patterns are respected in the representation in Fig. 2.29b.

High-resolution microscope observations clearly show that the corrensite layer cannot be subdivided (Beaufort et al. 1997). Chlorite and saponite "sublayers" as predicted in the models in Fig. 2.28 are not independent; they end strictly together at the level of dislocations or crystal endings (Fig. 1.33). So the crystal forms by growth of the 24 Å corrensite entity. Therefore, C/S mixed layers with chlorite content greater than 50% are the result of intergrowths of corrensite and chlorite.

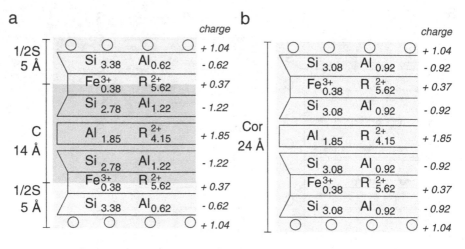

Fig. 2.29a,b. Schematic representations of the crystal structure of a corrensite (after Beaufort and Meunier 1994). **a)** Structure based on the fundamental particles model. Chlorite and saponite layers are separated. **b)** Structure of a homogenous phase. The chemical composition of 2:1 layers is homogenous and differs from that of chlorite or saponite

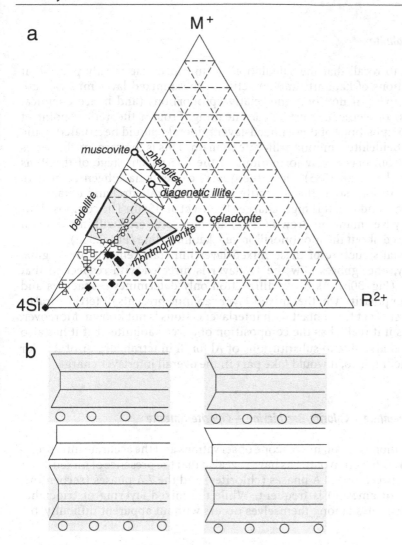

Fig. 2.30a,b. Kaolinite/smectite mixed layer minerals. **a)** Chemical composition of a series of kaolinite/smectite mixed layers from the formation of plastic clays of the Sparnacian stage of the Paris Basin. *Open squares* represent the composition of the bulk < 2 μm fraction (Brindley et al. 1983) and the corrected analyses of the quartz fraction (Proust et al. 1990). *Filled squares* correspond to compositions given by Newman and Brown (1987). **b)** The two possible crystal structures

2.2.3.3
Smectite/Kaolinite

It is useful to recall that the calculation of unit formulae is only possible if the proportions of kaolinite and smectite in these mixed-layer minerals are perfectly known. If not, only the relative proportions (and hence chemical projections) of elements have a true meaning. From a theoretical point of view, the compositions of these mixed-layer minerals should be situated in the kaolinite – beidellite – montmorillonite domain. This is the case for the series stemming from plastic clay formations of the Sparnacian stage of the Paris Basin (Brindley et al. 1983). Once the quartz contents have been corrected, these compositions show that smectite is mixed (tetrahedral and octahedral substitutions) and that it is high-charge (Proust et al. 1990). The compositions compiled by Newman and Brown (1987) are more scattered within a domain rather centred about the montmorillonite – kaolinite line (Fig. 2.30a).

The crystal structure of these mixed-layer minerals remains an enigma. Particularly, one ignores how the 1:1 layer is directed with respect to that of smectite (Fig. 30b). As X-ray diffraction only determines d-spacings and ignores layer polarity, whether mixed-layer kaolinite presents its tetrahedral or octahedral sheet at the contact with interlayer cations is not known. Moreover, one ignores if it really has the composition of a free kaolinite or if it has also an electric charge due to substitutions of Al for Si in tetrahedra or of R^{2+} for Al in octahedra. If so, it would take part in the overall interlayer charge.

2.2.3.4
Chlorite/Serpentine – Chlorite/Berthierine – Chlorite/Amesite

High-resolution electron microscope observations and the accurate interpretation of X-ray diffraction patterns have revealed that the presence of mixed-layer minerals between the 14 Å phases (chlorites) and the 7 Å phases (serpentine, berthierine or amesite) is frequent. While the mixed layering of trioctahedral phyllosilicates among themselves occurs without apparent difficulty, the

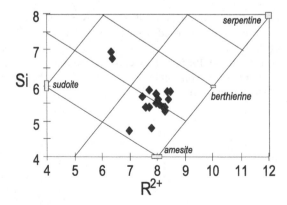

Fig. 2.31. Projection of the chemical composition of 14 Å chlorite/7 Å chlorite mixed layers known in nature. (Ahn and Peacor 1985; Hillier and Velde 1992; Ryan and Reynolds 1996) in the Wiewiora and Weiss system (1990)

chemical composition of these mixed-layer minerals known in nature (Ahn and Peacor 1985; Hillier and Velde 1992; Ryan and Reynolds 1996) shows that they are of type 14 Å chlorite/7 Å chlorite rather than 14 Å chlorite/serpentine (Fig. 2.31). The 7 Å structure is always characterised by the presence of Al^{3+} ions in tetrahedra and octahedra. The known mixed-layer minerals containing true serpentine are of talc/serpentine or saponite/serpentine types. In the present state of knowledge, the berthierine-to-14 Å chlorite transition is well demonstrated only for ferrous species (Fe-berthierine-chamosite) using high-resolution electron microscopy (Xu and Veblen 1996). This transition results from the solid-state transformation (cf. Sect. 1.2.5.1, Fig. 1.37c). Very recent data show that the 14 Å phase retains polytypes of the 7 Å phase. Nucleating in various points of the berthierine crystals, chamosite-like chlorite grows through the inversion process of tetrahedra. The junction of the different growth zones is represented by an odd number of 7 Å layers. Everything takes place as if these layers were antiphase walls (Billault 2002).

Suggested Readings

Crystal Structure and Chemical Composition of Clay Minerals

Bailey SW (1988) Hydrous phyllosilicates (exclusive of micas). Rev Mineral 19:725
Weaver CE, Pollard LD (1975) The chemistry of clay minerals. In: Developments in Sedimentology, 15. Elsevier, Amsterdam
Newman ACD (1987) Chemistry of clays and clay minerals. Mineralogical Society Monograph 6, Mineralogical Society, London, pp 480

Cation Exchange Capacity – Layer Charge

Sposito G (1989) The chemistry of soils. Oxford University Press, New York, 277 pp
McBride MB (1994) Environmental chemistry of soils. Oxford University Press, New York, 406 pp
Mermut AR (1994) Layer charge characteristics of 2:1 silicate clay minerals. CMS workshop lectures 6, 134 pp

Spectroscopic Methods

Farmer VC (1974) The infrared spectra of minerals. The Mineralogical Society, London
Hawthorne FC (1988) Spectroscopic methods in mineralogy and geology. Rev Mineral 18:698
Decarreau A (1990) Matériaux argileux. Structure, propriétés et applications. Soc Fr Miner Cristall, Paris, 586 pp
Wilson MJ (1994) Clay mineralogy: spectroscopic and chemical determinative methods. Chapman and Hall, London, 366 pp

CHAPTER 3

Energy Balances: Thermodynamics – Kinetics

The crystal structure and chemical composition of pure and interstratified clay species being familiar (see Chaps. 1 and 2), it is possible now to investigate a more subtle consequence: the distribution of energy inside the crystal lattice and its inner and outer surfaces. Indeed, if this is understood, then it becomes possible to relate the crystallochemical characteristics of clay minerals to the physicochemical conditions prevailing during their formation. This is a fascinating perspective for chemists, mineralogists and geologists who, for different but converging reasons, deal with mineral reactions in clay materials. Particularly, this makes it possible to study the effects for more or less long periods of time of chemical interactions in human environments (pollution, cropping, waste storage etc).

3.1
Thermodynamics of Equilibrium

3.1.1
Introduction

The characterisation of a clay, or of any other mineral species, requires the determination of its tridimensional structure and chemical composition. This amounts to a space representation of all the chemical bonds fixing cations and anions in a crystal lattice whose smallest entity is the crystal unit cell. Each type of bond (position of the ion in the structure) is characterised by a potential energy. Each ion vibrating or oscillating about this position adds a kinetic energy. The sum of the potential and kinetic energies form the enthalpy (H) of the crystal.

Entropy (S) is a parameter that describes quantitatively the degree of internal disorder of a substance. Vapour has a higher entropy than water, which itself has a higher entropy than ice. This concept is very important because it permits definition of the major parameter controlling chemical reactions: free energy (G):

$$G_{cryst} = H_{cryst} - TS_{cryst} \qquad (3.1)$$

Enthalpy and entropy increase with the mass or volume of the crystal; they are extensive variables. They are expressed in relation to the "molecule" rep-

Fig. 3.1. Respective domains of thermodynamics and kinetics

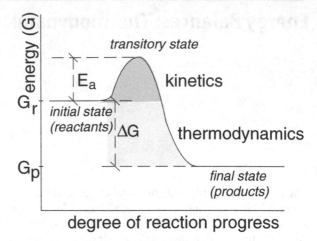

resentative of the crystal, namely the crystal unit cell: H in $J\,mol^{-1}$ and S in $J\,K^{-1}\,mol^{-1}$.

The first law of thermodynamics (conservation of energy) states that the total of energies is constant. The second law states that any spontaneous process leads to an increase in entropy. To establish the energy balance of a mineral reaction amounts to considering the difference (ΔG) between the free energies of products and those of reactants:

$$\Delta G = G_{\text{products}} - G_{\text{reactants}} \tag{3.2}$$

This balance, which only takes into account the final state and the initial state of a given system, is described by the thermodynamics of equilibrium. It does not help to determine how the system has changed from state 1 (reactants) to state 2 (products). In most of the chemical reactions, and specifically in those concerning clays, the process is triggered when an energy barrier is passed (activation energy, E_a). This process is described by kinetics (Fig. 3.1). In this chapter, although the SI unit of energy is the joule (J), energies will be deliberately expressed in kcal and kJ in order to become accustomed to the literature (1 cal = 4.184 J), in which both units are used.

3.1.2
Free Energy of Formation of Clay Minerals

The fundamentals of thermodynamics being presented in many books, it is proposed here to learn progressively how to calculate the enthalpy of formation of clay minerals using first simple methods and then a more refined procedure. The goal is to take into account as closely as possible the actual crystal structure of these minerals.

3.1.2.1
Experimental Measurement of the Solubility at Equilibrium of Simple Minerals

General Principle

Let's consider a reaction $aA + bB \rightarrow gC + dD$; the equilibrium constant K_{eq} is calculated as follows:

$$K_{eq} = \frac{a_C^g \cdot a_D^d}{a_A^a \cdot a_B^b} \tag{3.3}$$

The Gibbs free energy of the reaction ΔG_r per mole is given by the difference between the Gibbs free energies of the products (ΔG_p) and those of the reactants:

$$\Delta G_R = \Delta G_p - \Delta G_r \tag{3.4}$$

$$\Delta G_R = \Delta G_R^0 + RT \ln \left(\frac{a_C^g \cdot a_D^d}{a_A^a \cdot a_B^b} \right) \tag{3.5}$$

ΔG_R^0 is the standard free energy of the reaction, namely the change in free energy produced by the transformation of a moles from A and b moles from B in their standard states into g moles from C and d moles from D in their standard states. At equilibrium $\Delta G_R = 0$, so:

$$RT \ln \left(\frac{a_C^g \cdot a_D^d}{a_A^a \cdot a_B^b} \right) = -\Delta G_R^0 \tag{3.6}$$

hence

$$K_{eq} = \left(\frac{a_C^g \cdot a_D^d}{a_A^a \cdot a_B^b} \right) = \exp \left(\frac{-\Delta G_R^0}{RT} \right) \tag{3.7}$$

$$\ln K_{eq} = \frac{-\Delta G_R^0}{RT} \tag{3.8}$$

R perfect gas constant (1.987×10^{-3} kcal $^\circ K^{-1}$ mol^{-1} or 8.314×10^{-3} kJ $^\circ K$ mol^{-1})

T absolute temperature in degrees Kelvin ($^\circ K$)

at 25 $^\circ C$, $RT = 2.303 \times 1.987 \times 10^{-3} \times 298$ kcal $\cdot$ mol^{-1}

$$\ln K_{eq} = \frac{-\Delta G_R^0}{1.364} \tag{3.9}$$

The equilibrium constant can then be calculated if the standard free energies of formation of the reactants and products are known. This simple relation can be applied efficiently to natural systems, and particularly to weathered layers and to waters from those springs emerging from them.

Saturation, Undersaturation: Application to Serpentine

The chemical composition of spring waters in areas with crystalline basement (magmatic and metamorphic) is controlled by the equilibrium with some clay species. This is the case of water-kaolinite equilibria in granitic or gneissic areas (Feth et al. 1964) or water-brucite or water-serpentine equilibria in ultrabasic bodies (Wildman and Whittig 1971). The equilibrium concept can be illustrated using the example of ultrabasic rocks. The solubility equation of serpentine can be written as follows:

$$Si_2O_5Mg_3(OH)_{4\,(c)} + 6H^+_{(aq)} \rightarrow 2H_4SiO_{4\,(aq)} + 3Mg^{2+}_{(aq)} + H_2O_{(l)} \tag{3.10}$$

where (c), (aq) and (l) represent the various physical states crystal, aqueous solution and liquid, respectively

$$K_{serp} = \frac{a^2_{H_4SiO_4} \times a^3_{Mg^{2+}} \times a_{H_2O}}{a_{serp} \times a^6_{H^+}} \tag{3.11}$$

$$\log K_{serp} = 3\log\left(a^2_{Mg^{2+}}/a^2_{H^+}\right) + 2\log a_{H_4SiO_4} \tag{3.12}$$

$$\Delta G^0_R = [2(-312.6) + 3(-108.7) + (-56.7)] - [(-964.87)]$$
$$= -43.13\,kcal\,mol^{-1} \tag{3.13}$$

$$\log K_{serp} = -\Delta G^0_R/1.364 = +31.62 \tag{3.14}$$

$$\log\left(a^2_{Mg^{2+}}/a^2_{H^+}\right) = +10.54 - 2/3\log a_{H_4SiO_4} \tag{3.15}$$

The diagram for the system $\log(a^2_{Mg2+}/a^2_{H+})$ as a function of $\log a_{H_4SiO_4}$ shows the boundary between the undersaturated domain where serpentine is dissolved and the supersaturated domain where it precipitates (Fig. 3.2). If the activity of dissolved silicon is constant, dissolution can be triggered either by the decreasing activity of magnesium or by the decreasing pH. The supersaturation that brings about precipitation of serpentine is produced by opposite causes.

Solubility of Kaolinite

The unit formula of kaolinite is $Si_2O_5Al_2(OH)_4$. For greater convenience in writing dissolution equations, only the half-unit is considered: $SiO_{2.5}Al(OH)_2$. The chemical activity of pure phases kaolinite and gibbsite are theoretically equal to 1.

Dissolution of kaolinite:

$$SiO_{2.5}Al(OH)_{2\,(c)} + 3H^+_{(aq)} \rightarrow Al^{3+}_{(aq)} + SiO_{2\,(aq)} + 2.5\,H_2O_{(l)} \tag{3.16}$$

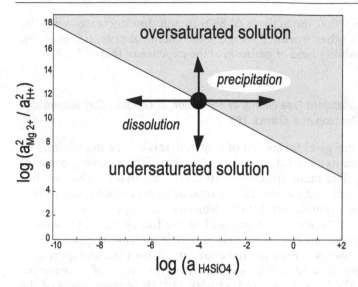

Fig. 3.2. Solubility of serpentine $Si_2O_5Mg_3(OH)_4$ at 25 °C, 1 atmosphere in water. Dissolution is triggered by undersaturation (decrease in $a_{H_4SiO_4}$ or in pH). Crystallisation is triggered by supersaturation (increase in $a_{Mg_{2+}}$ or in $a_{H_4SiO_4}$)

$$K_{sp.kaol} = \frac{a_{Al^{3+}} \times a_{SiO_2aq} \times a_{H_2O}^{2.5}}{a_{kaol} \times a_{H^+}^3} \tag{3.17}$$

$$\log \left(a_{Al^{3+}}/a_{3H^+} \right) = 0.5 \log K_{sp\text{-}kaol} - \log a_{SiO_2aq} - 2.5 \log a_{H_2O} \tag{3.18}$$

Dissolution of gibbsite:

$$Al(OH)_{3\,(c)} + 3\,H^+_{(aq)} \rightarrow Al^{3+}_{(aq)} + 3\,H_2O_{(l)} \tag{3.19}$$

$$K_{sp.kaol} = \frac{a_{Al^{3+}} \times a_{H_2O}^3}{a_{gib} \times a_{H^+}^3} \tag{3.20}$$

$$\log \left(a_{Al^{3+}}/a_{H^+}^3 \right) = \log K_{sp\text{-}gib} - 3 \log aH_2O \tag{3.21}$$

The following relation is inferred:

$$\log K_{sp.gib} - 3 \log a_{H_2O} = 0.5 \log K_{sp.kaol} - 2.5 \log a_{H_2O} - \log a_{SiO_2} \tag{3.22}$$

hence

$$\log a_{SiO_2} = 0.5 \log K_{sp.kaol} + 0.5 \log a_{H_2O} - \log K_{sp.gib} \tag{3.23}$$

Knowing the dissolution constants of gibbsite ($K_{sp.gib} = +8.205$) and kaolinite half-cell ($K_{sp.kaol} = +7.410$), the following is inferred:

$$\log a_{SiO_2\,aq} = -4.5 + 0.5 \log a_{H_2O} \tag{3.24}$$

This equation shows that the activity of SiO_2 in solution decreases with the activity of water. In other words, the higher the chemical potential of water, the greater is the stability field of gibbsite, at the expense of that of kaolinite.

3.1.2.2
Measurement of the Standard Free Energy of Formation of Complex Clay Minerals: Beavers Bend Illite (Routson and Kittrick 1971)

The standard free energy of formation of clay minerals can be determined by dissolution experiments aimed at measuring the value of the solubility constant at equilibrium K_{eq}. The main difficulty is to make sure that equilibrium has been reached. The best way, theoretically, would be to get close to equilibrium through two series of experiments in undersaturated and supersaturated states. Unfortunately, supersaturated solutions lead to the formation of metastable phases that impose intermediate equilibria. The only alternative proposed by the undersaturated way is to measure the apparent pK for various experiment durations. The method used will be explained by the study of an example extracted from works by Routson and Kittrick (1971): measurement of the standard free energy of formation of an illite (Beavers Bend).

The composition of the Beavers Bend illite is as follows: $[Si_{3.62}Al_{0.38}]O_{10}$ $(Al_{1.66}Fe^{3+}_{0.20}Mg_{0.13})(OH)_2K_{0.53}$. It has been dissolved in KOH solutions at pH 11.9. If ferric oxide (Fe_2O_3) and brucite $[Mg(OH)_2]$ are considered to be stable phases under the conditions of the experiment, then dissolution can be written as follows:

$$\log a_{SiO_2aq} = -4.5 + 0.5 \log a_{H_2O} \tag{3.25}$$

Considering that the activity of solid phases and H_2O is equal to 1, the solubility constant K_{sp} takes the following value:

$$K_{sp} = \left(K^+\right)^{0.53} \left(H_3SiO_4^-\right)^{3.62} \left[Al(OH)_4^-\right]^{2.04} \left(H^+\right)^{5.12} \tag{3.26}$$

or

$$pK_{sp} = 0.53p\left(K^+\right) + 3.62p\left(H_3SiO_4\right) + 2.04p\left[Al(OH)_4-\right] + 5.12pH \tag{3.27}$$

The K_{sp} values measured from the concentration of elements in solution (apparent solubility constant K_a) and plotted as a function of the inverse of the square root of time $(t^{-1/2})$ form a curve that can be extrapolated towards infinity. Extrapolation is valid only if the curve is continuous without any sudden change in slope. It is reliable if the curve is a straight line (Fig. 3.3). In this manner, one finds out that pK_{eq} equals 77.6.

The standard free energy of the reaction ΔG_r can be calculated:

$$\Delta G_r = -RT \ln K_{eq} = -1.364 \log K_{eq} = +105 \, \text{kcal} \, \text{g}^{-1} \tag{3.28}$$

Besides,

$$\Delta G_r = -RT \ln K_{eq} = -1.364 \log K_{eq} = +105 \, \text{kcal} \, \text{g}^{-1} \tag{3.29}$$

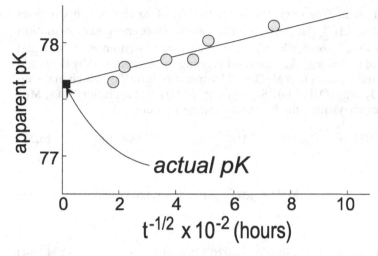

Fig. 3.3. Variation in apparent pK as a function of time. For an infinite time, the value of pK_{eq} at equilibrium is given by the intercept point at zero

from which the value of the standard free energy of formation of illite is inferred:

$$\Delta G_{illite} = [0.53 \times (-67.45)] + [2.04 \times (-310.02)] + [0.13 \times (-199.27)]$$
$$+ [3.63 \times (-299.7)] + [0.10 \times (-177.1)]$$
$$+ [5.12 \times (0)] - [10.20 \times (56.69)] - [105.8] \tag{3.30}$$

or

$$\Delta G_{illite} = -1,267.6 \, kcal \, mol^{-1} \tag{3.31}$$

3.1.2.3
Theoretical Calculation of Standard Free Energies of Formation

Calculation of Standard Free Energies of Formation of Oxides and Hydroxides Making up Silicates (Tardy and Garrels 1974)

As dissolution experiments are very delicate to perform, Tardy and Garrels (1974) have proposed a method of theoretical calculation enabling the values of the standard free energy of formation of clay minerals to be approached. This method rests on a simplifying assumption according to which each silicate can be represented by the sum of its elementary components in the form of oxides or hydroxides. These components have a free energy of formation within the silicate structure that is different from the one they have in the state of separate mineral phase. Therefore, it is first necessary to determine these free energies of formation within silicates (ΔG_{sil}), then to perform their sum in the stoichiometric proportions of the mineral considered.

The calculation of free energies of formation of oxides and hydroxides making up silicates ($\Delta G^0_{t,sil}$) is possible if the standard free energies of formation of these silicates are known. The number of equations written must be equal to the number of unknowns. The method is going to be explained by the study of an example: oxides SiO_2 and MgO, and hydroxide $Mg(OH)_2$, which make up chrysotile $Si_2 O_5 Mg_3 (OH)_4$, talc $Si_4 O_{10} Mg_3 (OH)_2$ and sepiolite $Si_3 O_6 Mg_2 (OH)_4$. Thus, for chrysotile, the following can be written:

$$2\Delta G^0_{t,sil}Mg(OH)_2 + \Delta G^0_{t,sil}MgO + 2\Delta G^0_{t,sil}SiO_2 = -965\,kcal\,mol^{-1} \tag{3.32}$$

for talc

$$\Delta G^0_{t,sil}Mg(OH)_2 + 2\Delta G^0_{t,sil}MgO + 4\Delta G^0_{t,sil}SiO_2 = -1,320\,kcal\,mol^{-1} \tag{3.33}$$

for sepiolite

$$2\Delta G^0_{t,sil}Mg(OH)_2 + 3\Delta G^0_{t,sil}SiO_2 = -1,020.5\,kcal\,mol^{-1} \tag{3.34}$$

The solution of these three equations yields:

$$\Delta G^0_{t,sil}Mg(OH)_2 = -203.3\,kcal\,mol^{-1} \tag{3.35}$$

$$\Delta G^0_{t,sil}MgO = -149.2\,kcal\,mol^{-1} \tag{3.36}$$

$$\Delta G^0_{t,sil}SiO_2 = -204.6\,kcal\,mol^{-1} \tag{3.37}$$

In the same manner, $\Delta G^0_{t,sil}$ Al_2O_3, $\Delta G^0_{t,sil}$ $Al(OH)_3$ and $\Delta G^0_{t,sil}$ H_2O have been calculated from the standard free energies of formation of kaolinite ($Si_2O_5Al_2(OH)_4$) and pyrophyllite ($Si_4O_{10}Al_2(OH)_2$):

$$\Delta G^0_{t,sil}Al_2O_3 = -382.4\,kcal\,mol^{-1} \tag{3.38}$$

$$\Delta G^0_{t,sil}Al(OH)_3 = -280.0\,kcal\,mol^{-1} \tag{3.39}$$

$$\Delta G^0_{t,sil}H_2O = -59.2\,kcal\,mol^{-1} \tag{3.40}$$

The values of $\Delta G^0_{t,sil}$ K_2O, $\Delta G^0_{t,sil}$ Na_2O, $\Delta G^0_{t,sil}$ $K(OH)$ and $\Delta G^0_{t,sil}$ $Na(OH)$ are determined using the standard free energies of formation of muscovite ($Si_3O_{10}Al_3(OH)_2K$) and paragonite ($Si_3O_{10}Al_3(OH)_2Na$):

$$\Delta G^0_{t,sil}K_2O = -188.0\,kcal\,mol^{-1} \tag{3.41}$$

$$\Delta G^0_{t,sil}Na_2O = -162.8\,kcal\,mol^{-1} \tag{3.42}$$

$$\Delta G^0_{t,sil}KOH = -123.6\,kcal\,mol^{-1} \tag{3.43}$$

$$\Delta G^0_{t,sil}NaOH = -111.0\,kcal\,mol^{-1} \tag{3.44}$$

Thus, gradually, the $\Delta G^0_{t,sil}$ of the major oxides and hydroxides are determined using the appropriate silicates, such as minnesotaite ($Si_4O_{10}Fe_3^{2+}(OH)_2$)

Table 3.1. Calculation of the standard free energy of formation of Beavers Bend illite using the method by Tardy and Garrels (1974). The resulting value is a little different from that measured by dissolution experiments (Routson and Kittrick 1971)

Component	Moles nb.	$\Delta G_f^\circ \, mol^{-1}$ $(kcal\,mol^{-1})$	ΔG_f° $(kcal\,mol^{-1})$	ΔG_f° dissol. $(kcal\,mol^{-1})$
K_2O_{sil}	0.26	−188	−48.88	
$SiO_{2\,sil}$	3.2	−204.6	−740.65	
$Al_2O_{3\,sil}$	1.02	−382.4	−390.05	
$Mg(OH)_{2\,sil}$	0.13	−203.3	−26.43	
H_2O_{sil}	0.87	−59.2	−51.5	
$Fe_2O_{3\,sil}$	0.10	−177.7	−17.77	
		Total	−1,275.28	−1,267.6

for FeO:

$$\Delta G_{t,sil}^0 FeO = -64.1\,kcal\,mol^{-1} \tag{3.45}$$

The value of $\Delta G_{t,sil}^0\, Fe_2O_3$ is considered equal to that of the free oxide phase, or $-177.7\,kcal\,mol^{-1}$.

Calculation of the Free Energy of Formation of the Beavers Bend illite (method by Tardy and Garrels 1974)

When the chemical composition of a phyllosilicate is known, it can be transformed into a sum of oxide and hydroxide molecules. Using the example of the Beavers Bend illite, the proportions indicated in Table 3.1 are obtained. Then the values of the free energies of formation of these oxides and hydroxides must be multiplied and added. The resulting value is an approximation of the free energy of formation of this illite.

Refinement of Calculations for the Beavers Bend Illite (Method by Vieillard 2000)

The foregoing theoretical calculations yield approximate values of the standard free energy of formation of clay minerals because of the drastic assumptions on which they rest. Particularly, the free energy of any oxide making up the silicate structure is supposed to have the same standard free energy of formation in all phyllosilicates. This method which is relatively efficient for phyllosilicates is definitely not applicable to other silicates or hydroxides. Consequently, empiric parameters have been searched to correct this defect (Tardy and Garrels 1976, 1977): ΔO^{2-} and $\Delta_{hydroxide}$.

The parameter ΔO^{2-} is defined as representing the difference between the free energy of formation of the elements in a crystal oxide and the free energy of formation of cations in aqueous solution:

$$\Delta O^{2-}M = \left(\Delta G_t^0 MO_x - \Delta G_{t,sil}^0 M^{2x+}\right)\,kcal\,mol^{-1} \tag{3.46}$$

M a given metallic cation

MO_x the oxide of this metal

M^{2x+} the corresponding cation of this metal in aqueous solution

x cation valency/2

For instance:

$$\Delta O^{2-} Mg^{2+} = \Delta G_t^0 MgO - \Delta G_t^0 Mg^{2+} \quad \text{or}$$
$$-136.10 + 108.7 \, \text{kcal mol}^{-1} = -27.40 \, \text{kcal mol}^{-1} \tag{3.47}$$
$$\Delta O^{2-} Li^+ = 2 \left(1/2 \Delta G_t^0 LiO_2 - \Delta G_t^0 Li^+ \right) \quad \text{or}$$
$$-134.29 + 140.44 \, \text{kcal mol}^{-1} = +6.11 \, \text{kcal mol}^{-1} \tag{3.48}$$
$$\Delta O^{2-} Al^{3+} = 2/3 \left(1/2 \Delta G_t^0 Al_2 O_3 - \Delta G_t^0 Al^{3+} \right) \quad \text{or}$$
$$2/3 \left(-189.1 + 116.0 / = -48.7 \, \text{kcal mol}^{-1} \right. \tag{3.49}$$

The endeavour to improve the accuracy of calculations of the free energy of formation of minerals has been continued by the integration of crystallographic and optic data, such as the refractive index, the molar volume, the mean distances between ions, the shortest cation-oxygen bond lengths in every site of the crystal structure. The calculation of the enthalpy of formation of oxides ($\Delta H_{t,oxides}^0$) is improved by the consideration of five parameters (Vieillard 1994, 2000):

1. Sites are distinguished according to whether they are occupied by one or several cations;

2. A new formulation of parameter $\Delta_H O^{2-}$ is used according to site occupancy;

3. Some polyhedra have extra-long cation-oxygen bonds;

4. Non-bridging oxygens between several neighbouring polyhedra are taken into account;

5. Introduction of the prediction of error as a function of the standard error made when measuring bond lengths.

The free enthalpy of formation of the Beavers Bend illite $\Delta G_{f\,(illite)}^\circ$ has been recalculated in order to illustrate the performances of this method. Uncorrected values of $\Delta G^\circ f$ of oxides (Table 3.2) have been compiled by Vieillard (2000):

$$K_{0.53} \left(Mg_{0.13} Fe_{0.2}^{3+} Al_{1.66} \right) \left(Si_{3.62} Al_{0.38} \right) O_{10} (OH)_2 \tag{3.50}$$
$$\rightarrow 0.53 \, K_2O + 1.02 \, Al_2O_3 + 0.1 \, Fe_2O_3 + 0.13 \, MgO + 3.48 \, SiO_2 + H_2O$$

Table 3.2. Uncorrected values of the free enthalpy of standard formation of oxides in a phyllosilicate frame ($\Delta G°$f oxides), compiled by Vieillard (2000)

Oxides	$\Delta G°$f oxides (kJ/mole)	Oxides	$\Delta G°$f oxides (kJ/mole)
K_2O	−322.10	Al_2O_3	−1,582.30
MgO	−569.30	SiO_2	−856.30
Fe_2O_3	−744.40	H_2O	−220.00

$$\Delta G_f^\circ \text{illite} = 0.53\Delta G_f^\circ \left(K_2O\right) + 1.02\Delta G_f^\circ \left(Al_2O_3\right)$$
$$+ 0.1\Delta G_f^\circ \left(Fe_2O_3\right) + 0.13\Delta G_f^\circ \left(MgO\right)$$
$$+ 3.48\Delta G_f^\circ \left(SiO_2\right)$$
$$+ \Delta G_f^\circ \left(H_2O\right) + \Delta G_{ox}^\circ \qquad (3.51)$$

hence

$$\Delta G_f^\circ \left(\text{illite}\right) = -5{,}170.17 + \Delta G_{ox}^\circ \qquad (3.52)$$

This approximate value can be improved if the energy of formation of oxide $[\Delta_G O = M^{z+} \text{(clay)}]$ is corrected by the effects of the electronegativity of cations in their respective crystal sites.

- Calculation of the numbers of oxygen atoms bonded to the various cations in the various sites (t: tetrahedral; o: octahedral; i: interlayer).

$0.265K_2O(l) + 0.83Al_2O_3(o) + 0.1Fe_2O_3(o) + 0.13MgO(o)$							
$+ 3.62SiO_2(t) + 0.19Al_2O_3(t) + H_2O(i)$							
Nb Oxyg.:	0.265	2.49	0.3	0.13	7.24	0.57	1=12
Mol. Fract.:	0.265/12	2.49/12	0.3/12	0.13/12	7.24/12	0.57/12	1/12

- Values of the parameter $\Delta_G O = M^{z+}$ (clay) used in the calculation of $\Delta G°_{ox}$ (Vieillard 2000)

	$K^+(l)$	$Mg^{2+}(o)$	$Al^{3+}(o)$	$Fe^{3+}(o)$	$Si^{4+}(t)$	$Al^{3+}(t)$	$H^+(arg)$
$\Delta_G O = M^{z+}$ (clay) (kJ/mole)	425.77	−112	−161.2	−164.1	−166.1	−197.31	−220

- Details of the calculation of $\Delta G°_{ox}$ (Table 3.3)

The following value of $\Delta G_{f(\text{illite})}^\circ$ is inferred $= -5{,}170.17-146.88$ or

$$\Delta G_{f\,(\text{illite})}^\circ = -5{,}317.05 \,\text{kJ mol}^{-1} \text{ or } -1{,}270.81 \,\text{kcal mole}^{-1} \qquad (3.53)$$

This value is closer to that measured by dissolution of illite (−1,262.6 kcal/mole) than the value given by the simple sum of oxides (−1,275.3 kcal/mole).

Table 3.3. Details of the calculation of the correction of electronegativity: $\Delta_G O = M^{z+}$ (clay) according to Vieillard's procedure (2000)

Interaction types	$12\,X_i.X_j$	$[\Delta_G O = M_i^{z+}$ (arg)$-\Delta_G O = M_j^{z+}$ (arg)] (kJ/mol)	$-12X_iX_j\,\{[\Delta_G O = M_i^{z+}$ (clay)$-\Delta_G O = M_j^{z+}$ (arg)]\}$ (kJ/mol)
		Within octahedral sites	
$Mg^{2+}(o)$–$Al^{3+}(o)$	0.0270	49.23	-1.33
$Mg^{2+}(o)$–$Fe^{3+}(o)$	0.0032	52.05	-0.17
$Al^{3+}(o)$–$Fe^{3+}(o)$	0.062	2.82	-0.18
		Within tetrahedral sites	
$Si^{4+}(t)$–$Al^{3+}(t)$	0.344	31.22	-10.74
		Between octahedral sites and OH	
$Mg^{2+}(o)$–$H(arg)$	0.0108	108	-1.17
$Al^{3+}(o)$–$H(arg)$	0.2079	58.77	-12.22
$Fe^{3+}(o)$–$H(arg)$	0.025	55.95	-1.40
		Between octahedral and tetrahedral sites	
$Mg^{2+}(o)$–$Si^{4+}(t)$	0.078	54.09	-4.24
$Mg^{2+}(o)$–$Al^{4+}(t)$	1.505	4.86	-7.32
$Al^{3+}(o)$–$Si^{4+}(t)$	0.181	2.04	-0.37
$Al^{3+}(o)$–$Al^{3+}(t)$	0.0062	85.31	-0.53
$Fe^{+3}(o)$–$Si^{4+}(t)$	0.1185	36.08	-4.28
$Fe^{+3}(o)$–$Al^{3+}(t)$	0.0143	33.26	-0.47
		Between the interlayer site and octahedral sites	
$K^+(l)$–$Si^{4+}(t)$	0.1599	591.86	-94.63
$K^+(l)$–$Al^{3+}(t)$	0.0126	623.08	-7.84
	$\Delta G_{ox.}^{\circ}$ (kJ/mole) =		-146.88

3.1.3
Equilibria Between Simple Minerals Without Solid Solutions

Knowing how to calculate the standard free energy of formation of clay species makes it possible to study the thermodynamics of mineral reactions involving these species. Why this is so important? For instance, this will help the geologists to determine some of the palaeo-conditions that prevailed in rocks at a given period of time. This is, of course, of great interest for oil or ore deposit prospecting.

3.1.3.1
Principle of Construction of Phase Diagrams

Fundamentals: The Phase Rule
Before undertaking the numerical construction of the phase diagrams applied to clays, it is first necessary to recall the basic principles to their design.

In a given chemical system, the maximum number of mineral phases forming stable assemblages depends on the number of variables considered. The Gibbs phase rule, generalised by Korzhinskii (1959), classifies variables into two classes and proposes a simple equation between variance, the maximum number of phases and the number of variables:

$$F = V_{ext} + V_{int} - p \qquad (3.54)$$

V_{ext} number of extensive variables. Extensive variables are the ones that depend on the system size. Recall that enthalpy, entropy, volume or mass are extensive variables. For a given chemical system at fixed pressure and temperature, the number of extensive variables is equal to the number of inert components. An inert component is a chemical component of which only the content plays a role and determines the quantity of each phase formed. Such components do not control the nature of phases. Note that, despite the qualifying term inert, they can be drained off (leaching) or concentrated in the given system.

V_{int} number of intensive variables. In natural systems, energy is transmitted – apart from the gravity field – by variations in temperature, pressure and chemical potential of the "mobile" component. Mobile components are elements whose chemical potential determines the nature of mineral phases and not their quantity. An element can be inert under certain circumstances and become mobile under others.

p maximum number of phases.

When pressure and temperature are fixed or do not vary enough to play their role of intensive variables, the phase rule is expressed as follows:

$$F = C_{mobile} + C_{inert} - p \qquad (3.55)$$

An Intensive Variable: the Chemical Potential of a Given Element (μ_x)

The chemical potential of a component x is expressed as follows: $\mu x = G°x + RT \ln a_x$ where $G°x$ and a_x are the standard free energy and the activity of this component respectively. In the case of ideality, this activity is equal to the concentration of the element x (C_x). For actual mixtures, $a_x = \gamma C_x$, where γ represents the activity coefficient. From these relations, one notes that μ_x varies like $\ln a_x$. The chemical potential is expressed in the same units as energy: kJ or kcal per moles.

Among all the chemical elements playing a role in those systems where clays are formed, liquid water holds a central place. The activity of liquid water changes as a function of its physical state: it is equal to 1 for free water and decreases as a function of the pore size and the fields of electric force at the surface of particles. Water exhibits a maximum organisation (that of ice) in the interlayer zones of smectites and vermiculites (Mercury 1997). The activity of water influences the activity of dissolved silica ($SiO_{2\,aq}$), and consequently the formation of phyllosilicates. Trolard and Tardy (1989) have shown the influence of the activity of water in the formation of minerals in lateritic alterations, notably the stability of goethite, hematite, gibbsite and kaolinite. It is the μ_{H_2O}–Al_2O_3–Fe_2O_3–SiO_2 system at 25 °C and 1 atmosphere.

Phase Diagrams

Several types of phase diagrams are commonly used for the study of assemblages of clay minerals according to whether the physicochemical systems in which they are formed are more or less open:

– *closed system*: no mobile component. The mineral assemblages are determined by the rock composition (X). If the simplified chemical system considered is reduced to three components or groups of components a–b–c, the

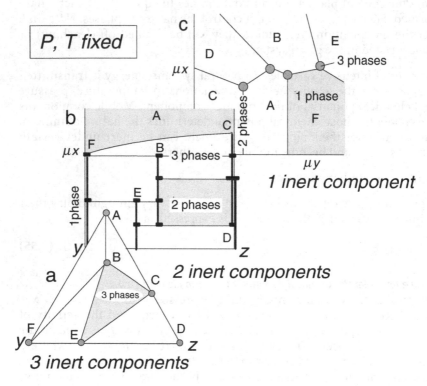

Fig. 3.4a–c. The three types of phase diagrams used to represent equilibria between mineral phases when pressure (P) and temperature (T) are fixed in a system with three components or groups of components. a) All components are inert (closed system). A triangular projection can be used. The maximum number of phases is three; three-phase assemblages are represented by *sub-triangles*. b) One component out of three is mobile (partially open system). The maximum number of phases is three. Three-phase assemblages are represented by *horizontal line segments* (chemical potential fixed). c) Two components out of three are mobiles (open system). The maximum number of phases is three. Three-phase assemblages are represented by *points* (intersection of three domains)

phase diagram can be constructed in a triangular projection (Fig. 3.4a). The phases represented by points are interconnected by tie lines. Three-phase assemblages delimit sub-triangles;

– *partially open system*: only one component out of three is mobile (a, for instance). The mineral assemblages depend on the chemical potential of this component in fluids and on the rock composition (μ_a–X_{b-c}). An orthogonal potential-composition representation is used (Fig. 3.4b). Phases are represented by vertical line segments and three-phase assemblages by horizontal line segments;

– *totally open system*: only one element remains inert, the two others are mobile (a and b, for instance). The mineral assemblages essentially depend on the chemical potential of the two mobile components (μ_a–μ_b–X_c). This last type of diagram is readily calculated using thermodynamic parameters determined for each mineral phase (Fig. 3.4c). Phases are represented by surfaces, and three-phase assemblages by points intersecting these surfaces.

3.1.3.2
SiO_2–Al_2O_3–K_2O–H_2O system at 25 °C, 1 atmosphere

The numerical construction of phase diagrams is relatively simple for minerals without solid solutions. Stability fields are straight lines whose slope is determined by the stoichiometry of the reaction and whose position is given by the numerical values of equilibrium constants, which are themselves determined by the free energies of formation of the various species. The SiO_2–Al_2O_3–K_2O–H_2O system at 25 °C, 1 atmosphere, is a classical example for learning the construction of potential-potential phase diagrams (open system).

In an open system, alkaline components and silica are mobile, only aluminium remains inert. The best representation of this system is an orthonormal diagram in which coordinates are the chemical potentials of alkaline elements and silica. The stability conditions of the simple mineral phases (gibbsite, kaolinite, muscovite, K-feldspar) have been studied by Hess (1966). Considering that there is no solid solution, each phase has the theoretical composition. The values of their free energies of formation (ΔG_f°) is given in Table 3.4.

Several mineral reactions are possible between the mineral phases of the SiO_2–Al_2O_3–K_2O–H_2O system. Each one of these reactions is characterised by an equilibrium constant K_{eq}. Solutions are assumed to be diluted, so $a_{H_2O} = 1$. The activities of pure bodies (solids) in their standard state are conventionally equal to 1 and $\Delta G_{H^+}^\circ = 0$. These assumptions permit calculation of the reaction constant $K_{reaction}$, which corresponds to the slope of the curve in coordinates $\log(a_{K^+}/a_{H^+})$ as a function of $\log(a_{H_4SiO_4})$.

(1) kaolinite → gibbsite

$$Si_2O_5Al_2(OH)_{4\,(c)} + 5\,H_2O_{(l)} \rightarrow 2\,Al(OH)_{3\,(c)} + 2\,H_4SiO_{4\,(aq)} \qquad (3.56)$$

Table 3.4. Composition and value of the free energies of formation (ΔG_f°) of the phases gibbsite – muscovite – kaolinite – pyrophyllite – K-feldspar characteristic of the SiO_2–Al_2O_3–K_2O–H_2O system ($kcal\,mol^{-1}$)

	Temperature (°C)		
Mineral	100	150	200
K-feldspar	−3,763.9	−3,777.7	−3,793.2
Muscovite	−5,615.8	−5,639.2	−5,656.8
Kaolinite	−3,806.6	−3,802.4	−3,836.3
K^+	− 290.4	− 295.8	− 301.7
H_2O_{liquid}	− 243.1	− 247.7	− 252.7

$$K_{eq} = \frac{a_{gib}^2 \times a_{H_4SiO_4}^2}{a_{kaol} \times a_{H_2O}^5} \tag{3.57}$$

$$K_{kaol\text{-}gib} = a_{H_4SiO_4}^2 \tag{3.58}$$

$$\Delta G_R^0 = [2(-276.2) + 2(-312.6)] - [(-905.6) + 5(-56.7)]$$
$$= 11.5\,kcal\,mol^{-1} \tag{3.59}$$

$$\log K_{kaol\text{-}gib} = 2\log a_{H_4SiO_4}$$
$$= -11.5/1.364\log a_{H_4SiO_4} = -4.2 \tag{3.60}$$

The straight line representing the reaction in a diagram $\log(a_{K^+}/a_{H^+})$ as a function of $\log(a_{H_4SiO_4})$ is parallel to the axis of ordinates and intercepts the axis of abscissas at −4.2 (Fig. 3.5).

(2) muscovite → kaolinite

$$2Si_3AlO_{10}Al_2(OH)_2K_{(c)} + 2H_{(aq)}^+ + 3H_2O_{(l)}$$
$$\rightarrow 3Si_2O_5Al_2(OH)_{4\,(c)} + 2K_{(aq)}^+ \tag{3.61}$$

$$K_{eq} = \frac{a_{kaol}^3 \times a_{K^+}^2}{a_{mus}^2 \times a_{H^+}^2 \times a_{H_2O}^3} \tag{3.62}$$

$$\log K_{mus\text{-}kaol} = 2\log\left(a_{K^+}/a_{H^+}\right) \tag{3.63}$$

$$\Delta G_R^0 = 2\left[1.5(-905.6) + (-67.5)\right] - \left[(-1336.3) + 1.5(-56.7)\right]$$
$$= -9.1\,kcal\,mol^{-1} \tag{3.64}$$

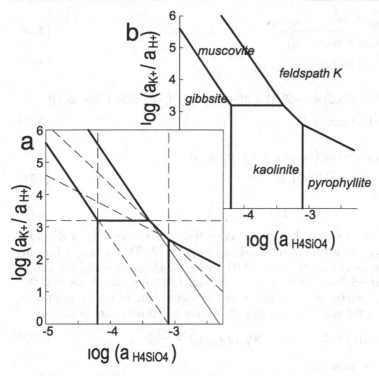

Fig. 3.5a,b. Phase diagram representing the relations between gibbsite, kaolinite, muscovite, pyrophyllite and K-feldspar under surface conditions (25 °C, 1 atmosphere). a) Numerical construction in the system a_{K+}/a_{H+} as a function of $a_{H_4SiO_4}$ (a represents the chemical activity). b) Phase diagram

$$\log K_{mus\text{-}kaol} = 2 \log (a_{K+}/a_{H+}) = -(-9.1/1.364) \text{ kcal mol}^{-1} \qquad (3.65)$$

Knowing that two muscovite molecules take part in the reaction, the stoichiometric coefficients must be divided by 2 to calculate the ΔG_R° for one molecule:

$$\log (a_{K+}/a_{H+}) = +3.3 \qquad (3.66)$$

The straight line representing the reaction in a diagram $\log(a_{K+}/a_{H+})$ as a function of $\log(a_{H_4SiO_4})$ is parallel to the axis of abscissas and intercepts the axis of ordinates at +3.3 (Fig. 3.5).

(3) muscovite → gibbsite

$$Si_3AlO_{10}Al_2(OH)2K_{(c)} + H^+_{(aq)} + 9H_2O_{(l)}$$
$$\rightarrow 3Al(OH)_{3\,(c)} + K^+_{(aq)} + 3H_4SiO_{4\,(aq)} \qquad (3.67)$$

$$K_{eq} = \frac{a_{gib}^3 \times a_{K^+} \times a_{H_4SiO_4}^3}{a_{mus} \times a_{H^+} \times a_{H_2O}^9} \tag{3.68}$$

$$K_{mus\text{-}gib} = \left(a_{K^+}/a_{H^+}\right) + \left(a_{H_4SiO_4}^3\right) \tag{3.69}$$

$$\Delta G_R^0 = [3(-276.2) + (-67.5) + 3(-312.6)] - [(-1336.3) + 9(-56.7)]$$
$$= +12.7\,\text{kcal mol}^{-1} \tag{3.70}$$

$$\log K_{mus\text{-}gib} = \log\left(a_{K^+}/a_{H^+}\right) + 3\log a_{H_4SiO_4}$$
$$= -\left(12.4/1.364\right) = -9.3 \tag{3.71}$$

$$\log\left(a_{K^+}/a_{H^+}\right) = -9.3 - 3\log\left(a_{H_4SiO_4}\right) \tag{3.72}$$

The slope of the straight line representing the reaction in a diagram $\log(a_{K+}/a_{H+})$ as a function of $\log(a_{H_4SiO_4})$ is -3. This straight line intercepts the axis of abscissas at -3.01 for $\log(a_{K^+}/a_{H^+}) = 0$ and the axis of ordinates at $+5.7$ for $\log a_{H_4SiO_4} = -5$ (Fig. 3.5). One can check that this straight line passes through the point of intersection of the straight lines representing the reactions (1) and (2) whose coordinates are as follows:

$$\log\left(a_{K^+}/a_{H^+}\right) = 3.3 \quad \text{and} \quad \log\left(a_{H_4SiO_4}\right) = -4.2 \tag{3.73}$$

(4) K-feldspar $\rightarrow$ kaolinite

$$2\,KAlSi_3O_{8\,(c)} + 2\,H_{(aq)}^+ + 9\,H_2O_{(l)}$$
$$\rightarrow Si_2O_5Al_2(OH)_{4\,(c)} + 2\,K_{(aq)}^+ + 4\,H_4SiO_{4\,(aq)} \tag{3.74}$$

$$K_{eq} = \frac{a_{kaol} \cdot a_{K^+}^2 \cdot a_{H_4SiO_4}^4}{a_{fels\,K}^2 \cdot a_{H^+}^2 \cdot a_{H_2O}^9} \tag{3.75}$$

$$K_{fels\,K\text{-}kaol} = \left(a_{K^+}/a_{H^+}\right)^2 \cdot \left(a_{H_4SiO_4}^4\right) \tag{3.76}$$

$$\log K_{fels\,K\text{-}kaol} = 2\log\left(a_{K^+}/a_{H^+}\right) + 4\log\left(a_{H_4SiO_4}\right) \tag{3.77}$$

$$\Delta G_R^0 = [(-905.6) + 2(-67.5) + 4(-312.6)] - [2(-895.4) + 9(-56.7)]$$
$$= +10.1\,\text{kcal mol}^{-1} \tag{3.78}$$

$$\log K_{fels\,K\text{-}kaol} = 2\log\left(a_{K^+}/a_{H^+}\right) + 4\log\left(a_{H_4SiO_4}\right)$$
$$= -10.1/1.364 = -7.40 \tag{3.79}$$

$$\log\left(a_{K^+}/a_{H^+}\right) = -3.7 + 2\log\left(a_{H_4SiO_4}\right) \tag{3.80}$$

The slope of the straight line representing the reaction in a diagram $\log(a_{K+}/a_{H+})$ as a function of $\log(a_{H_4SiO_4})$ is 2. This straight line intercepts the axis of ordinates at $+6.3$ for $\log(a_{H_4SiO_4}) = -5$ and the axis of abscissas at 1.85 for $\log(a_{K+}/a_{H+}) = 0$.

(5) K-feldspar → pyrophyllite

$$2KAlSi_3O_{8\,(c)} + 2H^+_{(aq)} + 4H_2O_{(l)}$$
$$\rightarrow Si_4O_{10}Al_2(OH)_{2\,(c)} + 2K^+_{(aq)} + 2H_4SiO_{4\,(aq)} \tag{3.81}$$

$$K_{eq} = \frac{a_{pyr} \cdot a_{K^+}^2 \cdot a_{H_4SiO_4}^2}{a_{felsK}^2 \cdot a_{H^+}^2 \cdot a_{H_2O}^4} \tag{3.82}$$

$$K_{fels\,K\text{-}pyr} = (a_{K^+}/a_{H^+})^2 \cdot (a_{H_4SiO_4}^2) \tag{3.83}$$

$$\log K_{fels\,K\text{-}pyr} = 2\log (a_{K^+}/a_{H^+}) + 2\log (a_{H_4SiO_4}) \tag{3.84}$$

$$\Delta G_R^0 = [(-1256) + 2(-67.5) + 2(-312.6)] - [2(-895.4) + 4(-56.7)]$$
$$= +1.4\,\text{kcal mol}^{-1} \tag{3.85}$$

$$\log K_{fels\,K\text{-}pyr} = 2\log (a_{K^+}/a_{H^+}) + 2\log (a_{H_4SiO_4})$$
$$= -1.4/1.364 = -1.03 \tag{3.86}$$

$$\log (a_{K^+}/a_{H^+}) = -0.56 + \log (a_{H_4SiO_4}) \tag{3.87}$$

The slope of the straight line representing the reaction in a diagram $\log(a_{K^+}/a_{H^+})$ as a function of $\log (a_{H_4SiO_4})$ is 2. This straight line intercepts the axis of abscissas at -0.56 for $\log(a_{K^+}/a_{H^+}) = 0$ and the axis of ordinates at -4.44 for $\log(a_{H_4SiO_4}) = -5$.

(6) K-feldspar → muscovite

$$3\,KAlSi_3O_{8\,(c)} + 2\,H^+_{(aq)} + 12\,H_2O_{(l)} \tag{3.88}$$
$$\rightarrow Si_3O_{10}Al_3(OH)_2\,K_{(c)} + 2\,K^+_{(aq)} + 6\,H_4SiO_{4\,(aq)} \tag{3.89}$$

$$K_{eq} = \frac{a_{mus} \cdot a_{K^+}^2 \cdot a_{H_4SiO_4}^6}{a_{felsK}^3 \cdot a_{H^+}^2 \cdot a_{H_2O}^{12}} \tag{3.90}$$

$$K_{fels\,K\text{-}mus} = (a_{K^+}/a_{H^+})^2 \cdot (a_{H_4SiO_4}^6) \tag{3.91}$$

$$\log K_{fels\,K\text{-}mus} = 2\log (a_{K^+}/a_{H^+}) + 6\log (a_{H_4SiO_4}) \tag{3.92}$$

$$\Delta G_R^0 = [(-1336.3) + 2(-67.5) + 6(-312.6)] - [3(-895.4) + 12(-56.7)]$$
$$= +19.7\,\text{kcal mol}^{-1} \tag{3.93}$$

$$\log K_{fels\,K\text{-}mus} = 2\log (a_{K^+}/a_{H^+}) + 6\log (a_{H_4SiO_4})$$
$$= -19.7/1.364 = -14.44 \tag{3.94}$$

$$\log (a_{K^+}/a_{H^+}) = -7.2 + 3\log (a_{H_4SiO_4}) \tag{3.95}$$

The slope of the straight line representing the reaction in a diagram $\log(a_{K^+}/a_{H^+})$ as a function of $(a_{H_4SiO_4})$ is 2. This straight line intercepts the axis of ordinates at $+7.8$ for $3\log(a_{H_4SiO_4}) = -5$ and the axis of abscissas at -2.4 for $\log(a_{K^+}/a_{H^+}) = 0$.

(7) pyrophyllite $\rightarrow$ kaolinite

$$Si_4O_{10}Al_2(OH)_{2\,(c)} + 5\,H_2O_{(l)} \rightarrow Si_2O_5Al_2(OH)_{4\,(c)} + 2\,H_4SiO_{4\,(aq)} \tag{3.96}$$

$$K_{eq} = \frac{a_{kaol} \cdot a_{H_4SiO_4}^2}{a_{pyr} \cdot a_{H_2O}^5} \tag{3.97}$$

$$K_{pyr-kaol} = \left(a_{H_4SiO_4}^2\right) \tag{3.98}$$

$$\log K_{pyr-kaol} = 2\log\left(a_{H_4SiO_4}\right) \tag{3.99}$$

$$\Delta G_R^0 = [(-905.6) + 2(-312.6)] - [(-1256) + 5(-56.7)] \tag{3.100}$$

$$= +8.7\,\text{kcal mol}^{-1} \tag{3.101}$$

$$\log K_{pyr-kaol} = 2\log\left(a_{H_4SiO_4}\right) = -8.7/1.364 = -6.38 \tag{3.102}$$

$$\log\left(a_{H_4SiO_4}\right) = -3.19 \tag{3.103}$$

The straight line representing the reaction in a diagram $\log(a_{K^+}/a_{H^+})$ as a function of $\log(a_{H_4SiO_4})$ does not intercept the axis of ordinates. It intercepts the axis of abscissas at -3.2.

3.1.3.3
The SiO_2–Al_2O_3–K_2O–H_2O System at Various Temperatures and Pressures

The thermodynamic parameters of minerals vary with temperature, thus displacing the limits between their respective stability fields. The example of the SiO_2–Al_2O_3–K_2O–H_2O system at 25, 100 and 200 °C permits illustration of these changes. Pressure plays a part too, but it is considered here as a variable not independent of temperature because it is controlled by the H_2O_{liquid}–H_2O_{vapour} equilibrium. Data bases are extracted from Fritz (1981).

The stability fields of various minerals move towards greater contents of H_4SiO_4 (Fig. 3.6). Recall that the solubility of quartz (and hence amorphous silica) increases with temperature. The saturation line of quartz is a useful reference point. Despite the observed displacement of stability fields, this reference always crosses the stability field of kaolinite, which nevertheless gets much narrower at 200 °C. The major change is the disappearance of the kaolinite-microcline tie line to the benefit of the muscovite-pyrophyllite tie line. This means that K-feldspars cannot transform directly into kaolinite at high temperatures. On the other hand, as the stability field of muscovite is very extended, this imposes practically that feldspars being altered yield micas (or illite or I/S).

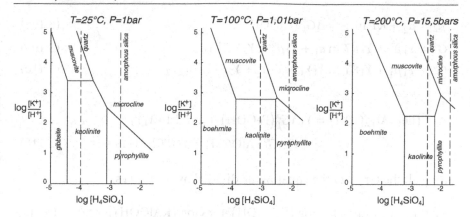

Fig. 3.6. Effect of temperature on the position of the stability fields of minerals in the SiO_2–Al_2O_3–K_2O–H_2O system (after Fritz 1981)

3.1.4
Equilibria Between Solid Solutions

3.1.4.1
Principle of H_2O – Solid Solutions Equilibria

In a solid solution, the chemical elements that substitute for each other in the same crystal sites (Fe^{2+} for Mg for instance) are characterised by a thermodynamic quantity: their chemical potential. Recall that the general expression of this potential for an element i is:

$$\mu_i = \mu_i^\circ + RT \ln a_i \tag{3.104}$$

where μ_i° is the reference chemical potential and a_i the activity of this element in the solid solution. When the solid solution is ideal, the activity is equal to the concentration: $a_i = X_i$.

Simple Minerals: Hydroxides Fe^{3+}–Al
How the thermodynamic formalism can be applied to solid solutions can best be understood by the consideration of a simple example (iron and aluminium hydroxides). The composition of the solid solution (Fe_{Y1} Al_{Y2} OOH) can be described by the mixture of two pure end members: Y_1 FeOOH $+Y_2$ AlOOH, whose activities are different from 1:

$$a_1 = \lambda_1 Y_1 \quad \text{and} \quad a_2 = \lambda_2 Y_2 \tag{3.105}$$

Therefore, the free energy of standard formation can be written as follows:

$$\bar{G}^\circ = Y_1\mu_1 + Y_2\mu_2 \tag{3.106}$$

$$\mu_1 = \mu_1^\circ + RT \ln a_1 = \Delta G_f^\circ(\text{FeOOH}) \tag{3.107}$$

$$\mu_2 = \mu_2^\circ + RT \ln a_2 = \Delta G_f^\circ \left(\text{AlOOH}\right) \tag{3.108}$$

$$\bar{G}^\circ = Y_1 \mu_1^\circ + Y_1 RT \ln a_1 + Y_2 \mu_2^\circ + Y_2 RT \ln a_2 \tag{3.109}$$

$$\bar{G}^\circ = Y_1 \mu_1^\circ + Y_1 RT \ln \lambda_1 Y_1 + Y_2 \mu_2^\circ + Y_2 RT \ln \lambda_2 Y_2 \tag{3.110}$$

$$\Delta G_f^\circ \left(\text{Fe}_{Y1}\text{Al}_{Y2}\text{OOH}\right) = Y_1 \Delta G_f^\circ \left(\text{FeOOH}\right) + Y_1 RT \ln \lambda_1 Y_1$$
$$+ Y_2 \Delta G_f^\circ \left(\text{AlOOH}\right) + Y_2 RT \ln \lambda_2 Y_2 \tag{3.111}$$

Letting all the terms of the equation be divided by $-RT$ gives:

$$\text{Log} K_{\text{Fe}_{Y1}\text{Al}_{Y2}\text{OOH}} = \left[Y_1 \log \left(\text{KFeOOH}\right) + Y_2 \log \left(\text{KAlOOH}\right) \right] \tag{3.112}$$
$$+ \left[Y_1 \log Y_1 + Y_2 \log Y_2 \right] + \left[Y_1 \log \lambda_1 + Y_2 \log \lambda_2 \right]$$

The three terms of the equation correspond to the energy of the mechanical mixture, the entropy and the excess energy from the mixture, respectively. In the case of ideal solid solutions, $\lambda_1 = \lambda_2 = 1$. The term excess energy from the mixture becomes zero and the activities are equivalent to concentrations. The equilibrium constant in aqueous solution can be written in the following simplified way:

$$K_{\text{Fe}_{Y1}\text{Al}_{Y2}\text{OOH}} = \frac{[\text{FeOOH}]^{Y1} [\text{AlOOH}]^{Y2}}{[\text{Fe}_{Y1}\text{Al}_{Y2}\text{OOH}]} \tag{3.113}$$

Phyllosilicates: the Iron-Bearing Kaolinite

Aluminium in kaolinite can be partially replaced by iron in a ferric state (refer to Sect. 2.1.3.2). Therefore, the iron-bearing kaolinite can be considered as the result of the mixture in limited proportions of the aluminous end member $Si_2O_5Al_2(OH)_2$ (kaolinite) with the Fe^{3+}-bearing end member $Si_2O_5Fe_2^{3+}(OH)_2$ (ferrikaolinite). The half unit formula can then be written as follows: $SiO_{2.5}(Fe_{Z1}^{3+}Al_{Z2})(OH)_2$ with $Z_1 + Z_2 = 1$. The solid-solution equilibrium is described by two partial equilibria:

Fe^{3+}-bearing end member Z_1 of the iron-bearing kaolinite

$$\text{SiO}_{2.5} \left(\text{Fe}^{3+}\right) \left(\text{OH}\right)_2 + 3\,\text{H}^+ \rightarrow \text{Al}^{3+} + \text{SiO}_2\,(\text{aq}) + 2.5\,\text{H}_2\text{O} \tag{3.114}$$

$$K_{\text{sp.Fekaol}} = \frac{a_{\text{Al}^{3+}} \times a_{\text{SiO}_2\,\text{aq}} \times a_{\text{H}_2\text{O}}^{2,5}}{a_{\text{Fekaol}} \times a_{\text{H}^+}^3} \tag{3.115}$$

$$\log \left(a_{\text{Fe}^{3+}}/a_{\text{H}^+}^3\right) = \log K_{\text{sp-Fekaol}} - \log a_{\text{SiO}_2(\text{aq})}$$
$$- 2.5 \log a_{\text{H}_2\text{O}} + \log a_{\text{Fekaol}} \tag{3.116}$$

Aluminous end member Z_2 of kaolinite

$$SiO_{2.5}(Al^{3+})(OH)_2 + 3H^+ \rightarrow Al^{3+} + SiO_{2\,(aq)} + 2.5\,H_2O \tag{3.117}$$

$$K_{sp.kaol} = \frac{a_{Al^{3+}} \times a_{SiO_2\,aq} \times a_{H_2O}^{2.5}}{a_{kaol} \times a_{H^+}^3} \tag{3.118}$$

$$\log(a_{Al^{3+}}/a_{H^+}^3) = \log K_{sp\text{-}kaol} - \log a_{SiO_2\,aq} - 2.5 \log a_{H_2O} + \log a_{kaol} \tag{3.119}$$

Ideality allows for similarity between activity and concentration; so $a_{Fekaol} = Z_1$ and $a_{kaol} = Z_2$.

3.1.4.2
Ideal Solid Solutions

How the composition of a mineral with a solid solution is fixed by its relations with the fluids and minerals that crystallise with it remains a major question. In other words, how can the partitioning of the chemical components between different minerals be calculated? This complex problem has been addressed in the study of the partitioning between the iron-bearing kaolinite and the oxides and iron hydroxides in laterites (Trolard and Tardy 1989). These tropical soils essentially composed of oxides, hydroxides and kaolinite can be described in the Fe_2O_3–Al_2O_3–SiO_2–H_2O system. In order to simplify the construction of phase diagrams, the μ_{H_2O}–Fe_2O_3–Al_2O_3 system will be considered for fixed values of μ_{SiO_2} and a total absence of iron in the kaolinite lattice will be assumed. The solid solutions are related to the presence of aluminium in goethite and in hematite.

Aluminous Goethite
It is considered as an ideal solid solution between the pure member FeOOH (Y_1) and the diaspore AlOOH (Y_2): $(Fe_{Y1}\,Al_{Y2})OOH$ with $Y_1 + Y_2 = 1$. The amounts of Al^{3+} and Fe^{3+} depend on the local conditions imposed by the rock composition:

$$b\,(Fe_{Y1}Al_{Y2})\,OOH + f\,(Fe_{Z1}Al_{Z2})\,SiO_{2.5}(OH)_2 \tag{3.120}$$

The equilibrium between Al-goethite and the solutions is reached when the partial equilibria with both composition end members is reached.
Aluminous end member (Y_2) of goethite

$$AlOOH_{(c)} + 3H^+_{(aq)} \rightarrow Al^{3+}_{(aq)} + 2H_2O_{(l)} \tag{3.121}$$

$$K_{sp.dia} = \frac{a_{Al^{3+}} \times a_{H_2O}^2}{a_{dia} \times a_{H^+}^3} \tag{3.122}$$

$$\log(a_{Al^{3+}}/a_{H^+}^3) = \log K_{sp\text{-}dia} - 2 \log a_{H_2O} + \log Y_2 \tag{3.123}$$

Ferric end member (Y_1) of goethite

$$FeOOH_{(c)} + 3H^+_{(aq)} \rightarrow Fe^{3+}_{(aq)} + 2H_2O_{(l)} \tag{3.124}$$

$$K_{sp.dia} = \frac{a_{Fe^{3+}} \cdot a^2_{H_2O}}{a_{goe} \cdot a^3_{H^+}} \tag{3.125}$$

$$\log\left(a_{Fe^{3+}}/a^3_{H^+}\right) = \log K_{sp\text{-}goe} - 2\log a_{H_2O} + \log Y_1 \tag{3.126}$$

Combining the equation of the aluminous end member with that of the dissolution of the iron-bearing kaolinite gives:

$$Y_2 = \frac{Z_2 K^{1/2}_{sp.dia}}{K_{sp.dia} \cdot a^{1/2}_{H_2O} \times a_{SiO_2aq}} \tag{3.127}$$

$$Z_2 = \frac{K_{sp.dia}\left[K^{1/2}_{sp.Fekaol} - K_{sp.goe} \cdot a_{SiO_2aq} \times a^{1/2}_{H_2O}\right]}{K_{sp.dia} \cdot K^{1/2}_{sp.Fekaol} - K_{sp.goe} \cdot K^{1/2}_{sp.kaol}} \tag{3.128}$$

At equilibrium, the chemical balance imposes that the total amounts of iron (Fe_t), aluminium (Al_t) and silicon (Si_t) confirm the following relations:

$$Fe_t = bY_1 + Si_t Z_1 \tag{3.129}$$

$$Al_t = bY_2 + Si_t Z_2 \tag{3.130}$$

$$Al_t > Si_t Z_2 \tag{3.131}$$

Calculation of the Chemical Potential-Composition Diagram for the μ_{H_2O}–Fe^{3+}–Al system
The formation of minerals in arenae (friable weathered rocks), i.e., clays and oxides in weathering profiles, is perfectly described in potential-composition phase diagrams. Under surface conditions, temperature and pressure can be considered as fixed. The weathered rock has a high porosity and a permeability much greater than that of fresh rock. Therefore, the chemical potential of elements in solution is controlled by the external environment of the system or by the activity of water. Only the less soluble elements accumulate in the rock (inert components). In a potential–composition system (P and T fixed), two chemical components (or groups of components) y and z are inert (2 intensive variables), the third x being mobile (1 intensive variable). The phase rule applied to this μ_x–y–z system is expressed as follows: $F = 2 + 1 - p$. This rule states that a maximum of three phases co-exist when $F = 0$. Zero degree of freedom means that the intensive variable no longer varies. This may occur for some of its values (Fig. 3.4b). When the intensive variable "varies", $F = 1$; assemblages are two-phased maximum.

The aluminous goethite is considered as an ideal solid solution between the pure member FeOOH (Y_1) and the diaspore AlOOH (Y_2): (Fe$_{Y1}$ Al$_{Y2}$)OOH with $Y1 + Y_2 = 1$. Hematite is considered as an ideal solid solution between the pure member FeO$_{1,5}$ (X_1) and corundum AlO$_{1,5}$ (X_2): (Fe$_{X1}$Al$_{X2}$)O$_{1,5}$ with

$X_1 + X_2 = 1$. The amounts of Al^{3+} and Fe^{3+} are controlled by the rock composition, or more accurately by the composition of the microenvironments in which the mineral reactions take place. The activity of water depends on its physical state in the rock porosity (capillary water, double layer).

3.1.4.3
Al-goethite – Kaolinite

The equilibrium between goethite and the solutions will be reached when the partial equilibria with both composition end members is reached.
   Aluminous end member (Y_2)

$$AlOOH_{(c)} + 3H^+_{(aq)} \rightarrow Al^{3+}_{(aq)} + 2H_2O_{(l)} \tag{3.132}$$

$$K_{sp.dia} = \frac{a_{Al^{3+}} \cdot a^2_{H_2O}}{a_{dia} \cdot a^3_{H^+}} \tag{3.133}$$

$$\log\left(a_{Al^{3+}}/a^3_{H^+}\right) = \log K_{sp\text{-}dia} - 2\log a_{H_2O} + \log Y_2 \tag{3.134}$$

Ferric end member (Y_1)

$$FeOOH_{(c)} + 3H^+_{(aq)} \rightarrow Fe^{3+}_{(aq)} + 2H_2O_{(l)} \tag{3.135}$$

$$K_{sp.dia} = \frac{a_{Fe^{3+}} \cdot a^2_{H_2O}}{a_{goe} \cdot a^3_{H^+}} \tag{3.136}$$

$$\log\left(a_{Fe^{3+}}/a^3_{H^+}\right) = \log K_{sp\text{-}goe} - 2\log a_{H_2O} + \log Y_1 \tag{3.137}$$

Combining the equation of the aluminous end member with that of the dissolution of kaolinite gives:

$$Y_2 = \frac{K^{-1/2}_{sp.kaol}}{K_{sp.dia} \cdot a^{1/2}_{H_2O} \cdot a_{SiO_2aq}} \tag{3.138}$$

As $Y_2 \le 1$, it yields:

$$1/2\log a_{H_2O} + \log a_{SiO_2aq} > 1/2\log K_{sp.kaol} - \log K_{sp.dia} \tag{3.139}$$

When this equation is not respected, only the aluminous goethite is formed; kaolinite is not formed. When these two minerals co-exist, the total amounts of iron (Fe_t) and aluminium (Al_t) of the rock are distributed as follows:

$$b\left(Fe_{Y1}Al_{Y2}\right)OOH + eSiAlO_{2,5}(OH)_2 \tag{3.140}$$

where b represents the number of aluminous goethite moles and e the number of kaolinite moles. For one Fe_2O_3 mole, $Fe_t = 2 = bY_1$ and $Al_t = bY_2 + e$. We know that $Y_1 + Y_2 = 1$, so: $b = 2/(1 - Y_2)$ and $e = Al_t - 2Y_2$. This shows that

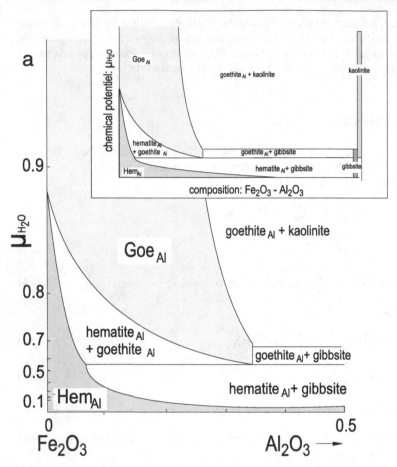

Fig. 3.7a,b. Potential-composition phase diagram for the μ_{H_2O}–Fe_2O_3–Al_2O_3 system. Pressure, temperature and μ_{SiO_2} are fixed. **a)** Details of the relationships between phases near the end member Fe (after Trolard and Tardy 1989). **b)** Overall phase diagram

the rock Al_t content must be greater than a minimum $Al_{t\,min}$ for kaolinite to be formed ($e > 0$): $Al_{t\,min} = 2Y_2/(1 - Y_2)$.

In this system, when goethite and kaolinite co-exist, the Al content in goethite and the amount of Al-goethite are determined (Fig. 3.7a). Consequently, hematite is not stable with kaolinite. Two- and three-phase assemblages in the whole system are described in Fig. 3.7b.

3.1.4.4
Al-hematite – Kaolinite

If one considers that hematite can be stable with kaolinite, goethite-gibbsite and hematite-gibbsite assemblages are forbidden (Fig. 3.7b). The dissolution reaction of the aluminous hematite ($Fe_{X1} Al_{X2})O_{1,5}$, where X_1 represents the number of moles of the end member hematite and X_2 that of the end member corundum, can be written with the two following equations:

$$0.5Fe_2O_3 + 3H^+ \rightarrow Fe^{3+} + 1.5H_2O \tag{3.141}$$

$$0.5Al_2O_3 + 3H^+ \rightarrow Al^{3+} + 1.5H_2O \tag{3.142}$$

The following equations are inferred:

$$\log\left(a_{Al^{3+}}/a_{H^+}^3\right) = \log K_{\text{sp-cor}} - 1.5 \log a_{H_2O} + \log X_2 \tag{3.143}$$

$$\log\left(a_{Fe^{3+}}/a_{H^+}^3\right) = \log K_{\text{sp-hem}} - 1.5 \log a_{H_2O} + \log X_1 \tag{3.144}$$

Writing again the dissolution equation of kaolinite gives:

$$X_2 = \frac{K_{\text{sp.kaol}}^{1/2}}{K_{\text{sp.cor}} \cdot a_{H_2O} \cdot a_{SiO_2aq}} \tag{3.145}$$

As $X_2 \leq 1$, the following is inferred:

$$\log a_{H_2O} + \log a_{SiO_2aq} > 1/2 \log K_{\text{sp.kaol}} - 1/2 \log K_{\text{sp.cor}} \tag{3.146}$$

When the aluminous hematite and kaolinite co-exist, the total amounts of iron (Fe_t) and aluminium (Al_t) in the rock are distributed as follows:

$$a\left(Fe_{X1} Al_{X2}\right)_2 O_3 + eSiAlO_{2,5}(OH)_2 \tag{3.147}$$

where a represents the number of aluminous hematite moles and e the number of kaolinite moles. The minimum total amount of aluminium ($Al_{t\,min}$) in the rock is equal to: $Al_{t\,min} = 2X_2/(1 - X_2)$. For one Fe_2O_3 mole ($Fe_t = 2$), $a = 2/(1 - X_2)$ and $e = Al_t - 2X_2/(1 - X_2)$. This shows that the composition and the amount of aluminous hematite are controlled by the rock composition (Fig. 3.7b).

3.1.5
Qualitative Construction of Phase Diagrams

Calculated phase diagrams are very useful tools but their construction is time consuming . In some cases, for petrological or chemical reasoning, it is better to use simplified ones that can be easily and quickly established. For instance, this is particularly the case in weathered rocks where the compositions of clays are highly variable due to their dependence to local chemical compositions (microsystems, see Sect. 6.1.2.4).

3.1.5.1
Qualitative Method by Graphical Derivation

Although the dissolution equations of the various composition end members can be written in a simple way thanks to the definition of the ideal solid solutions, their solution becomes rapidly a cumbersome numerical exercise that goes well beyond the scope of this book. However, a simpler graphical method enables the relations between phases when solid solutions are involved to be qualitatively solved: the derivation of phase diagrams based on the plotting of perpendiculars to the tie lines (Garrels and Mackenzie 1971). The problem of the thermodynamic status of the illite/montmorillonite mixed layers is now a classical example for illustrating this method. Indeed, one of the major questions raised by these minerals is to determine whether they behave like a mixture of two phases illite and montmorillonite or whether they belong to a continuous solid solution between an end member illite and an end member montmorillonite.

The first step is the drawing of the phase diagrams that meet both assumptions, using the chemical composition of the mineral phases forming in the $SiO_2-Al_2O_3-K_2O-H_2O$ system: gibbsite, kaolinite, muscovite, illite, montmorillonite, pyrophyllite, K-feldspar (Garrels 1984). Plotting the perpendiculars to the tie lines allows the construction of phase diagrams in coordinates $\log a_{K+}/a_{H+} - \log(a_{H_4SiO_4})$ corresponding to the mixture assumption (Fig. 3.8a) or to the solid solution assumption (Fig. 3.8b).

3.1.5.2
Construction of Phase Diagrams from Petrographic Observations

Determination of Parageneses of Clay Minerals

Clay minerals have a common characteristic: their small size. They form assemblages whose determination by petrographic observation is delicate. Despite these obstacles, observation remains the only way to know whether these minerals have formed together (paragenesis) or in the course of successive events in the rock history. In no way does an X-ray diffraction analysis of the $< 2\,\mu m$ fraction extracted from a rock powder make possible the determination of these assemblages. Indeed, minerals from early episodes and clays from recent neogenesis are very commonly observed together in the same diffraction pattern.

What are the petrographical properties of assemblages? There is no absolute criterion and traps are numerous. Theoretically, those clay minerals forming parageneses are associated in the same microsites. Unfortunately, these microsites form porous zones that are constantly visited by fluids. The latter bring about dissolution – precipitation reactions locally, showing new species that modify the existing assemblages. The only way to handle it is to observe a great number of identical microsites and to determine their clay content. In this manner, mineralogical sequences appear, enabling recent and early assemblages to be identified.

A Procedure for Petrographical Analyses of Microsystems

Two techniques are essential for determining clay minerals in their formation site: X-ray diffraction applied to microsites and chemical microanalyses. The petrographic examination of an altered rock for instance permits classification of the numerous microsites in which clay minerals appear in a limited

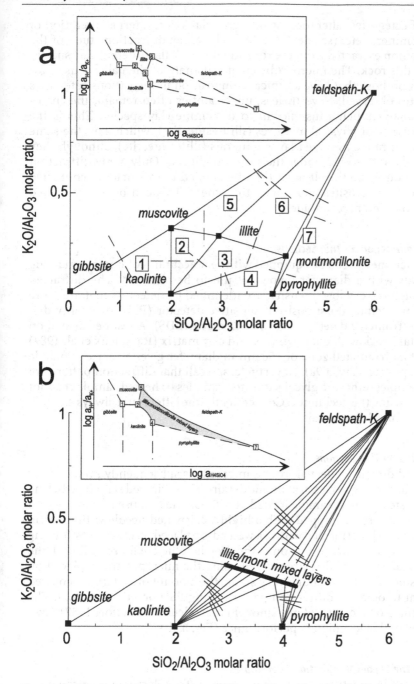

Fig. 3.8a,b. The thermodynamic status of illite/montmorillonite mixed layers treated by the derivation method by Garrels and Mackenzie (1971). **a)** The mixed layers are considered as a physical mixture of two phases. **b)** The mixed layers are considered as a continuous solid solution between two end members

number of categories: alteration at intergranular joint, internal alteration of
a primary mineral etc. (see Sect. 6.1.2.4). The X-ray diffraction study of the
< 2 μm fraction extracted from the rock identifies all the different clay species
present in this rock. The microprobe chemical analyses then allows us to lo-
calise most of those that have a typical composition in the various categories
of microsites observed. Nevertheless, uncertainties often remain, the chemical microanalyses being insufficient to determine clay species. This is the
case for instance of physical mixtures (illite+smectite), which have the same
compositional range as mixed layer minerals (illite/smectite), although they
are obviously not formed under the same conditions. Only X-ray diffraction
can remove uncertainties. It must then be applied to the material contained
in problematic microsites and not in the others. This can be accomplished
through two different techniques.

Direct Microdiffraction on Thin Sections

Some diffractometers can be equipped with very accurate collimators enabling
X-ray beams with a diameter below 100 μm to be obtained. The diffracted
rays, although of very low intensity, are identified by detectors that are more
sensitive than scintillators: position sensitive detector (PSD) or energy dis-
persive spectrometry detector with a Si-Li diode (EDS). A test performed on
a sedimentary rock with quartz grains and clay matrix (Rassineux et al. 1987)
shows that an irradiated zone of 150 μm in diameter gives a very exploitable
diffraction pattern (Fig. 3.9a). Nevertheless, recall that diffraction patterns of
oriented samples, ethylene glycol saturated samples or heated samples cannot
be obtained using this technique. Consequently, the latter is not always efficient
for determining clay minerals.

Microsampling on Thin Sections

The material directly collected from a microsite is subsequently prepared for
diffraction according to classical methods: air-dried oriented sample, ethylene
glycol saturated sample, sample heated at 550 °C. The tool used for extraction
is an ultrasound rod equipped with a highly sharpened needle with a 50 μm
head. Extraction is performed within a water drop placed on the microsite using
a micropipette. The surface tension of the drop is sufficient for retaining all the
extracted material. The drop, recovered using the micropipette, is placed on
a glass or silicon plate. Although very small, the amount of material recovered
is sufficient to obtain a diffraction pattern using PSD or EDS detectors. The
test performed on a crystal of metamorphic vermiculite (Beaufort 1987) gives
a good idea of this technique's performance (Fig. 3.9b).

Selection of the Chemical Projection Coordinates

The first stage in constructing phase diagrams is the selection of the chemical
system enabling the composition of the observed mineral phases to be represented. We have seen the complexity of natural minerals that may be composed
of up to ten major elements. Nevertheless, some simplifications can be made

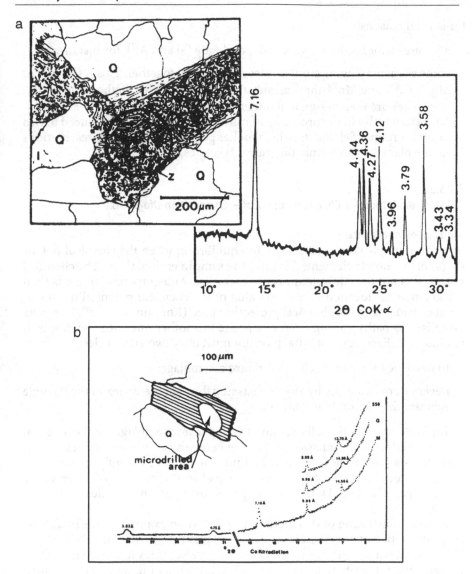

Fig. 3.9a,b. X-ray diffraction determination of clays in microsites. **a)** Direct diffraction on thin section of a sedimentary rock with quartz grains and clay matrix consisting of illite and chlorite. The *circle* corresponds to the irradiated zone (Rassineux et al. 1987). **b)** Microsampling of material within a metavermiculite crystal and X-ray diffraction analysis of this material in an air-dried oriented sample, an ethylene glycol saturated sample and a sample heated at 550 °C (Beaufort 1987)

for several reasons:

- all these elements do not vary independently (Si and Al^{IV} for instance);

- some elements playing similar roles are grouped together. This is the case of Mg^{2+}, Fe^{2+}, and Mn^{2+} that can substitute for each other in the same crystal sites. They are then designated under the same term R^{2+}. The chemical systems truly active in the processes of mineral reactions can be reduced to two or three groups of elements, which makes possible graphical representations in one plane (orthonormal diagrams, triangles).

3.1.5.3
A Graphical Method for Chemical Potential – Composition Diagrams

The Equipotential Concept

By definition, phases are said to be in equilibrium when the chemical potential(s) of the mobile element(s) is (are) the same in each of them. The chemical potential is not the equivalent of concentration. An equipotential line is then the geometrical location of a fixed value of the chemical potential of an element considered in the chemical projection used (Korzhinskii 1959). It may be visualised as paths leading from one phase to another one with which it is in equilibrium. Each equipotential plotting must obey two strict rules:

1. to describe only once each three-phase assemblage;

2. never be crossed twice by any line passing through the vertex of the triangle representing the mobile elements.

A third rule requires that all two- and three-phase assemblages of the diagram be described without omission. A tie line between two phases of a triangular diagram is a portion of equipotential line. The number of equipotential lines necessary for describing any system is equal to the number of three-phase assemblages plus two. Details of the proof are given in Meunier and Velde (1989).

A better illustration of this method is given by an example in the literature: serpentinisation of peridotites described by Fyfe et al. (1978) in the H_2O–MgO–SiO_2 system when H_2O, from inert component, becomes mobile. The system contains ten three-phase assemblages (Fig. 3.10a). Its full description is then given by $10+2 = 12$ different equipotential lines (Fig. 3.10b). The equipotential lines are plotted in order of decreasing value of the chemical potential μ_{H_2O}, from the highest (1) to the lowest (12).

Derivation of the Potential – Composition Diagram

Once the equipotential lines are plotted, a potential-chemical composition diagram for the μ_{H_2O}–$X_{MgO-SiO_2}$ system can be constructed qualitatively (Fig. 3.10c). Inert elements form the MgO–SiO_2 binary system in which the compositions of the mineral phases are reported (black squares in Fig. 3.10a). The mineral phases described by each equipotential line μ_{H_2O} in Fig. 3.11b

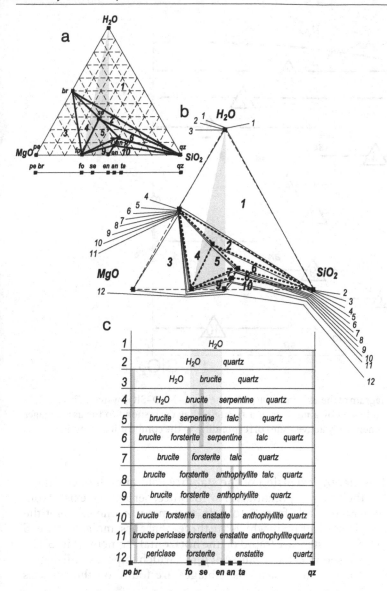

Fig. 3.10a–c. Derivation method of phase diagrams based on the equipotential lines (Korzhinskii 1959; Meunier and Velde 1989) applied to the serpentinisation of peridotites (after Fyfe et al. 1978). **a)** Phase diagram in the H_2O–MgO–SiO_2 system in which all components are inert. **b)** H_2O becomes a mobile component: 12 equipotential lines of μ_{H_2O} are necessary for describing all the mono-, two- and three-phase assemblages of the system. **c)** Construction of the potential-composition diagram for the μ_{H_2O}–MgO–SiO_2 system; *br*: brucite; *se*: serpentine; *ta*: talc; *an*: anthophyllite; *pe*: periclase; *fo*: forsterite; *en*: enstatite; *qz*: quartz. The chemically active zone corresponding to the composition of peridotites is *shaded*

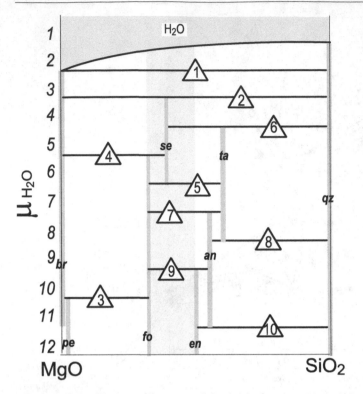

Fig. 3.11. Phase diagram of the serpentinisation in the μ_{H_2O}–MgO–SiO$_2$ system. Three-phase assemblages are indicated by a number in a *triangle*; they correspond to the assemblages in Fig. 3.10a. The chemically active zone corresponding to the composition of peridotites is *shaded*

are mentioned so that the extension of the stability field of each one of them can be plotted in the order of potentials. For instance, serpentine exists from potential 4 to potential 6, talc from 5 to 8. In this manner, one can plot the μ_{H_2O}–MgO–SiO$_2$ phase diagram in which the three-phase assemblages defined in the H$_2$O–MgO–SiO$_2$ system are found on the horizontal segments (Fig. 3.11). These segments correspond to fixed values of μ_{H_2O}, which imposes a nil degree of freedom ($F = 0$). The two-phase assemblages are found on the surfaces ($F = 1$).

3.1.5.4
Equilibria Between Solid Solutions: Representation Modes

Clay minerals form complex solid solutions (see Chap. 2). In natural environments where conditions remain stable (P, T constant), like under weathering conditions, they form mineral assemblages between which the chemical elements contained in the rock are distributed. The concept of equilibrium between coexisting phases is then fundamental. Knowing how to represent these equilibria in phase diagrams is essential to the petrographic study of weathered rocks, a field in which great advances have been made with the advent of the microprobe.

Let's consider a system with three components a, b and c (P, T fixed) in which five phases (solid solutions) precipitate and form four three-phase assemblages (Fig. 3.12a). The two-phase fields are indicated by the paths of the equipotential lines. The equilibrium between the solid solutions O and P is detailed in Fig. 3.12c; it is symbolised by a pencil of tie lines. The potential-composition diagram (Fig. 3.12b) shows the effects of the change in the thermodynamic status of the component a when it becomes mobile (intensive variable). The equilibrium between the phases O and P is represented by a pencil of tie lines (Fig. 3.12c–d). In both types of diagrams, the definition of equilibrium is respected: the chemical potential μ_a is identical in each one of the phases at equilibrium.

Under natural conditions, equilibria are seldom perfectly reached and the chemical compositions of the various phases are rarely controlled by a distribution coefficient. In the case of clay minerals formed in weathered rocks for instance, the phases observed are actually those corresponding to the conditions imposed, but their compositions depend on the local availability of the chemical elements and on the reaction rate. Compositions are often scattered within a domain whereas they should be controlled by tie lines. Nevertheless, despite these interfering effects, the trend towards equilibrium can sometimes be detected when a same phase exhibits different chemical composition domains depending on the phase with which it forms an assemblage. In this manner, illite that is formed in granitic arenae contains more Al when it coexists with kaolinite rather than with smectites or vermiculites (see Sect. 6.1.3.1).

3.2
Kinetics of Mineral Reactions

How long a bentonite smectite may "survive" when aggressed by cement alkaline solutions or granitic waters? What is the effect of heat transfer in the smectite-to-illite conversion? As the prediction of barrier stability is a fundamental parameter of any decision for waste storage, the kinetic studies take a great importance. This is why it is necessary to understand how kinetic constants can be determined through experiments. We shall see further how they can also be determined using natural systems such as diagenetic series or contact metamorphosed clayey rocks (Sect. 8.1.2.2 and Sect. 9.2.1.2)

3.2.1
Introduction

Considering that the kinetics of a mineral reaction with a given activation energy depends on temperature, the measurement of a duration necessitates the control of temperature too. Time measuring is possible in some cases: (1) experiments; (2) particular natural settings in which dating is possible (soils

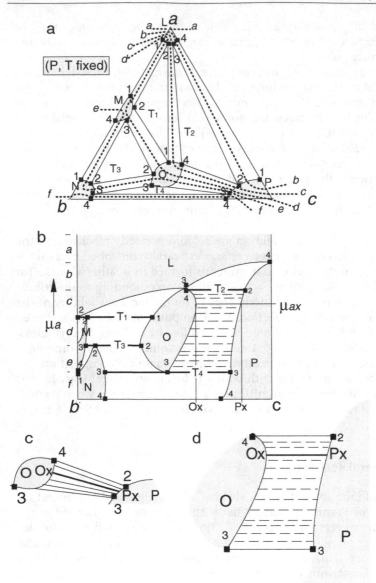

Fig. 3.12a–d. Representation of equilibria between solid solutions in phase diagrams. **a)** System without mobile component. The equilibria between solid solutions correspond to the two-phase fields covered by the various equipotential lines. **b)** System with one mobile component: a. The equilibria between solid solutions are defined by the chemical potential μ_a. **c)** Detail of Figure a representing the equilibrium between solid phases for the phases O and P in the system without mobile component. **d** Detail of the equilibrium between the same phases in the system with 1 mobile component

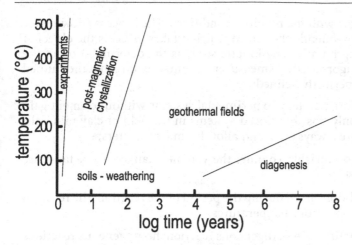

Fig. 3.13. Schematic representation of the domains *time × temperature* (*t × T*), in which clay minerals are formed or transformed

formed under equivalent climato-pedogenetic conditions during historical periods or glacier recoil). It remains difficult to do in hydrothermal or diagenetic systems where the unique possibility depends on the fit of measurement series using kinetic models. Despite these difficulties, the factor *time × temperature* (*t × T*)–basic element of the kinetics of mineral reactions – is characteristic of the main environments in which clays are formed or transformed (Fig. 3.13). This factor will be addressed in this book for each one of the fields – either experimental or natural – where coherent data have been published.

This section, based on the experimental data acquired in the study of the smectite-to-illite transformation, is aimed at recalling the elementary laws of kinetics. This transformation reaction is the best documented today because of its significance in the field of diagenesis (search for petroleum) as well as in the field of the prediction of the sustainability of clay barriers planned for the confinement of radioactive waste storage.

3.2.2
Fundamental Laws of Kinetics

The fundamentals will be summarised here in order to "refresh" the mind. Detailed explanations will be found in many books dealing with chemistry. Here, the goal is to present as simply as possible the principles that are applied to clay synthesis experiments.

3.2.2.1
Theoretical Basis of Kinetic Laws

Mineral reactions, like any chemical reaction, are a response to changes in the system conditions. They make possible the change from a disequilibrium state to an equilibrium state with the required conditions. The process allowing for this change, and consequently the rate at which it takes place, is the object of kinetics. The starting point of any kinetic study is the determination of the reaction rate. To be rigorous, these measurements first require that the studied chemical system be perfectly defined:

- open system or closed system: experimental systems without element gain nor loss are the only possible closed systems in the field of clay minerals. Natural systems are always open and allow for mass transfers;

- constant volume or variable volume: the volume changes if reactants are added continuously;

- temperature conditions fixed during the period of the reaction: the reaction rate is measured for various temperatures;

- homogeneous reaction or heterogeneous reaction: homogeneous reactions are those occurring totally in one phase. A reaction that requires the presence of at least two phases to take place at the observed rate is a heterogeneous reaction. This is the case of the reactions described in this book.

Let's consider, in a closed system with constant volume, a reaction of type

$$aA + bB \rightarrow xX + yY \tag{3.148}$$

where $a, b, c,$ and d are the concentrations of the minerals A, B, C and D, respectively. To simplify, the square brackets symbols of the concentration will not be used, $[a]=a$. The rate of the mineral reaction is measured by the variation in the concentration of the reactants (A or B) or of the products (X or Y) as a function of time. It is expressed as follows:

$$-\frac{da}{dt}ou - \frac{db}{dt}ou + \frac{dx}{dt}ou + \frac{dy}{dt} \tag{3.149}$$

The sign is imposed by the necessity for giving a positive value to the numerical expression of the rate. Units are then concentrations divided per time: $mol\,dm^{-3}\,s^{-1}$ or $mol\,litre^{-1}\,s^{-1}$ or $M\,s^{-1}$. The reaction rate has several mathematical expressions according to whether it is constant or varies with time. These expressions have been summarised by Lasaga (1981); they are graphically presented in Fig. 3.14.

Zero-order reaction. Concentration varies proportionally to the reaction time:

$$-\frac{da}{dt} = k \text{ or } a = a_0 - kt \tag{3.150}$$

First-order reaction. The decrease in concentration becomes slower with time. Concentration tends asymptotically towards zero. The rate for any value of time is given by the negative slope of the curve concentration as function of time:

$$-\frac{da}{dt} = ka \tag{3.151}$$

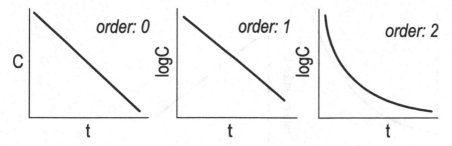

Fig. 3.14. Variation in concentration as a function of time for order reactions $n = 0$, $n = 1$ and $n > 1$

Rearranging and integrating the equation of time $t = 0$, for which the initial concentration is a_0, into the time t, for which the initial concentration is a, yields:

$$-\frac{da}{a} = k\,dt \tag{3.152}$$

$$-\int_{a0}^{a} \frac{da}{a} = \int_{0}^{t} k\,dt \tag{3.153}$$

$$\ln a - \ln a_0 = -kt \tag{3.154}$$

$$\ln \frac{a_0}{a} = kt \tag{3.155}$$

$$a = a_0 e^{-kt} \tag{3.156}$$

A first-order reaction is recognisable from an exponential decrease in concentration as a function of time.

Order reaction ($n > 1$). The order reactions ($n > 1$) are expressed as follows:

$$\frac{da}{dt} = -ka^n \tag{3.157}$$

$$\frac{1}{a^{n-1}} - \frac{1}{a_0^{n-1}} = (n - 1/kt \tag{3.158}$$

Temperature and reaction rate. The rate of most chemical reactions, and notably those relating to clays in rocks, increases with temperature (Fig. 3.15). S.A. Arrhenius, a 19th-century Swedish chemist, discovered the relationships between the rate constant k and the temperature through the following rule:

$$k = A e^{-\frac{E_a}{RT}} \tag{3.159}$$

E_a activation energy of the reaction ($J\,mol^{-1}$)

R perfect gas constant ($8.314\,J\,mol^{-1}\,K^{-1}$)

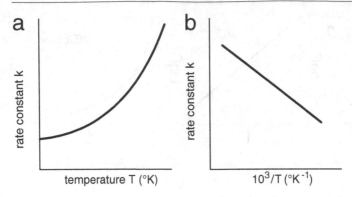

Fig. 3.15a,b. Variation of the rate constant k as a function of temperature T expressed in degrees Kelvin (**a**) and its equivalent expressed as a function of the inverse of temperature $1/K$ (**b**)

A pre-exponential factor (same units as k, which depends on the order of the reaction involved)

The constants A and E_a are characteristic of the reaction considered; they change from one reaction to another. The concept of activation energy rests on the following observation: to pass from a stable state having a high energy level, to another stable state having a lower energy level, the reaction has to get over a transitory energy barrier called the activated complex state (Fig. 3.1).

3.2.2.2
An Example: the Experimental Synthesis of Illite from Glasses

The experimental synthesis has been used for more than a century to determine the stability fields of the constituent minerals of natural rocks. The example selected here rests on the study by Eberl and Hower (1976) dedicated to the progressive formation of illite from the transformation of a glass of known chemical composition under controlled conditions of temperature and duration. The glass is made from a stoichiometry of beidellite: $[Si_{3,66} \ Al_{0,34}]$ O_{10} (Al_2) $(OH)_2$ $K_{0,34}$ or $Na_{0,34}$. Experimental data are given in Table 3.5.

The experiments carried out with the K-glass show the following sequence of mineral reactions:

1. glass $\rightarrow$ 100% expandable smectite

2. smectite $\rightarrow$ illite-smectite mixed layers+kaolinite (or pyrophyllite) + quartz + feldspar

The formation of illite layers in the second reaction can be described by a first-order kinetic equation with the following integration:

$$\ln \frac{a}{a - x} = kt \tag{3.160}$$

Table 3.5. Kinetic data of the experimental transformation from smectite into illite-smectite mixed layers (after Eberl and Hower 1976). Kaol.: kaolinite; Qz: quartz; Felds.: alkaline feldspar; Pyr: pyrophyllite

Temperature (°C)	Duration (days)	Smectite (%)	Other products
152	101	No reaction	
260	99	85	Kaol.
260	266	65	Kaol.
300	31	85	Kaol. + qz + felds.
300	88	70	Kaol. + felds.
343	5	90	–
343	23	80	Kaol.
343	88	35	Qz + felds.
343	99	25	Qz?
393	3	70	Qz
393	14	70	Qz? + pyr.?
393	23	35	Qz + pyr.? + felds.
393	169	15	Qz + pyr.

a initial concentration of smectite (100%)

x percentage of smectite layers transformed into illite.

Therefore, $a-x$ represents the percentage of smectite layers measured by the degree of expansion on the X-ray diffraction patterns X (column smectite % in Table 3.3), so

$$\ln \frac{100}{\% \text{ smectite}} = kt \qquad (3.161)$$

The rate constant k is expressed in days^{-1} here. If the order of the reaction is one, the Eq. (3.161) shall be represented by a straight line with slope k in a diagram ln (100% smectite) as a function of time for each one of the temperatures selected for the experiment (Fig. 3.16).

The authors have shown that the values of k increase with increasing temperature (Table 3.6). The activation energy of the reaction is calculated from the Arrhenius equation:

$$E_a = \frac{\Delta \log_{10} k}{\Delta \frac{1}{T}} \times 2.303R \qquad (3.162)$$

The equation above is represented by a straight line in a diagram $\log_{10} k$ as a function of the inverse of temperature (Fig. 3.17); its slope is equal to: $\frac{E_a}{2.303R}$. From this diagram is inferred the value of E_a for the illitisation reaction of smectite under experimental conditions: $E_a = 81.9 \pm 14.6 \, \text{kJ mol}^{-1}$.

What does the value of E_a represent? In reality, the mechanism enabling smectite to disappear to the benefit of illite is not well known. Several explanations may be proposed. For instance, one can imagine that each smectite layer

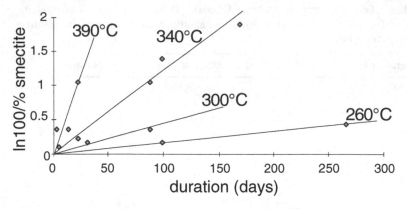

Fig. 3.16. Graphical representation of Eq. (3.161): $\ln \frac{100}{\%\text{smectite}} = kt$. The rate constant for each temperature corresponds to the slope of the various straight lines

Table 3.6. Values of the rate constants measured experimentally (slope of the straight lines in Fig. 3.9)

Temperature (°C)	k (day^{-1})
260	$1.2 \pm 0.5 \times 10^{-3}$
300	$4.3 \pm 1.2 \times 10^{-3}$
340	$13.0 \pm 3.0 \times 10^{-3}$
390	$45.0 \pm 7.0 \times 10^{-3}$

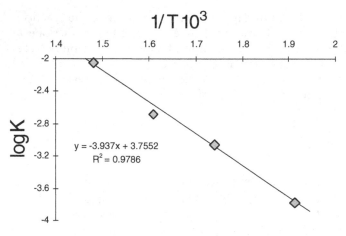

Fig. 3.17. Arrhenius diagram using the values of k given in Table 3.6.. The calculated activation energy is: $Ea = 81.9 \pm 14.6\,\text{kJ mol}^{-1}$

is transformed individually into illite. In this case, E_a represents the energy necessary for breaking the Si–O bond in tetrahedra so that Al may substitute for Si. One can also imagine that the illite layers grow on certain smectite layers by heterogeneous nucleation and "feed" on the chemical elements resulting from the dissolution of the other smectite layers. In this case, two mechanisms take place in parallel: dissolution and growth. The slower one controls the rate of the global reaction. If it is dissolution, E_a then represents the energy necessary for breaking all the chemical bonds within the smectite crystals. By contrast, if it is growth, E_a represents the energy necessary for each ion or group of ions to find its place in the growth steps, i.e., crystal incorporation (see Sect. 1.2.1.3).

3.2.3
Kinetics of the Montmorillonite → Illite Reaction

The transformation reaction of montmorillonite into illite has been widely studied because of its vital importance for the calculation of clay barrier durability or for prospecting potential zones of oil reservoirs in diagenetic series. The reaction may be written qualitatively as follows:

$$\text{montmorillonite} + \text{K} \rightarrow \text{illite} + \left(\text{Fe}^{2+}, Mg^{2+}\right) + \text{SiO}_2 \qquad (3.163)$$

Two facts clearly appear: (1) the reaction depends on the concentration of the dissolved potassium; (2) it is a heterogeneous reaction that produces several different mineral phases: quartz and one ferro-magnesian phyllosilicate (saponite or chlorite at low or high temperatures respectively). Ignoring *a priori* the order of this reaction, the general kinetic equation may be written as follows (Pytte 1982; Pytte and Reynolds 1989; Huang et al. 1993):

$$-\frac{dS}{dt} = k \left[K^+\right]^a S^b \qquad (3.164)$$

S molar fraction (smectite %) of smectite in the illite-smectite mixed layer

K^+ concentration of the dissolved potassium

k rate constant

a and b constants defining the order of the reaction.

If the concentration of potassium $[K^+]$ is constant, $k[K^+] = k'$ is constant; the equation above becomes:

$$-\frac{dS}{dt} = k'S^b \qquad (3.165)$$

This equation shows that the smectite ratio in illite-smectite mixed layers decreases with time according to a power rule of the molar fraction of smectite in mixed layer minerals (smectite %). If $b = 2$, the order of the reaction kinetics is 2 and the equation becomes:

$$-\frac{dS}{dt} = k'S^2 \tag{3.166}$$

or

$$-\frac{dS}{S^2} = k'\,dt \tag{3.167}$$

The integral of the equation above gives: $\frac{1}{S} = k'y + I$ where I is an integration constant. At the limit conditions $S = 1$ and $t = 0$, one obtains: $\frac{1}{S} = k't + 1$. This shows that the quantity $\frac{1}{S}$ is a linear function of t. This has been verified by experimental data from Huang et al. (1993), as shown in Fig. 3.18a.

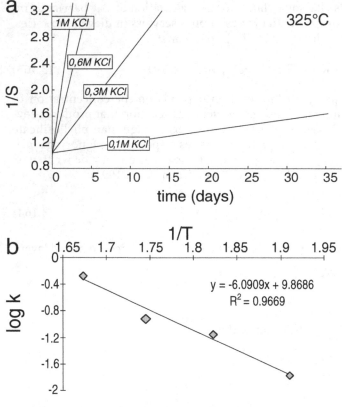

Fig. 3.18a,b. Kinetics of the smectite → illite reaction according to the experimental data from Huang et al. (1993). **a)** Progress of the reaction as a function of time (S being the smectite content in I/S). **b)** Arrhenius diagram showing that the experimental data are consistent with a second-order kinetics

Table 3.7. Values of k' for various temperatures according to the experimental data from Huang et al. (1993)

Temperatures (°C)	$\log k$ (days^{-1})
250	−1.8
275	−1.13
300	−0.91
325	−0.25

We know that $k' = k[K^+]$, so $\log k' = \log k + a \log[K^+]$. The value of a can then be obtained by the slope of the curves in a diagram $\log k'$ as a function of $\log[K^+]$. Huang et al. (1993) have shown that $a \approx 1$. This suggests that the reaction is of first order for the concentration of potassium ($k' = k[K^+]$) and of second order for the molar fraction of smectite (smectite %) in illite –smectite mixed layers. Equation

$$-\frac{dS}{dt} = k[K^+]^a S^b \tag{3.168}$$

is then written as follows:

$$-\frac{dS}{dt} = k[K^+] S^2 \tag{3.169}$$

k' is determined in diagram $\log$ (duration in days) as a function of $\log[K^+]$ for the value of $\log[K^+] = 0$. When $\log[K^+] = 0$, $k' = k$. The results obtained by Huang et al. (1993) are given in Table 3.7. The activation energy E_a can then be obtained from these data exploited in an Arrhenius diagram according to the following equation:

$$k = A \exp^{-\frac{E_a}{RT}} \tag{3.170}$$

hence

$$\ln k = -\left(\frac{E_a}{R}\right)\left(\frac{1}{T}\right) + \ln A \tag{3.171}$$

The results obtained (Fig. 3.18b) give $E_a = 117 \, \text{kJ} \, \text{mol}^{-1}$ and $A = 8.08 \times 10^{-4}$ $\text{s}^{-1} \, \text{mol}^{-1} \, \text{l}$.

The smectite $\rightarrow$ illite reaction has been widely studied either by experimental reproduction or by modelling of the I/S series observed in natural deposits. The values of the activation energy so determined vary greatly. They must be used as indicators depending on the conditions in which the reaction has occurred and mostly corresponding to stages in the transformation process (for instance, low-charge smectite $\rightarrow$ high-charge smectite $\rightarrow$ vermiculite $\rightarrow$ illite). The main data available in the literature are as follows:

Experiment

natural smectite → illite

4 kcal   Howard and Roy (1985)
18 kcal   Whitney and Northrop (1986)
28 kcal   Huang et al. (1993)
30 kcal   Robertson and Lahann (1981)

synthetic beidellite → illite

20 kcal   Eberl and Hower (1976)

kaolinite → muscovite

30 kcal   Chermarck (1989)
28 kcal   Small (1993)

Estimation Through the Modeling of Natural I/S Series

Illitisation of bentonite by thermal effect

25 kcal   Pusch and Madsen (1995)
27 and 7 kcal   Esposito and Whitney (1995)
25 kcal   Pytte and Reynolds (1989)

Illitisation of shales in a geothermal field

28 kcal   Velde and Lanson (1992)

Illitisation of shales by diagenesis

27 and 7 kcal   Velde and Vasseur (1992)
20 kcal   Bethke and Altaner (1986)
30 kcal   Elliott et al. (1991)

Suggestions for Additional Reading

Fundamentals

Garrels RM, Christ LL (1965) Solutions, minerals and equilibria. Freeman, Cooper, San Francisco, 450 pp
Drever JI (1982) The geochemistry of natural waters. Prentice Hall, Englewood Cliffs, NJ, 388 pp

Advanced Studies

Lasaga AC, Kirkpatrick RJ (1981) Kinetics of geochemical processes. Rev Mineral 8:398
Lagache M (1984) Thermométrie et barométrie géologique. Société Française de Minéralogie et de Cristallographie, 663 pp

Isotopic Composition of Clay Minerals

Geologists are always faced with many questions when studying natural systems. Among them, some can be answered using clay minerals if one considers a particular aspect of their chemical composition: isotopes. The goal of this chapter is to give the basics of the stable and radioactive isotope chemistry through the description of some geological examples. This is simply an introduction to a broad field of investigation. More detailed presentations may be found in the suggested reading listed at the end of the chapter.

4.1
Stable Isotopes

Introduction

Oxygen exhibits three stable isotopes whose mean contents in the Earth's crust are as follows: $^{16}O = 99.762\%$, $^{17}O = 0.038\%$ and $^{18}O = 0.200\%$. Hydrogen exhibits two stable isotopes: $^{1}H = 99.985\%$ and ^{2}H (or deuterium, D) $= 0.015\%$. Phyllosilicates contain O and H in fixed proportions (or almost fixed). For a unit formula with four tetrahedral sites, these proportions are as follows:

Type of phyllosilicate	Number of oxygens	Number of hydrogens
1:1	18	8
2:1	12	2
2:1:1	18	8

Knowing the proportions between heavy and light isotopes is an approach to the conditions in which phyllosilicates have crystallised. It is an attempt to answer three fundamental questions: What was the fluid origin? What was the fluid/rock ratio? What was the formation temperature?

4.1.1
Isotopic Fractionation

4.1.1.1
Fundamentals

The Isotopic Ratio R and the δ (Delta) Notation

The relative proportion of the various isotopes is expressed by the ratio R of the number of atoms of the heavy isotope to the number of atoms of the light isotope. For oxygen $R = {}^{18}O/{}^{16}O$, and for hydrogen $R = D/H$. The ratio R varies as a function of the conditions of formation of minerals. Nevertheless, these variations are too small to be reliably determined as an absolute value. It is much easier to measure with great accuracy the relative difference δ between the ratio in the studied mineral and the ratio in some standard. The isotopic compositions used correspond to these differences in ‰: $\delta^{18}O = 10^3(\frac{R_X - R_{STD}}{R_{STD}})$ and $\delta D = 10^3(\frac{R_X - R_{STD}}{R_{STD}})$. A clay is said to have a $\delta^{18}O$ of $+15‰$ if its heavy isotope content is 15 per thousand greater than that of the standard. Conversely, its δD is of $-60‰$ if its heavy isotope content is 60 per thousand lower than that of the standard. The international standard SMOW (standard mean ocean water) has been defined by Craig (1961). Therefore, the $\delta^{18}O$ and δD values of the standard SMOW are equal to 0‰.

Isotopic Composition of Natural Waters

The minerals precipitating from aqueous solutions concentrate ${}^{18}O$. Therefore, the $\delta^{18}O$ will always be positive if the mineral precipitates in the present-day

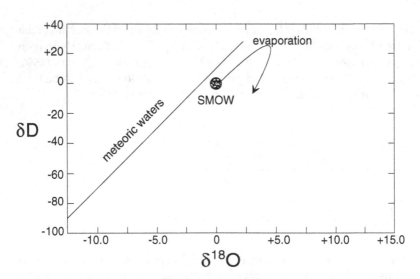

Fig. 4.1. Isotopic composition of meteoric waters (Craig 1961). Evaporation of seawater first causes D enrichment followed by depletion, whereas ${}^{18}O$ increases constantly

seawater, and *a fortiori* in meteoric waters that are always depleted in comparison with seawater (evaporation – condensation process). Two parameters control the partitioning (fractionation) of ^{18}O between the forming crystal and the solution in which it is formed: the $\delta^{18}O$ of the solution and temperature. For example, for a same temperature, the kaolinite formed at the latitude of France will have a lower $\delta^{18}O$ than that formed in Morocco.

The water-mineral fractionation coefficients cannot be determined if the isotopic composition of natural waters is not known. This composition varies with latitude (Craig 1961). The $\delta^{18}O$ and δD values are controlled by a simple relationship (Fig. 4.1):

$$\delta D = 8\delta^{18}O + 10 \tag{4.1}$$

The isotopic composition of seawater or of evaporated fresh water is located to the right of the meteoric water line. Compositions of those waters percolating through the rocks in diagenetic or hydrothermal series are also located to the right because they exchange oxygen with the rocks.

4.1.1.2
The Mineral-Water Isotopic Fractionation

Mechanism
The energy of the chemical bonds involving oxygen or hydrogen slightly varies with the mass of the isotope: it is lower with light isotopes and greater with heavy isotopes. Some physical processes such as evaporation bring about an isotopic fractionation: vapour does not have the same $\delta^{18}O$ and δD as the liquid phase. The fractionation between two phases A and B is expressed by the fractionation coefficient α_{A-B}:

$$\alpha_{A-B} = \frac{\left(^{18}O/^{16}O\right)_A}{\left(^{18}O/^{16}O\right)_B} \tag{4.2}$$

α as a function of δ can be expressed as follows:

$$\alpha_{A-B} = \frac{1 + \frac{\delta^{18}O_A}{1000}}{1 + \frac{\delta^{18}O_B}{1000}} \approx 1 + \frac{\delta^{18}O_A - \delta^{18}O_B}{1000} \tag{4.3}$$

When $\delta^{18}O_A - \delta^{18}O_B < 10\%_0$, the following logarithmic expression is inferred:

$$1000 \ln \alpha_{A-B} \approx \delta^{18}O_A - \delta^{18}O_B \tag{4.4}$$

This approximation is acceptable for high-temperature fractionations, which remain relatively weak. It is no longer acceptable at low temperature where the difference is greater than $10\%_0$. Under natural conditions of low temperature, α values are close to 1.

Empirical Calculation of the Fractionation Equation as a Function of Temperature
The fractionation coefficient α is similar to the equilibrium constant K. Its value equally depends on temperature. Urey (1942) has shown that, for perfect gas, $\ln \alpha$ varies proportionally to $1/T^2$ under high temperature conditions and to $1/T$ under low temperature conditions. Experimental works relating to oxygen isotopes in minerals show that the relationship is of type:

$$10^3 \ln \alpha_{\text{mineral water}} = A\, 10^6/T^2 + B \tag{4.5}$$

where A and B are constants.

For some minerals, the relationship is of type $10^3/T$ or $10^9/T^3$ or $10^{12}/T^4$.

These equations can be established empirically when two parameters are known:

1. the proportion of cation – O or cation – (OH) chemical bonds in the structure of minerals;

2. the fractionation characteristic of each bond.

Then, fractionations that can be attributed to each type of bond just have to be added to obtain the isotopic fractionation equation of the mineral. Let's take the example of kaolinite. How can the proportion of each type of bond in the structure be calculated? Kaolinite is a 1:1 mineral. Therefore, the external surfaces of the layers are formed by the basal oxygens of the tetrahedral sheet and the external $(OH)^-$ of the octahedral sheet (Fig. 4.2a). The half unit formula can be written as follows: $Si_2 O_5 Al_2 (OH)_4$. It contains 3 basal oxygens, 2 apical

Table 4.1. Isotopic fraction equation of oxygen for the main types of chemical bonds characterising clay minerals. M represents the trivalent elements other than Al. The bonds established by interlayer cations are not taken into account

Bond	Isotopic fractionation equation of oxygen	Source
Si–O–Si	$10^3 \ln \alpha = 3.34 \times 10^6/T^2 - 3.31$	Matsuhisa et al. (1979)
Si–O–Al	$10^3 \ln \alpha = -11.6 \times 10^3/T + 5.40 \times 10^6/T^2 + 6.50$	Matsuhisa et al. (1979)
Al–O–Al	$10^3 \ln \alpha = -23.14 \times 10^3/T + 7.455 \times 10^6/T^2 + 16.31$	Matsuhisa et al. (1979)
M–O–M	$10^3 \ln \alpha = -45.4 \times 10^3/T + 24.3 \times 10^6/T^2 - 5.61 \times 10^9/T^{-3} + 0.504 \times 10^{12}/T^{-4} + 22.38$	Becker (1971), Friedman and O'Neil (1977)
Si–O–M	$10^3 \ln \alpha = -22.7 \times 10^3/T + 13.82 \times 10^6/T^2 - 2.81 \times 10^9/T^{-3} + 0.252 \times 10^{12}/T^{-4} + 9.54$	Savin and Lee (1988)
Al–O–M	$10^3 \ln \alpha = -34.3 \times 10^3/T + 15.88 \times 10^6/T^2 - 2.81 \times 10^9/T^{-3} + 0.252 \times 10^{12}/T^{-4} + 19.33$	Savin and Lee (1988)
Al–OH	$10^3 \ln \alpha = 29.6 \times 10^3/T - 4.25 \times 10^6/T^2 - 35.28$	Savin and Lee (1988)

Fig. 4.2a,b. Calculation of
the proportion of bonds
in kaolinite. a) Crystal
structure of the unit cell.
b) Counting of the bonds
of O^{2-} or (OH) with Si^{4+} or
Al^{3+} cations

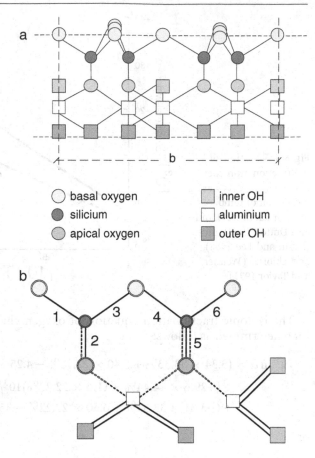

○ basal oxygen   ◻ inner OH

● silicium   ▢ aluminium

◌ apical oxygen   ▩ outer OH

oxygens and 4 radicals $(OH)^-$. Each one of the 3 basal oxygens is bonded to
two Si^{4+} ions, the two apical oxygens are bonded to 1 Si^{4+} ion and 1 Al^{3+} ion,
and finally each one of the 4 $(OH)^-$ is bonded to one of the free valencies of
the Al^{3+} ions (Fig. 4.2b). In total, out of the 9 bonds contained in the half unit
formula, 3 are of type Si–O–Si (33.33%), 2 of type Si–O Al (22.22%) and 4 of
type Al–OH (44.44%).

The isotopic fractionation equations of oxygen have been established empirically for the main types of chemical bonds occurring in the crystal structure of
clay minerals. In reality, only those strong bonds forming the layer superstructure are taken into account, and the weak bonds in the interlayer are ignored.
These calculations rest on a number of simplifying assumptions (Savin and
Lee 1988). The values obtained are given in Table 4.1.

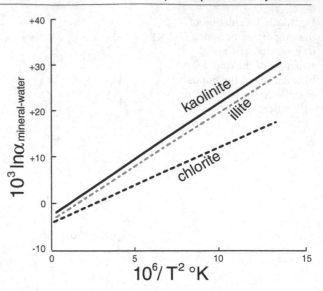

Fig. 4.3. Variation of the fractionation coefficient $\alpha_{mineral-water}$ as a function of temperature for kaolinite (Land and Dutton 1978), illite (Savin and Lee 1988) and chlorite (Wenner and Taylor 1971)

The isotopic fractionation equation of oxygen characterising kaolinite is then determined as follows:

$$10^3 \ln \alpha = (3.34 \times 33.33\% + 5.40 \times 22.22\% - 4.25 \times 44.44\%) \, 10^6/T^2$$
$$+ (+29.6 \times 44.44\% - 11.6 \times 22.22\%)10^3/T$$
$$+ (-3.31 \times 33.33\% + 6.50 \times 22.22\% - 35.28 \times 44.44\%) \quad (4.6)$$

or

$$10^3 \ln \alpha = +10.577 10^3/T + 0.424 10^6/T^2 - 15.337 \quad (4.7)$$

The clay-water isotopic fractionation coefficient of oxygen regularly decreases as temperature rises. It becomes very low at high temperature (Fig. 4.3). It can also greatly vary with the chemical composition of clay minerals. This is essential in the case of extended solid solutions (chlorites) or pseudo-solid solutions (illite/smectite mixed layers). A few isotopic fractionation equations are shown in Tables 4.2 and 4.3.

4.1.1.3
The Mineral-Mineral Isotopic Fractionation

Can the formation temperature of a mineral be determined when the isotopic composition of the fluid in which it was formed is not known? This question is fundamental because it is in fact very rare to be able to collect fluids in which the minerals have been formed. When fluids are collected (in active geothermal systems in particular), it is even frequent to see that they are not in isotopic equilibrium with the minerals in the rock or in the fractures.

Table 4.2. Isotopic fractionation equation of oxygen for a few species of clay minerals. $\alpha_{\text{mineral-water}}$: fractionation coefficient

Mineral	Isotopic fractionation equation of oxygen	Source
Kaolinite	$10^3 \ln \alpha = 10.6 \times 10^3/T + 0.42 \times 10^6/T^2$ $-15.337 \ (< 200 \,^\circ\text{C})$	Savin and Lee (1988)
	$10^3 \ln \alpha = 2.50 \times 10^6/T^2 - 2.87$ $(< 200 \,^\circ\text{C})$	Land and Dutton (1978), Eslinger (1971)
	$10^3 \ln \alpha = 2.76 \times 10^6/T^2 - 6.75$	Sheppard and Gilg (1996)
Di. smectite	$10^3 \ln \alpha = 2.58 \times 10^6/T^2 - 4.19$	Savin and Lee (1988), Yeh (1974), Yeh and Savin (1977)
	$10^3 \ln \alpha = 2.55 \times 10^6/T^2 - 4.05$	Sheppard and Gilg (1996)
Tri. smectite	$10^3 \ln \alpha = 3.31 \times 10^6/T^2 - 4.82$	Escande (1983)
Illite	$10^3 \ln \alpha = 2.39 \times 10^6/T^2 - 4.19$	Savin and Lee (1988)
	$10^3 \ln \alpha = 2.43 \times 10^6/T^2 - 3.78$	Glassman et al. (1989)
	$10^3 \ln \alpha = 1.83 \times 10^6/T^2$ $+ 0.0614 \times (10^6/T^2)^2$ $- 0.00115 \times (10^6/T^2)^3 - 2.87$	Lee (1984)
	$10^3 \ln \alpha = 2.39 \times 10^6/T^2 - 3.76$	Sheppard and Gilg (1996)
Illite/Smectite	$10^3 \ln \alpha = (2.58 - 0.19 \times \text{I}) \times 10^6/T^2$ $- 4.19 \ (\text{I=illite \%})$	Savin and Lee (1988)
Mg Chlorite	$10^3 \ln \alpha = 1.56 \times 10^6/T^2 - 4.70$	Wenner and Taylor (1971)
Fe Chlorite	$10^3 \ln \alpha = 3.72 \times 10^3/T + 2.50 \times 10^6/T^2$ $- 0.312 \times 10^9/T^3 + 0.028 \times 10^{12}/T^4 - 12.62$	Savin and Lee (1988)
Al Chlorite	$10^3 \ln \alpha = 6.78 \times 10^3/T + 1.19 \times 10^6/T^2$ $- 13.68$	Savin and Lee (1988)
Bertierine	$10^3 \ln \alpha = 4.375 \times 10^3/T + 2.602 \times 10^6/T^2$ $- 0.422 \times 10^9/T^3 + 0.038 \times 10^{12}/T^4 - 12.95$	Longstaffe et al. (1992)
Serpentine	$10^3 \ln \alpha = 1.56 \times 10^6/T^2 - 4.70$	Wenner and Taylor (1971)
	$10^3 \ln \alpha = 4.61 \times 10^3/T + 3.08 \times 10^6/T^2$ $- 0.623 \times 10^9/T^3 + 0.056 \times 10^{12}/T^4 - 14.43$	Savin and Lee (1988)
Talc	$10^3 \ln \alpha = -3.94 \times 10^3/T + 5.86 \times 10^6/T^2$ $- 0.934 \times 10^9/T^3 + 0.084 \times 10^{12}/T^4 - 4.27$	Savin and Lee (1988)

The only procedure is to analyse the isotopic composition of two types of minerals formed at the same time in the same fluid. Fortunately, the precipitation of clay minerals with quartz or carbonates, zeolites etc. is very commonly observed. If the slopes of the fractionation curves are sufficiently different, a geothermometer may be calculated by subtracting one function from the other. That is the case for the quartz-illite and quartz-calcite pairs (Fig. 4.4).

The slopes of the fractionation curves of the various types of clay minerals are too close for clay-clay geothermometers to be defined. On the other hand,

Table 4.3. Isotopic fraction equation of hydrogen for a few species of clay minerals. $\alpha_{mineral-water}$: fractionation coefficient

Mineral	Isotopic fractionation equation of hydrogen	Source
Kaolinite	$10^3 \ln \alpha = -4.53 \times 10^6/T^2$ $+ 19.4\ (< 200\,°C)$	Lambert and Epstein (1980)
	$10^3 \ln \alpha = -42.2 \times 10^6/T^2 - 7.7$	Sheppard and Gilg (1996)
Illite/Smectite	$10^3 \ln \alpha = -45.3 \times 10^3/T$ $+ 94.7\ (0–150\,°C)$	Capuano (1992)
	$10^3 \ln \alpha = -19.6 \times 10^3/T$ $+ 25\ (30–120\,°C)$	Yeh (1980)
Serpentine	$10^3 \ln \alpha = 2.75 \times 10^7/T^2 - 7.69 \times 10^4/T$ $+ 40.8$	Sakai and Tutsumi (1978)

clay minerals can be used with quartz, calcite or any other mineral to determine the temperature or the $\delta^{18}O$ composition of the formation fluid. One example is the use of the pair quartz-illite/smectite mixed layers to calculate the temperature at the maximum burial depth in diagenesis. It is assumed that the equilibrium between I/S and solutions contained in the pore rocks is continuously maintained during the diagenetic process by a reaction of type

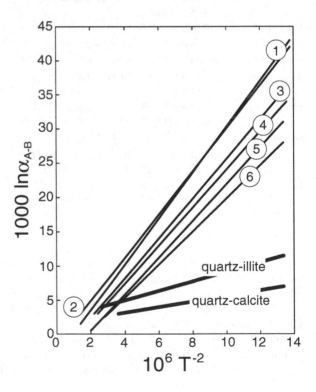

Fig. 4.4. Mineral-water fractionation curves plotted according to the equations in Table 4b. 1: quartz-water (Clayton et al. 1972); 2: dolomite-water (Northrop and Clayton 1966); 3: alkaline feldspar-water (O'Neil and Taylor 1967); 4: calcite-water (O'Neil et al. 1969); smectite-water (Yeh and Savin 1977); illite-water (Eslinger and Savin 1973). The mineral-mineral fractionation curves have been obtained by subtracting the corresponding mineral-water equations. T: temperature in K

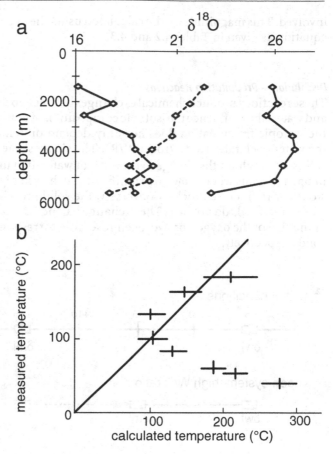

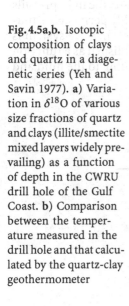

Fig. 4.5a,b. Isotopic composition of clays and quartz in a diagenetic series (Yeh and Savin 1977). **a)** Variation in $\delta^{18}O$ of various size fractions of quartz and clays (illite/smectite mixed layers widely prevailing) as a function of depth in the CWRU drill hole of the Gulf Coast. **b)** Comparison between the temperature measured in the drill hole and that calculated by the quartz-clay geothermometer

smectite → illite + quartz (Yeh and Savin 1977). Figure 4.5 shows that the temperatures determined by the $\delta^{18}O$ of the pair quartz-clays are abnormally high, up to 85 °C or so. Above this threshold, measured temperatures are consistent with calculated temperatures. This means that quartz and clays are not in isotopic equilibrium beneath 85 °C because of the influence of the sedimentary inheritance (see Sect. 8.1.2.1). Above this temperature, both types of minerals are formed together and fractionate the ^{18}O of the same fluid.

4.1.2
Water – Clay Mineral Interactions

4.1.2.1
Isotope Balance Principle (from Girard and Fouillac 1995)

Precipitation from a Solution
The isotopic fractionation magnitude at equilibrium between a mineral and water only depends on temperature, and not on the volumes of liquid and solid

involved. This magnitude can be calculated using the appropriate fractionation equation as given in Tables 4.2 and 4.3.

Dissolution – Precipitation Reactions

These reactions produce chemical exchanges between rocks or minerals (solids) and water; they influence the isotopic composition of both water and rock. Then the isotopic fractionation does not only depend on temperature but also on the water/rock ratio (water/rock: W/R). The slightest the amount of dissolved solids, the slightest the change in the $\delta^{18}O$ of water. In nature, the fluid prevails in open systems where flows are significant, such as in highly permeable sandstones or in fractured rocks. Conversely, the solid prevails when the system is closed and fluids do not flow. The exchange balance is established as a function of the size of the oxygen or hydrogen reservoir corresponding to the fluid and solid, respectively.

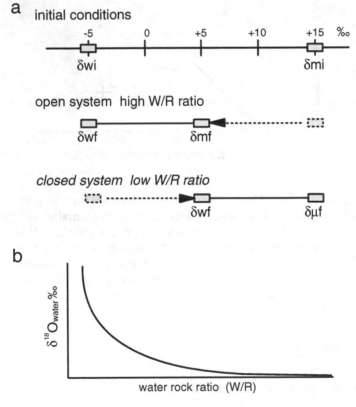

Fig. 4.6a,b. Water – mineral interactions. **a)** Schematic representation of the isotope balance of exchanges between water and the mineral in closed or open systems. **b)** Variation in the $\delta^{18}O$ of water as a function of the water-rock ratio (W/R) from Girard and Fouillac (1995)

In the case of a closed system, exchanges between fluid and solid occur without modification in the amounts of each one of the isotopes. A mass balance can then be established between the initial stage and the final stage (Fig. 4.6a):

$$M\delta m_i + W\delta w_i = M\delta m_f + W\delta w_f \quad \text{with} \quad M + W = 1 \tag{4.8}$$

M mole fraction of oxygen of the mineral

W mole fraction of oxygen of water

δm_i and δm_f initial and final δ of the mineral

δw_i and δw_f initial and final δ of water

From this equation is inferred an expression of the W/R ratio by introducing the oxygen contents of water (Cw_i) and of the mineral (Cm_i):

$$W/M = \left[\delta m_f - \delta m_i/\delta w_i - \delta w_f\right] Cm_i/Cw_i \tag{4.9}$$

The lower the water/rock ratio, the more the isotopic composition of water is affected and the less is that of rock (Fig. 4.6b).

4.1.2.2
Closed System

One example of a closed system is given by the experimental alterations of dioctahedral smectites (montmorillonite) into illite (Whitney and Northrop 1989). The progress of the mineral reaction is measured by the smectite content at the end of each test. This content decreases as a function of time and temperature, the totally expandable layers (17 Å) of montmorillonite being replaced by layers collapsed at 10 Å when K-saturated (true illite and high-charge layers). The isotopic analyses of the products (Table 4.4) are distributed according to two linear functions of the smectite content (Fig. 4.7): one corresponds to the formation of randomly ordered illite/smectite mixed layers ($R0$), the other to ordered mixed-layer minerals ($R1$).

An isotopic fractionation coefficient between water and clay minerals has been calculated in a simple way: $\alpha = 100 \; (\delta^{18}O_f - \delta^{18}O_m)/\delta^{18}O_f$. Its variation as a function of the smectite content determines two straight lines (Fig. 4.8):

1. the line corresponding to randomly ordered mixed-layer minerals, extrapolated to 0% smectite, shows that the isotopic equilibrium is obtained at 65% only.

2. the line corresponding to regularly ordered mixed-layer minerals approaches equilibrium (extrapolation to 100%).

This has been interpreted by the authors as a transformation of smectite into illite controlled by two processes:

Table 4.4. Analyses of the run products of the experimental alteration of montmorillonite into illite, from Whitney and Northrop (1989)

Temp. (°C)	Time (days)	Stacking order	Smectite (%)	$\delta^{18}O$ Fluid	$\delta^{18}O$ Mineral
	0		100	−19.7	19.2
250	7	R0	83	−17.4	14.9
250	14	R0	78	−16.7	13.9
250	30	R0	57	−15.8	13.5
250	60	R0	52		12.2
250	120	R0 + R1	45	−14.9	11
250	220	R0 + R1	44	−14.9	9.6
300	1	R0	70	−16.6	13.7
300	7	R0 + R1	58	−14.4	9.8
300	14	R1 + R0	50		8.7
300	30	R1 + R0	45		7.5
300	120	R1	35	−12.1	5.3
300	220	R1	35	−11.4	4.2
350	1	R1 + R0	47	−13.6	8.1
350	7	R1 + R0	41		6
350	14	R1	34	−12.1	5
350	30	R1	27		4
350	60	R1	24		1.8
350	120	R1	21	− 9.9	1.4
400	14	R1	23	− 9.1	0.6
450	60	R1	4	− 6.8	

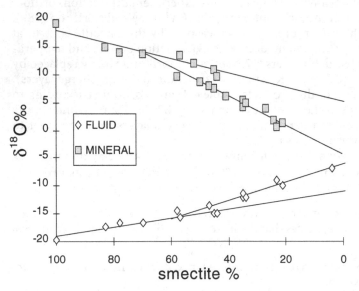

Fig. 4.7. Variation in the isotopic composition of water and clays in the transformation process of montmorillonite into illite, from experiments by Whitney and Northrop (1989)

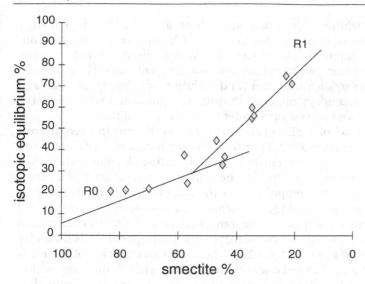

Fig. 4.8. Variation in the isotopic fractionation coefficient of oxygen between water and the montmorillonite illitised under experimental conditions (after Whitney and Northop 1989)

1. a first reaction transforms smectite into R0 I/S through an incomplete dissolution process (65%). Consequently, the crystal structure of smectite is partially "re-arranged" in the solid state;

2. a second reaction in which R0 I/S are transformed into ordered mixed-layer minerals (R1 I/S) through a dissolution – precipitation mechanism resulting in a complete isotopic equilibrium between fluid and solid.

Even if the solid-state transformation is not the only possible solution (Meunier et al. 1998), the action of two different processes is unquestionable.

4.1.3
The isotopic Composition of Clays of the Weathering – Sedimentation Cycle

4.1.3.1
Weathering Clays

Clays formed by weathering result from the interaction between rocks and meteoric water. The isotopic composition of those minerals forming igneous or metamorphic rocks has been acquired under pressure – temperature conditions that greatly differ from the surface conditions. Consequently, disequilibrium does exist. Similarly, sedimentary rocks, even if they contain the same clay minerals as those forming under surface conditions (kaolinite, smectite ...), have an isotopic composition characteristic of exchanges with sedimentary fluids (seawater, connate waters) or diagenetic fluids. Again here, disequilibrium does exist. Surface weathering in soils and alterite mantles brings into contact minerals whose $^{18}O/^{16}O$ ratio stems from a more or less long history with rainwater, the isotopic composition of which depends on climatic conditions. From a theoretical point of view, the isotopic disequilibrium between minerals and rainwater can be reduced in two different ways: (1) solid state isotopic exchanges by diffusion, and (2) dissolution of primary minerals and crystallisation of secondary minerals in isotopic equilibrium with rainwater. The second process widely prevails under weathering conditions.

Rocks at the surface of the Earth are affected by weathering in a very narrow range of temperatures (0–30 °C). Therefore, the fractionation coefficient varies very slightly. The isotopic composition of the resulting clay minerals is then mainly controlled by three factors: (1) the isotopic composition of the parent minerals, (2) the isotopic composition of the local meteoric waters (varying according to the latitude), and (3) the exchange rate between these waters and the parent minerals that depend on the water/rock ratio. This type of alteration takes place through a multitude of microsystems that impose locally fluid-clay minerals disequilibria (see Sect. 6.1.2.4). These microsystems are of three different types. In the most closed ones (contact microsystems), the composition and the number of clay phases are controlled by the parent minerals. In plasmic microsystems (partially open systems), the number of phases decreases and the exchange rate with fluids increases. Finally, in totally open microsystems (fissures, high permeability), only one clay phase crystallises (generally kaolinite); the influence of parent minerals is zero.

The study of biotites in a lateritic profile in Tchad (Clauer et al. 1982) shows that their $\delta^{18}O$ and δD change suddenly right at the beginning of alteration. No secondary phase can be determined yet at this stage, only the water content increases. The sudden change in isotopic composition is related to the dissolution of mica and to the formation of neogenetic minerals, whose presence could not be detected because of their too small size and amount. The authors have interpreted these facts as evidence of the absence of exchange between meteoric water and biotites. This confirms the observations performed on unaltered relic micas in soils, which show that their isotopic composition (O and H) is identical to that of primary micas (Lawrence and Taylor 1972).

Kaolinite is generally the weathering final product. It results from an intensive leaching of profiles, hence from water-rock interactions where water amounts are significant. Lawrence and Taylor (1971) have shown that kaolinite bears witness to the isotopic composition of meteoric waters, which varies according to latitude (Fig. 4.9). Clay minerals formed in contact or plasmic microsystems have an isotopic composition intermediate between that of the parent mineral they replace and that of water at the latitude considered.

Although it is difficult to be sure that clays formed in soils are in isotopic equilibrium (O and H) with their environment, the analyses show that δD varies linearly with $\delta^{18}O$, at least in the low-temperature range. This is how the δD of the kaolinites formed in soils in the presence of meteoric waters of different isotopic compositions vary linearly with their $\delta^{18}O$ (Fig. 4.10a). Note that, for kaolinite, the fractionation equation of oxygen inferred from the measurements on natural samples is different from that calculated theoretically

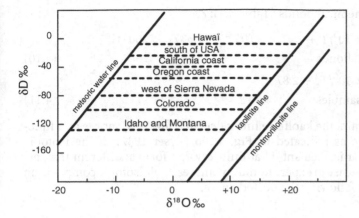

Fig. 4.9. Relationship between the isotopic compositions of kaolinite samples from soil profiles of North America – Hawaii and those of the corresponding meteoric waters (after Lawrence and Taylor 1971)

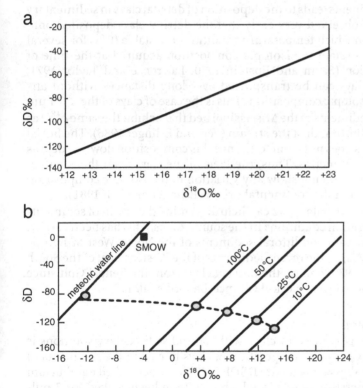

Fig. 4.10a,b. Isotopic composition of kaolinite in a $\delta D–\delta^{18}O$ diagram. **a)** The *"line"* of the weathering kaolinites has been determined by the distribution of $\delta^{18}O$ and δD measured from the crystals formed under various latitudes. **b)** Variation in the position of the *kaolinite line* as a function of temperature, from Kyser (1987)

by the balance of chemical bonds (Table 4.1):

$$10^3 \ln \alpha = 10.6 \times 10^3/T + 0.42 \times 10^6/T^2 - 15.337 \quad (< 200\,^\circ\text{C})$$

Chemical bonds   (4.10)

$$10^3 \ln \alpha = 2.50 \times 10^6/T^2 - 2.87$$

Natural samples   (4.11)

In a δD–$\delta^{18}O$ diagram, the kaolinite line calculated for 25 °C moves as a function of temperature as indicated in Fig. 4.10b (Kyser 1987). If the isotopic composition of water is fixed ant if kaolinite crystals form at different temperatures, as may be the case in experimental syntheses, their isotopic composition varies theoretically following the dotted curve.

4.1.3.2
Sedimentary Clays

Detrital clays
The erosion of continents leads to the deposition of detrital clays in sedimentary basins. It has been observed very early that the detrital clays deposited onto the ocean floors – in which temperature conditions are stable (0 °C for several millions of years) – retain the isotopic composition acquired at the time of continental alteration (Savin and Epstein 1970; Lawrence and Taylor 1971, 1972). Moreover, clays can be transported over long distances without any alteration of their isotopic composition. This is the case of clays of the < 0.1 µm fraction from the sediments of the Mississippi bed that exhibit the same $\delta^{18}O$ as those deposited at the mouth of the stream (Yeh and Eslinger 1986). The burial of marine sediments does not seem to change this composition down to depths of several hundreds of metres. Thus, the Pleistocene clays from the Aleutian trench buried at 500 m do not show any variation in the isotopic composition O and H inherited from the continental origin (Eslinger and Yeh 1981).

The $\delta^{18}O$ and δD of detrital clays can help determine the origin of sediments and specify the alteration conditions in the source zones. This has been carried out on the recent stream and littoral sediments of the North-West of Europe (Salomons et al. 1975), on the recent sediments of the Western part of the South Atlantic (Lawrence 1979) and on the detrital clays from the Bengal Gulf since the beginning of the Miocene period (France-Lanord et al. 1993).

Sediment-Fluid Exchanges
The isotopic exchange process of clays with natural sedimentary solutions is very slow. Thus, the isotopic composition of clays buried in the sediments of the Bellingshausen abyssal plain (site DSDP 3223) has not reached equilibrium with the marine interstitial fluids in which they have been soaked for 3 millions of years (Yeh and Eslinger 1986). This disequilibrium is obvious when the measured isotopic compositions are compared with those calculated for equilibrium at the given temperature (Table 4.5). Nevertheless, the difference at equilibrium decreases with depth (Fig. 4.11).

Table 4.5. Value of $\delta^{18}O$ of Bellingshausen clays (DSDP 323) as a function of depth (Yeh and Eslinger 1986)

Depth (m)	$\delta^{18}O\%{}_{clay}$	Estimated temp. (°C)	$\delta^{18}O\%{}_{pore\ water}$	$\delta^{18}O\%{}_{equilibrium}$
80	19.00	3.20	−0.70	30.00
165	18.90	6.60	−1.10	28.70
260	21.80	10.40	−0.70	28.10
365	21.00	14.60	−1.04	26.70
410	19.50	16.40	−2.70	24.80
460	17.90	18.40	−2.50	24.40
510	20.70	20.40	−2.50	23.90
555	18.80	22.20	−2.70	23.60
600	16.80	24.00	−2.50	23.20

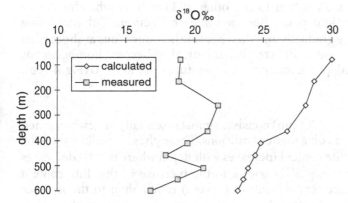

Fig. 4.11. Variation in $\delta^{18}O$ of Bellingshausen clays (DSDP 323) as a function of depth (from Yeh and Eslinger 1986). The measured values are always lower than the values calculated at isotopic equilibrium of solutions. The difference at equilibrium decreases with depth

The fact that isotopic equilibrium between clay minerals and fluids is not reached in sedimentary environments does not mean that sediment remains totally inert. Since reactions are very slow, the mass of the rock that has reacted is low. The study of isotopic compositions of several size fractions of sediments from ocean floors of the North Pacific <2.7 m.y. in age has shown significant differences. Clays with a size > 0.1 μm have retained the original continental isotopic signature for oxygen and hydrogen. On the other hand, the $\delta^{18}O$ and δD of < 0.1 μm size fractions are always high. The question is to know whether these modifications are due to diffusion or to crystallisation of new particles. In the first case, the differences would be the result of isotope exchanges ranging from 10 to 30%. Nevertheless, the simultaneous variations in $\delta^{18}O$ and δD are not consistent with a diffusion mechanism. The modifications of the isotopic composition of the < 0.1 μm fraction are due to the low-temperature crystallisation of newly formed clays.

4.1.4
Isotopic Composition of Clays Under Diagenetic Conditions

4.1.4.1
Evolution of the Isotopic Composition of I/S During Illitisation

$\delta^{18}O$ of Clays and Burial of Sediments

The isotopic composition of sedimentary clays changes when burial is greater than 1 km for a geothermal gradient of 20–30 °C km^{-1}. In order to measure the effect of exchanges between fluids and sediments, clays of the Tertiary series of the Gulf Coast (USA) have been divided into four size fractions (1–2, 0.5–1, 0.1–0.5 and < 0.1 μm) then analysed (Yeh and Savin 1977; Yeh 1980). Their mineralogy is dominated by illite/smectite mixed layers whose illite content progressively increases with depth (Table 4.6). The $\delta^{18}O$ and δD are very different from one fraction to the other at 1 km in depth. This is due to the various proportions of detrital clays in each fraction. With increasing depth, the $\delta^{18}O$ change gradually and then become homogeneous at about 5 km (Fig. 4.12a). The δD show a different behaviour: they become homogeneous suddenly at 2.5 km in depth, namely at a temperature close to 70 °C (Fig. 4.12b).

$\delta^{18}O$ and Illite %

The finest size fraction (< 0.1 μm) consists almost essentially of newly-formed phases that crystallise in diagenetic conditions. These phases are illite/smectite mixed layers whose illite content increases with depth whereas $\delta^{18}O$ decreases (Fig. 4.13). The decreasing $\delta^{18}O$ is due to the increase in the illite content in I/S with temperature (crystallisation process) rather than to the isotopic equilibrium reaction of detrital illites (diffusion process). Nevertheless, $\delta^{18}O$ does not vary regularly as a function of the illite content: it keeps decreasing whereas the illite content remains quite constant at 80% (Fig. 4.13). This shows

Table 4.6. Evolution of the $\delta^{18}O$ and δD of illite/smectite mixed layers in the Tertiary clays from the Gulf Coast as a function of depth (drill hole CWRU 6) and size of the particles analysed (Yeh and Savin 1977; Yeh 1980)

Depth	$\delta^{18}O$ (Yeh and Savin 1977)				Illite %	δ (Yeh 1980)			
(m)	2–1 μm	1–0.5 μm	0.5–0.1 μm	< 0.1 μm	< 0.1 μm	2–1 μm	1–0.5 μm	0.5–0.1 μm	< 0.1 μm
1,402	17.64	19.58	19.64	22.45	87	−45	−54	−54	−56
2,073	18.4	20.68	22.13	21.72	79	−44	−48	−54	−53
2,561	18.29	20.93	22.22	20.94	64	−43	−47		−53
3,354	20.67	19.28	18.74	20.78	42	−37	−37	−34	−35
3,902	18.67	19.65	18.36	19.78	20	−35	−38	−41	−38
4,512	20.14	19.28	19.15	18.62	20		−36	−38	−39
5,122	18.99	19.35	19.03	19.63	20	−40	−40	−38	−36
5,610	18.58	18.13	18.2	17.66	20	−38	−40	−36	−35

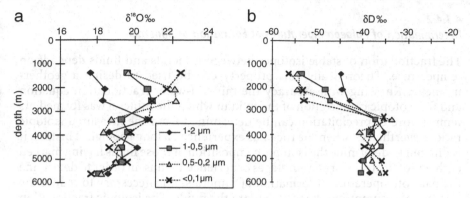

Fig. 4.12a,b. Progressive homogenisation of $\delta^{18}O$ (a) and δD (b) of the various size fractions of clays from the Tertiary series of the Gulf Coast (USA) as a function of depth (Yeh and Savin 1977; Yeh 1980). These clays are dominated by illite/smectite mixed layers whose illite content increases with depth

that the clay material keeps forming probably by the increase in the particle size in the *a-b* plane. This lateral growth takes place from a "nucleus" that retains smectite layers (see Sect. 8.1.2.3). This is accompanied by the growth of zoned quartz grains: their periphery is enriched with ^{18}O in comparison with the centre.

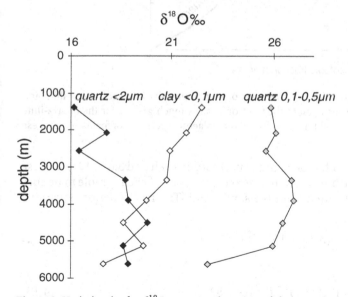

Fig. 4.13. Variation in the $\delta^{18}O$ rate as a function of the smectite content in illite/smectite mixed layers (according to data from Yeh and Savin 1977)

4.1.4.2
Determination of Palaeotemperatures of Formation of Minerals

The fractionation of stable isotopes between minerals and fluids depends on temperature. Theoretically, this property can be used to design a geothermometer. Knowing with accuracy the mineral-water fractionation equation and the isotopic composition of the fluid in which the mineral was formed, its temperature of precipitation can be determined from the measured isotopic ratio. Nevertheless, when the rock has experienced a long geological history, it is difficult to determine the isotopic composition of those fluids having marked each one of the transformation stages experienced. Thus, in order to determine the palaeotemperature of formation of minerals, it is necessary to reduce the number of assumptions, hence to ignore the fluids. The isotopic fractionation between cogenetic minerals can be used in the temperature range where its variation is significant. This is not the case for instance for the kaolinite-illite pair, as shown in Fig. 4.14.

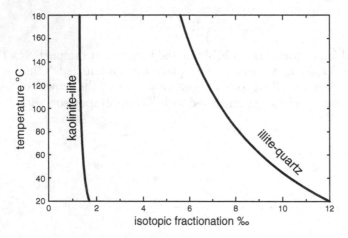

Fig. 4.14. Isotopic fractionation for the pairs of cogenetic minerals kaolinite – illite and quartz – illite. Since the variation of the fractionation is much greater for the quartz-illite pair, the latter will be selected for determining the palaeotemperature of formation of these minerals

The illite-quartz pair has been used successfully (Yeh and Savin 1977). These authors have used the finest clay and quartz fractions of each sample to be sure to analyse minerals formed from the same fluid. The fractionation equations used are as follows:

$$10^3 \ln \alpha_{\text{quartz-water}} = \left(3.38 \times 10^6 T^{-2}\right) - 3.40 \tag{4.12}$$

Clayton et al. (1972)

$$10^3 \ln \alpha_{\text{illite-water}} = \left(2.43 \times 10^6 T^{-2}\right) - 4.82 \tag{4.13}$$

Eslinger and Savin (1973).

The result (Fig. 4.14) shows that the slope of the curve is sufficient to determine accurately the temperature corresponding to the mineral-mineral fractionation observed.

4.1.4.3
Determination of the Origin of Fluids

Searching for the origin of those fluids flowing in a basin throughout their diagenetic history is essential in oil exploration. Indeed, during the sedimentary stage of this long history, fluids may be of meteoric or marine origin. During the burial period, these fluids are located in the rock porosity. Their chemical composition varies as a function of the dissolution of detrital minerals and as a function of the precipitation of new minerals (clays, carbonates, quartz for instance). Therefore, their isotopic composition changes over time: they get richer in heavy isotopes until burial has reached a maximum. If burial gets interrupted by tectonic movements uplifting the sedimentary layers towards the surface through faults, rocks are invaded again by meteoric waters. The isotopic composition of fluids located in the porosity is then modified again. During the erosion and cooling periods, fluids keep getting poorer in heavy isotope down to the value measured in present-day fluids. The reconstruction of the isotopic palaeocompositions of fluids is possible only if the temperature conditions are determined at each stage by another method. Most of the time, information extracted from the analysis of fluid inclusions observed in growth aureoles of quartz, carbonates or sulphates are used. Indeed, knowing the temperature, the isotopic composition of the fluid can be calculated using the mineral-water fractionation equation (Fig. 4.15). Note that uncertainty is great.

This method as been used to study the variation in the isotopic composition of fluids in diagenetic environments. The example selected is that of the Viking

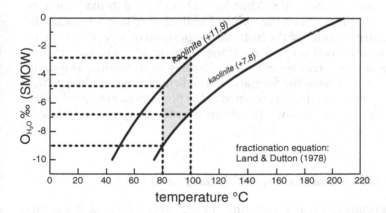

Fig. 4.15. Principle of determination of the composition of the fluids contained in the porosity of diagenetic rocks from the evaluation of temperatures via other methods such as the fluid inclusion analysis for instance

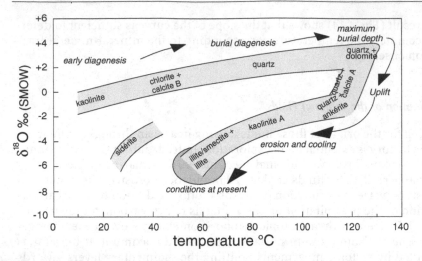

Fig. 4.16. Variation in the isotopic composition of fluids and in the temperatures during the various episodes of the geological history of the Viking Formation (Alberta, Canada) (after Longstaffe and Ayalon 1987). Details of the procedure are exposed in the text

Formation (Longstaffe and Ayalon 1987). The evolution curve of the isotopic composition of fluids filling the porosity as a function of temperature is divided into five stages, each one being marked by precipitating minerals (Fig. 4.16). The first stage is characterised by poral fluids close to seawater. Kaolinite and siderite are formed in a shallow environment in which meteoric waters can enter owing to temporary emersions. This explains the low $\delta^{18}O$ values of siderite. This stage (early diagenesis) precedes the progressive burial marked by the enrichment in ^{18}O of poral fluids. Quartz precipitates in the form of overgrowths on detrital grains. The petrographic analysis shows that chlorite+calcite B precipitate first, followed by dolomite. When burial has reached its maximum, or just afterwards, tectonics causes the uplift of sedimentary layers towards the surface. The sudden variation in the isotopic composition of poral fluids can be explained by the introduction of meteoric waters owing to faults. Calcite then ankerite precipitates whereas quartz keeps forming overgrowths. The uplift period triggers an erosion of the formations and brings about cooling, which leads to the composition of fluids contained in the porosity as measured today. A new generation of kaolinite then illite/smectite mixed layers precipitate in the porosity.

4.1.4.4
Fluid Flows in Sandstone Beds (Whitney and Northrop 1987)

Fluid flows in porous formations deposited in sedimentary basins can sometimes be revealed by the isotopic composition of clays. This is the case of the sandstones from the Morrison Formation (late Jurassic) in the San Juan Basin in New Mexico (Whitney and Northrop 1987). The sandstone bar 100 m thick is inserted between two more or less impermeable silt-clay beds. The clay fraction of sandstones is composed essentially of chlorite and illite/smectite mixed layers whose composition varies between 90 and 0% smectite. This variation determines a double zonation inside the sandstone bar (Fig. 4.17a):

– vertical zonation: I/S mixed layers get richer in illite upwards and downwards in the vicinity of impermeable levels, and chlorite content increases,

– horizontal zonation: I/S mixed layers are richer in illite towards the centre of the basin and the chlorite content increases.

The mineralogical zonation is accompanied by a zonation of the isotopic composition of I/S mixed layers. As the illite content increases, the $\delta^{18}O$ decreases by 15 to 8‰ whereas the δD increases by -116 to -77‰ (Fig. 4.17b–c). This shows that hot fluids ($\approx 130\,°C$) stemming from the bottom of the basin and going up towards zones of lesser depth have caused the smectite-to-chlorite + illite transformation via a sequence of I/S mixed layers. These fluids, whose $\delta^{18}O$ are close to 3‰ and δD to -50‰, are believed to correspond to very advanced formation waters, namely waters resulting from intense exchanges with the rock at the bottom of the basin.

4.1.5
Isotopic Composition of Clays in Active Geothermal Systems

4.1.5.1
Continental Geothermal Systems

Active geothermal fields offer the possibility to measure the present-day temperature of fluids in drill holes. This has enabled Eslinger and Savin (1973) to show that the isotopic fractionation of oxygen between solutions and illite or illite/smectite mixed layers depends on the measured temperature. Indeed, they have observed a decrease in the $\delta^{18}O$ of the bulk rock (12 to 6‰), quartz (9 to 4‰) and clays of I/S or illite type (6 to 1‰), more or less steady with depth, namely with the increase in temperature (Table 4.7; Fig. 4.18). Illite (or I/S mixed layers) and quartz are formed co-genetically.

4.1.5.2
Oceanic Hydrothermal Systems

The isotopic composition of clay minerals produced by the submarine hydrothermal alteration of oceanic basalts (saponite, chlorite and celadonite) depends essentially on the temperature and on the water/rock ratio at the time when the reactions produced them. Since water has an infinite volume in comparison with that of rock, its isotopic composition does not change: 0‰. Therefore, the $\delta^{18}O$ value of those clays formed by low-temperature alteration shall range between 0‰ and $+5.7 \pm 0.2$‰ (composition of fresh basalts). This value can greatly exceed the value of fresh basalts if alteration takes place at

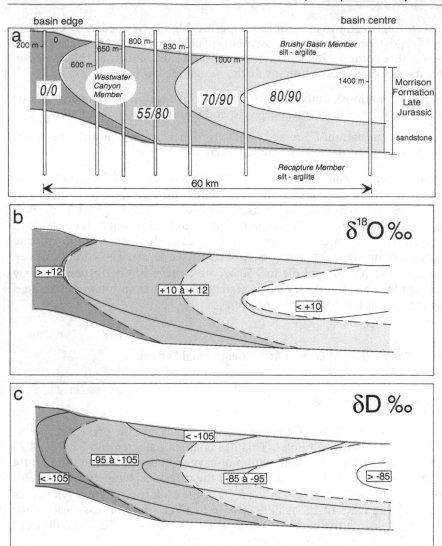

Fig. 4.17a–c. Spatial variation of the isotopic composition of clay minerals (dominated by illite/smectite mixed layers) in the sandstones from the Morrison Formation of the San Juan Basin, United States (Whitney and Northrop 1987). **a)** Sampling location (< 1 µm fraction); illite contents of I/S mixed layers are indicated by *two digits*: mean content to the *left* and maximum content to the *right*. **b)** Variation in $\delta^{18}O$ of the < 1 µm fraction. **c)** Variation in δD of the < 1 µm fraction

Table 4.7. $\delta^{18}O$ values of the bulk rock, quartz and clays (I/S mixed layers and illite) as a function of depth in the geothermal field of Oaki Broadland, New-Zealand (Yeh and Savin 1973)

Depth (m)	Measured temp. (°C)	$\delta^{18}O$ Rock	$\delta^{18}O$ Quartz	$\delta^{18}O$ Clays
80	70	9.37		
152	150	6.67	9.86	3.71
306	160		11.87	5.24
363	164	9.06		
345	158		11.19	5.05
380	157		11.77	6.29
489	140		10.01	4.68
619	126	6.61		
708	179	5.71	9.05	
850	236	5.47		
896	241		7.9	3.6
993	251	4.58	7.43	2.06
1,175	259		6.37	2.04
1,358	270	3.83	6.04	2.94

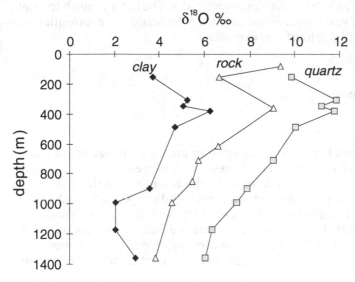

Fig. 4.18. Variation in the $\delta^{18}O$ of the bulk rock (*triangles*), quartz (*squares*) and clays (*lozenges*) as a function of depth in the geothermal field of Oaki Broadland, New-Zealand (Yeh and Savin 1973)

higher temperature. Thus, saponite (formed under very high conditions of water/rock ratio > 50:1) crystallises in a wide range of temperatures: from 30 to 200 °C. Consequently, $\delta^{18}O$ values vary in a wide range: 2 to 24‰. (Lawrence and Drever 1981; Stakes and O'Neil 1982; Alt et al. 1986; Proust et al. 1992).

Another example helps to understand how the origin of clays can be determined using their isotopic composition in an oceanic environment. Iron-bearing dioctahedral clays are ubiquist: they are formed both on continents by meteoric alteration (see Sects. 6.1.3.2 and 6.1.3.3) and in oceans by reactions of detrital clays with seawater (see Sect. 7.2.2.2) or by direct precipitation from hydrothermal fluids (see Chap. 9). This question has been raised concerning the origin of nontronites discovered about the hydrothermal sites of the Galapagos Islands (McMurtry et al. 1983). Far away from any continent, these green clays are probably authigenic, even if they contain a small amount of quartz and kaolinite. Their $\delta^{18}O$ value is high and varies from 21 to 25‰. If the solutions in which they were formed are considered to have the isotopic composition of the seawater ($\delta^{18}O = 0$‰), the calculated crystallisation temperatures are contained between 27 and 39 °C. These temperatures are then 10 to 32 °C greater than those measured today in the poral solutions of the green clay deposit. The "isotopic" temperatures are consistent with those of the hydrothermal fluids that enter the system today. The latter is much less hot than it was 130,000 years ago. Several periods of formation of nontronites are recorded in the stratigraphy of clay deposits.

4.2
Radioactive Isotopes

Introduction

To date clay minerals is potentially a mean to measure the age or the duration of some important geological processes such as weathering, diagenesis, hydrothermal alteration. Several datation methods are commonly used: K–Ar, Ar–Ar and Rb–Sr. The procedures of acquisition of data specific to each one of these methods are described in several books (Hoefs 1980; Faure 1986; Clauer and Chaudhuri 1995). The goal here is to give the basic principles on which the datation methods rest. To illustrate the demonstrations, a few examples are selected in the literature, which has been very abundant for thirty years or so.

4.2.1
Datation Principle: Closed System

4.2.1.1
Method of Calculation

All datation methods rest on the spontaneous transformation of unstable radioactive atoms into stable atoms by one or several successive reactions. These reactions are accompanied by the emission of α, β, and γ particles, positrons

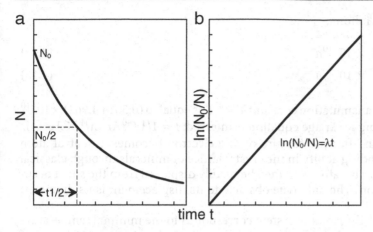

Fig. 4.19a,b. Representation of the decrease in the amount of atoms of the mother isotope as a function of time. **a)** The negative exponential enables the half life of the mother isotope to be readily determined. **b)** A first-order reaction is represented by a linear function in a semi-logarithmic plot. The slope of the *straight line* corresponds to λ, the transmutation constant

and heat. The reaction rate of a radioactive species is proportional to the number of atoms (N) present at instant t: $-\frac{dN}{dt} = \lambda N$ $t = \frac{1}{\lambda} \ln\left(1 + \frac{D}{N}\right)$ where D represents the number of daughter atoms formed at instant t (Fig. 4.19). The half-life of a radioisotope is the time required for the disintegration of $N_0/2$ mother atoms ($N = N_0/2$) hence $\ln(2/1) = \lambda t_{1/2}$. Therefore, $t_{1/2} = 0.6931/\lambda$.

For any radioactive element whose disintegration sequence is perfectly known, a datation method can be devised if four conditions are met:

1. an accurate definition of the transmutation constant;

2. a measurement as accurate as possible of the mother and daughter isotope amounts;

3. the conservation of the mother and daughter isotope amounts without loss nor gain (closed system);

4. equality of the initial isotopic compositions at the instant $t = 0$ of the system closure.

4.2.1.2
The K-Ar Method

During nuclear transmutation, the $^{40}K^+$ ion is transformed either into $^{40}Ca + \beta$ particle (89.9%) or into $^{40}Ar^+$ ion. The $^{40}Ar^+$ ion, in turn, is transformed into a neutral atom by electron capture: Ca. The values of the transmutation constants are as follows:

$$\lambda_{K-Ar} = 0.581 \times 10^{-10} a^{-1} \qquad\qquad (4.14)$$

$$\lambda_{K-Ca} = 4.962 \times 10^{-10} a^{-1} \qquad\qquad (4.15)$$

Therefore, the transmutation constant λ of ^{40}K is equal to $(0.581 + 4.962) \times 10^{-10}$ y^{-1}. The following K–Ar age equation is inferred: $t = 1/\lambda$ ($^{40}Ar^* \lambda/^{40}K \lambda_{K-Ar}$).

The $^{40}Ar^+$ ion after the capture of one electron becomes a neutral atom that loses its bonding ability in the crystal lattice of minerals through classical chemical bonds. This atom will then be readily displaced from the site it occupies by any cation. The only true obstacle to its displacement is its diameter: about 1.9 Å.

The closure of the isotopic system corresponds to the minimal temperature below which the daughter isotope (^{40}Ar) stays trapped in the crystal lattice. For clay minerals, the closure temperature of the radiogenic argon is of the order of $260 \pm 30\,^{\circ}C$ (Purdy and Jäger 1976). Closure can be considered instantaneous when the time required for mineral formation is lower than the experimental error ($< 2\sigma$). In this case, the age measured is the age of the mineral formation. By contrast, if the system closure is very long ($> 2\sigma$) – because the mineral keeps growing – the age measured will not be the actual age of the mineral formation but, rather, "an integrated age" whose value depends on the duration of the growth process.

The outflow of argon from crystal lattices is controlled by two different processes: (1) destruction of the lattice (dissolution or fusion), and (2) diffusion within lattices. Conversely, the adsorption of argon bears witness to the conditions at the instant t of the trapping:

– the isotopic composition of this argon is equal to that of the atmospheric argon ($^{40}Ar/^{36}Ar = 295.5$);

– the isotopic composition of argon is more or less enriched in ^{40}Ar. Any impoverishment indicates that the sample measured is a mixture of mineral phases.

Two graphical representations are commonly used (Fig. 4.20): diagrams ^{40}Ar as a function of K_2O (Harper 1970) and $^{40}Ar/^{36}Ar$ as a function of $^{40}K/^{36}Ar$ (Roddick and Farrar 1971).

The two types of diagram are considered to be significant of an age (the straight regression lines are then isochrons) if they yield the same result (Shafiqullah and Damon 1974). The results given by both tests are valid if the MSDW (mean squares weighted deviate) values are low (below 3). The results are expressed with an experimental error σ calculated following classical procedures. From a theoretical standpoint, the measured age truly corresponds to the age of crystallisation of the studied minerals when the maximum accumulation rate is acquired during a period shorter than the experimental error 2σ (Fig. 4.21).

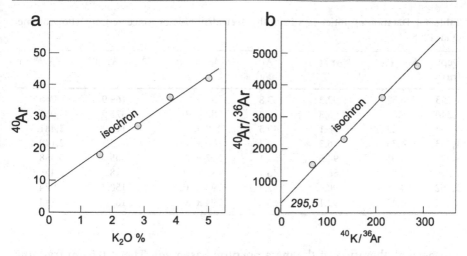

Fig. 4.20a,b. Principle of the definition of isochrons with diagrams: **a)** ^{40}Ar as a function of K$_2$O (Harper 1970), and **b)** ^{40}Ar/^{36}Ar as a function of ^{40}K/^{36}Ar (Roddick and Farrar 1971)

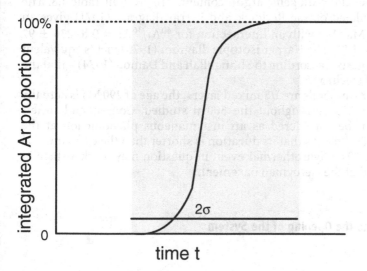

Fig. 4.21. Schematic representation of the daughter isotope accumulation (^{40}Ar) during an instantaneous event. Accumulation takes place during a period shorter than 2σ (*shaded area*)

4.2.1.3
Application: Datation of Clay Veins in a Basement (Sancerre-Couy)

The deep drill hole at Sancerre-Couy, Southern Paris Basin (Géologie Profonde de la France, French Scientific Program) has permitted the collection of clay minerals whose crystallisation in veins took place during an episode of hy-

Table 4.8. Datation of illitic clays from the deep drill hole of Sancerre-Couy (Bonhomme et al. 1995)

Depth (m)	K_2O (%)	$^{40}Ar^*$ (%)	$^{40}Ar^*$ (nl/g)	t ($My \pm 1\sigma$)	$^{40}K/^{36}Ar \cdot 10^3$	$^{40}Ar/^{36}Ar$
1,153	6.52	92.2	63.8	280 ± 7	169.9	3,766
1,240	4.89	95.8	48.2	282 ± 8	375.9	6,965
1,264	5.25	88.1	49.3	270 ± 8	129.2	2,481.2
1,353	4.73	86.3	46.3	280 ± 9	105.3	2,149.4
1,373	5.51	94.5	51.1	267 ± 7	305	5,388
1,390	4.59	88	46.4	289 ± 10	118.7	2,454.6
1,462	4.17	90.4	42.8	293 ± 10	150.4	3,074
1,652	6.42	92.1	66.2	294 ± 8	185.6	3,744

drothermal alteration of the metamorphic basement. The $< 0.5\,\mu m$ fraction of the samples essentially consists of ordered illite/smectite mixed layers associated to chlorite and calcite. Corrensite has been identified at $-1,353$ and $-1,390\,m$. The K_2O and radiogenic argon contents are given in Table 4.8. The isotopic line calculated on the Roddick and Farrar diagram (1971) gives an age $t = 290 \pm 12$ Ma (1σ) with an intersection for $^{40}Ar/^{36}Ar = 0$ to 236 ± 97 (1σ) and MSDW$= 1.33$. The Harper isotopic diagram (1970) yields equivalent results, which suggests – according to Shafiqullah and Damon (1974) – that the isotopic line is an isochron.

Since the only K-minerals are I/S mixed layers, the age of 290 Ma is actually that of their formation throughout the 500 m studied. Consequently, their crystallisation can be considered as an instantaneous phenomenon at the geological scale. This means that its duration is shorter than the experimental error σ (Fig. 4.22). The hydrothermal event in question may mark a stage in the cooling period of the Hercynian basement.

4.2.2
Disturbances Due to the Opening of the System

4.2.2.1
Theoretical Aspect

The opening of the system disturbs the relationship between the mother and daughter isotope amounts, and consequently distorts the calculation of the age of formation of minerals as given by the equation: $t = 1/\lambda$ ($^{40}Ar^* \lambda/^{40}K$ λ_{K-Ar}). Figure 4.23a shows the four possibilities of modification of mother and daughter isotope amounts, and their effect on the slope of the isotopic line. In reality, this slope measures the age of the opening of the system and erases any information about the age of formation of the minerals (Fig. 4.23b).

The opening of the system in the stability field of clay minerals is generally due to the temperature rise that causes the loss of argon. If the loss is complete,

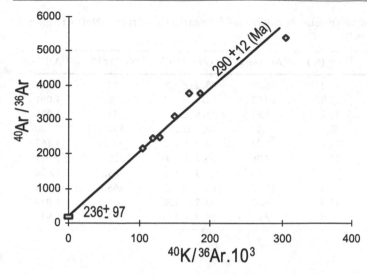

Fig. 4.22. Determination of the age of the clay veins crossing the metamorphic basement of the Paris Basin at Sancerre-Couy. In view of the experimental error, the regression line is considered as an isochron at 290 ± 12 Ma, because it crosses the axis of ordinates at the value close to that of the ^{40}Ar/^{36}Ar ratio of atmosphere (295.5)

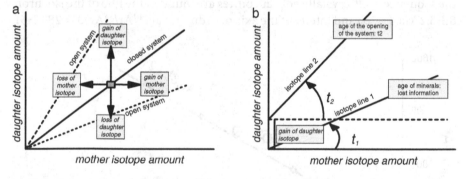

Fig. 4.23a,b. Representation in Harper's diagram (1970) of the influence of the opening of the system on the isotopic composition of minerals. **a)** The opening of the system can increase the slope of the isotopic line either by gain of daughter isotope or by loss of mother isotope. It can decrease the slope either by loss of daughter isotope or by gain of mother isotope. **b)** In the case where the opening of the system at an instant t_2 leads to a gain of daughter isotope, the isotopic line permits measurement of the age of the opening of the system, and not of the age of the formation of minerals

the measured age t_2 will be that of the new closure of the system following its opening (cooling period). If it is partial, the measured age will be a weighted mean between the accumulation rate of argon during the formation of the mineral and that produced by the new closure of the system. The interpretation of this age becomes impossible in terms of geological history.

Table 4.9. Datation of the Ypresian-Bartonian sands from the Cassel region, Northern France (Bonhomme et al. 1995)

Sample	K_2O (%)	$^{40}Ar^*$ (%)	$^{40}Ar^*$ (nl/g)	t ($My \pm 1\sigma$)	$^{40}K/^{36}Ar.10^3$	$^{40}Ar/^{36}Ar$
G 898	7.28	79.2	10.27	43.2 ± 1.2	441	1,421
G 899	7.28	72.9	10.02	42.1 ± 1.2	320.4	1,091
G 900	6.7	71.2	9.62	43.9 ± 1.3	281.9	1,025
G 901	5.66	61.1	7.84	42.5 ± 1.5	185.8	760
G 902	7.03	74.7	9.95	43.3 ± 1.2	341.3	1,167
G 903	5.74	71	9.99	53.2 ± 1.6	229.7	1,017
G 904	6.56	72	9.68	45.1 ± 1.3	284.9	1,054
G 905	5.38	76.4	10.77	61.0 ± 1.7	265.3	1,254
G 906	5.57	71.3	9.61	52.7 ± 1.6	235.6	1,029
G 907	4.61	68.6	7.6	50.4 ± 1.6	217	941

4.2.2.2
Example: Glauconitic Sands

The Ypresian- Bartonian sands from the Cassel region (Northern France) contain glauconites. In a sequence of vertical sampling, the extracted glauconites do not exhibit the same isotopic composition (Table 4.9; Fig. 4.24). At the top of the sequence, well-crystallised glauconites are found in the line of the isochron 8 ± 3.2 Ma. ($\pm 2\sigma$) and intercept the axis of ordinates at $(^{40}Ar/^{36}Ar)0 = 283 \pm 61$

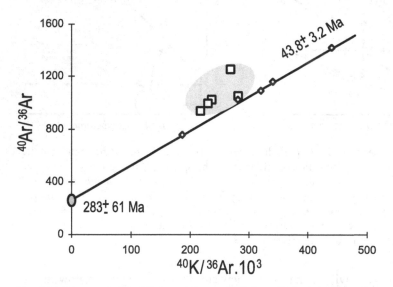

Fig. 4.24. Diagram of Roddick and Farrar (1971) applied to the Ypresian-Bartonian glauconites from the Cassel region (Northern France). The glauconites of the upper part of the deposit are found in the *line of an isochron*; those of the lower part (*squares in the shaded area*) are too "old"

($\pm 2\sigma$). At the bottom of the sequence, they are enriched in ^{40}Ar and would therefore be older than the deposit. In the absence of a rearrangement of older glauconitic layers, this age makes no geological sense. It can only be explained by the incorporation of argon (open system) or by the formation of glauconites on K-minerals of detrital origin.

4.2.3
Mixtures of Phases

The clay fraction of soils – diagenetic or hydrothermal rocks – is very seldom monophased. It mostly consists of newly formed minerals mixed with debris of older minerals. One of the most commonly occurring problem is the ambiguity of the "illite" phase, which can be formed via a series of illite/smectite mixed layers, or which can result from the progressive destruction of magmatic or metamorphic micas. A K–Ar datation of such a mixture cannot be exploited in geological terms. Fortunately, if the analysed clay fractions can be shown to consist of only two K-bearing phases (newly-formed illite and detrital illite), the study of their mixture in different proportions enables the ages of both components to be determined by simple extrapolation.

This approach has been successfully tried by Pevear (1992) for the shales of the Albian – Aptian and the Turonian of Arkansas (USA). The K–Ar datation

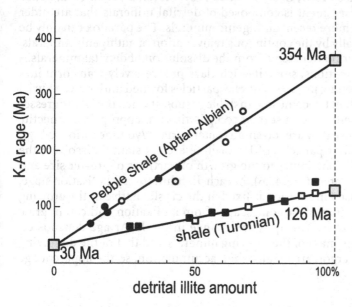

Fig. 4.25. K–Ar datation of samples of diagenetic shales from two series and composed of a mixture of detrital illite and newly-formed illite whose proportions have been determined by X-ray diffraction (from Pevear 1992). *Points, circles, black squares* and *white squares* correspond to different size fractions. The *two straight lines* have been obtained by linear regression

has been performed on several size fractions of the shales (2.0–0.2; 0.2–0.02; < 0.02 µm). The detrital illite fraction has been measured by X-ray diffraction using the calculation code NEWMOD (Reynolds 1985). The results (Fig. 4.25) show that both series of analyses can be represented by two straight lines (simple linear regression) that converge at 0% detrital illite at about the same age: 30 my. This age corresponds to that of purely neogenetic illites extracted from bentonites intercalated in the Albo-Turonian series. This is a diagenetic age, as opposed to ages 354 and 126 Ma respectively, which are detrital ages. Moreover, Pevear has concluded that the detrital input has changed between the Albian-Aptian and the Turonian because of major tectonic movements at that time.

4.2.4
Datation and Crystal Growth

4.2.4.1
Theory

A paradox appears in the datation of clays from diagenetic environments: the older the bed of the buried sediment, the younger the illites and illite/smectite mixed layers they contain (Aronson and Hower 1976; Morton 1985; Rinkenbach 1988; Burley and Flisch 1989; Glasman et al. 1989; Mossman et al. 1992; Renac 1994). Generally, a sediment is composed of detrital minerals that are older than the bed and of more recent authigenic minerals. The paradox can only be explained theoretically by the continuous rejuvenation of authigenic minerals, even if a part of their matter stems from the dissolution of detrital minerals.

In diagenetic formations, smectite-rich clays progressively transform into illite following a ripening process. The clay particles formed under given burial conditions at an instant t become too unstable to subsist when burial progresses or when t increases, either because their composition is inappropriate (smectite % too high) or because they are too small (outer surface /volume ratio too big). Therefore, the smallest particles of I/S or illite are constantly dissolved; the matter thus released contributes to the growth of particles of greater size and with a greater illite content (Fig. 4.26). At each dissolution-crystallisation stage, the potassium released in solution is fixed in the crystal lattice of the growing particles. But the radiogenic argon escapes from all fixation and can migrate out of the reaction zone. The apparent change in the K–Ar age depends on the ratio between the mass of the growing minerals and that of the dissolving minerals. Losses of argon are greater than accumulation, so the apparent age decreases.

4.2.4.2
Example: I/S Under Diagenetic Conditions

Burley and Flisch (1989) have shown that the I/S mixed layers of the mudstones and sandstones of the Kimmeridgian have increasing illite contents (hence

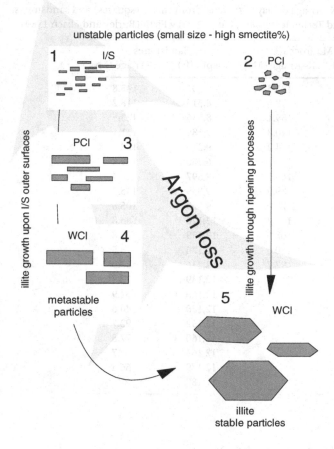

Fig. 4.26. The growth of I/S or illite particles and the increase in the illite content depend on a dissolution process of the unstable particles whose elements are used for the growth of illite layers. PCI: poorly crystallized illite; WCI: well crystallized illite.

increasing K_2O contents) with the burial depth. In the same way, the K–Ar age of these minerals becomes younger as depth increases (Table 4.10; Fig. 4.27). This is due to the fact that the fixation of potassium increases with depth, thus causing the formation of illite over a very long period, greater than the experimental error. This potassium is provided by the dissolution of I/S crystals that are richer in smectite or of smaller size. The dissolution-recrystallisation rate increases with depth, releasing more argon. Consequently, the greater the burial depth, the more recent the calculated age.

Since potassium is an element readily displaced in geological environments by interactions of rocks and fluids, the accuracy of K-Ar datations is quite often questionable. This drawback has been solved by basing the datations on very little soluble elements, like certain rare earth elements such as samarium and

Table 4.10. Values of the K–Ar age of clays extracted from shales (squares) and sandstones (lozenges) of the Piper and Tartan formation, Outer Moray Firth (Burley and Flisch 1989)

Mudrocks		Sandstones	
Depth (ft)	K–Ar age (My)	Depth (ft)	K–Ar age (My)
8,482	125.3	8,506	143.8
8,608	128.5	8,515	118.2
9,285	139.1	8,536	123.5
9,948	140.8	9,189	131.1
10,391	154.2	9,210	122.3
10,444	141.2	6,260	129.2
11,671	160.7	9,897	134.4
12,195	89.9	9,899	143.4
12,286	72.6	9,933	105.8
12,355	102.2	12,120	86.8
12,490	68.5	12,139	76.8
13,039	140.6	12,170	73.5
13,980	97.7	12,236	65.9
		12,139	77.7
		12,166	62.9
		12,180	40.6
		12,152	92.7
		12,160	77.8
		12,169	67.7
		12,179	66.4
		12,189	66

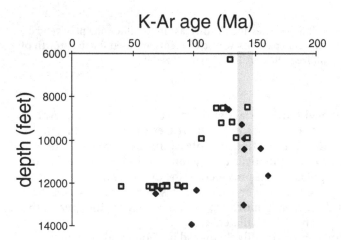

Fig. 4.27. Variation in the K–Ar age of clays extracted from shales (*squares*) and sandstones (*lozenges*) of the Piper and Tartan formation, Outer Moray Firth (Burley and Flisch 1989). The *shaded area* represents the stratigraphic age of this formation (Kimmeridgian)

neodymium. The application to clay minerals of these methods widely used for magmatic and metamorphic rocks presents technical difficulties. Clauer (1997) has shown how to bypass these difficulties with specific preparations. His article presents a very useful synthesis of these methods.

Suggested Readings

Dalrymple GB, Lanphere MA (1969) Potassium argon dating principles, techniques and application to geochronology. Freemann, San Francisco, 258 pp

Fritz P, Fontes JC (1980) Handbook of environmental isotope geochemistry, vol 1. Elsevier, Amsterdam, 328 pp

Faure G (1982) Principles of isotope geology, 2nd edn. Wiley, New York, 589 pp

Clauer N, Chaudhuri S (1995) Clays in crustal environments. Isotope dating and tracing. Springer, Berlin Heidelberg New York, 359 pp

CHAPTER 5

Surface Properties – Behaviour Rules – Microtextures

We saw in Chaps. 1 to 4 how to "read the genetic memory" of clay minerals using their crystallochemical properties. The scale of investigation was the crystal and the conditions of formation are deduced from the analyses of the chemical composition and the stacking sequence of layers. However, clays have another particular characteristic: the huge development of their outer or inner crystal surfaces. This property governs their chemical exchanges with the ambient solutions and the way their crystals are associated in three-dimensional structures, i. e. the "texture" of clays. Textures record the mechanical or rheological history of the natural rock or the artificial material. How surface chemical properties and textures are related is the theme of this chapter.

5.1
Chemical Properties

Introduction

Surface properties of clay minerals are defined according to the investigation scale. Indeed, several organisation levels from the nanometer to the centimetre can be distinguished: layers, crystallites, particles, aggregates and clay materials (rocks, muds, suspensions). Water and ions can be exchanged at all levels through physical or chemical processes, the energy of which is greater as the scale is small.

Clay materials have a physical organisation that varies according to the constituent crystal species (kaolinite, smectite, illite ...) and according to their water content. The presence of "swelling" minerals such as smectites introduces an additional factor: the organisation also depends on the nature of those cations saturating the interlayer spaces. All these parameters determine the physical properties of the material: mechanical strength, drying shrinkage, viscosity of suspensions etc.

The organisation of the material, at the scale of small crystals (crystallites), involves complex interactions between chemical and electrical forces. This chapter proposes an introduction to these problems, further developments of which are to be searched for in specialised books (van Olphen 1977; Sposito 1984; Stumm 1992; McBride 1994).

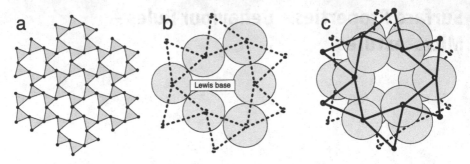

Fig. 5.1a–c. Surfaces of tetrahedral layers. **a)** Trigonal lattice formed by tetrahedra. **b)** The ditrigonal cavity behaves like a weak base site (Lewis base). **c)** Coordination polyhedron with 12 vertices formed by the oxygens of the ditrigonal cavities of two consecutive layers. In smectites, the more stable stacking sequence is obtained by rotation of $\pi/3$

5.1.1
Structure of Clay Minerals at Various Scales

5.1.1.1
Layer Surfaces

The external surfaces of 2:1 layers are essentially formed by their tetrahedral sheets. The latter form "siloxane" surfaces. The SiO_4 tetrahedra are associated into a trigonal lattice (Fig. 5.1a) and undergo a rotation about their apex (tilt angle α) and about an axis in the a-b plane (surface corrugation) which results in the deformation of the surface of siloxanes. A height difference of 0.02 nm separates two consecutive oxygens. The cavities of the deformed hexagons are about 0.26 nm in diameter. The configuration of the electron orbitals of the 6 oxygens gives these cavities a character of Lewis bases (Fig. 5.1b). If no isomorphic substitution of ions occur in the tetrahedral layer (Al^{3+} for Si^{4+}) or in the octahedral layer (R^{2+} for R^{3+}), the basic behaviour, namely of electron donor, remains very low. It is merely sufficient to complex dipolar molecules such as H_2O for instance (the H^+ ion is located on a line orthogonal to the siloxane surface at the centre of a ditrigonal cavity). These complexes are then poorly stable and can be destroyed by low-energy processes.

The isomorphous substitutions of a bivalent cation for a trivalent one in the octahedral layer introduce a positive charge deficiency in the replaced octahedron. The resulting negative charge is distributed between the 10 oxygens of the 4 tetrahedra bonded to the deficient octahedron. Therefore, the Lewis base behaviour of the ditrigonal cavity is reinforced and becomes sufficient to form complexes with dipolar molecules and cations with their hydration sphere.

Isomorphous substitutions in tetrahedral layers also lead to the occurrence of negative charges in the replaced tetrahedra. These charges are distributed essentially between the three surface oxygens (the fourth one being bonded to an octahedron). Therefore, much more stable complexes are formed with cations or dipole-molecules. Let's consider the stacking sequence of two layers

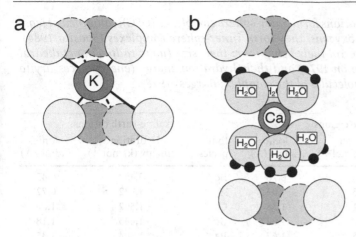

Fig. 5.2a,b. Bonding of interlayer cations. **a** Interlayer cations like potassium are bonded by 6 strong bonds and 6 weak bonds. **b** Those cations less strongly bonded are accompanied by their hydration sphere. The thickness of the interlayer space increases

in a smectite crystallite. The ditrigonal cavities are superposed after a rotation of 60° (Fig. 5.1c), thus determining a polyhedron with 12 vertices, 6 of them being closer to the cation. Depending on the negative charge value inside each cavity, this polyhedron houses cations with or without their hydration sphere.

The 12 oxygens of the siloxane surfaces form the inner sphere of K^+ ions (Fig. 5.2a). The geometrical configuration is almost perfect because the ionic diameters of O^{2-} and K^+ ions are very close. Nevertheless, on a chemical bonding basis, owing to the degeneration of the hexagonal symmetry into ditrigonal symmetry, each K^+ ion is strongly bonded to every second oxygen, as shown in Fig. 1.5b. This is the case of micas.

When the negative charges in the ditrigonal cavities are low (0.3 to 0.6 per 4 Si), cations such as Ca^{2+} are bonded to the water molecules of their hydration sphere (Fig. 5.2b). Weakly bonded to the silicate layers, these molecules are released by low-energy processes (temperature rise between 80 and 120 °C). They organise themselves on the siloxane surface (Fig. 2.7) into a structure with thermodynamic properties close to those of ice (see Sect. 5.1.2.1). This is typically the case of smectites.

Minerals composed of 1:1 layers such as kaolinite, serpentines or berthierines exhibit two types of outer surfaces: a siloxane surface and a surface formed by (OH)⁻ groups. The layers are electrically neutral, thus preventing the adsorption of ions or molecules in the interlayer spaces.

Interactions Between Interlayer Cations and the Surface of Layers

Cations are housed at the centre of the coordination polyhedra whose vertices are either water molecules or oxygens from the ditrigonal cavities, or both. Sur-

rounded by water, cations form outer-sphere complexes (OSC); combined at least in part with basal oxygens, they form inner-sphere complexes (Sposito 1984). Two parameters control their behaviour: their size (ionic radius in octahedral coordination; Shannon 1976) and their hydration energy (energy necessary to release the water molecules of the complex; Burgess 1978).

Ionic species	Alkaline cations		Ionic species	Alkaline-earth cations	
	Hydration enthalpy (kJ mol^{-1})	Ionic radius (Å)		Hydration enthalpy (kJ mol^{-1})	Ionic radius (Å)
Li^+	−515	0.76	Be^{2+}	−2,487	0.45
Na^+	−405	1.02	Mg^{2+}	−1,922	0.72
K^+	−321	1.38	Ca^{2+}	−1,592	1.00
Rb^+	−296	1.52	Sr^{2+}	−1,445	1.18
Cs^+	−263	1.67	Ba^{2+}	−1,304	1.35

K^+, Rb^+ and Cs^+ alkaline cations readily lose their water molecule "shell". They form complexes with basal oxygens because their hydration enthalpy is low. Since their diameter is greater than the size of the hexagonal cavity of tetrahedral sheets (> 1.4 Å), they can occupy two positions according to the origin of the layer charge: either perpendicular to the basal oxygens of the tetrahedron in which Al replaces Si, or partially engaged in the hexagonal cavity when the charge is octahedral

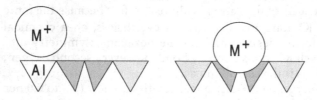

Ba^{2+}, Na^+ and Li^+ cations have greater hydration energy. Due to their smaller ionic diameter, they can enter farther into the hexagonal cavity. The Li^+ ion is small enough to enter with a water molecule. Mg^{2+}, Ca^{2+} and Sr^{2+} cations have very small hydration enthalpies and ionic radii even smaller. They remain strongly bonded to the water molecules forming the complex (hydration sphere). These water molecules are the ones bonded to basal oxygens through weak bonds (hydrogen bonding). The interlayer spaces saturated by these bivalent cations exhibit 1 or 2 water layers depending on the partial water pressure (relative humidity).

5.1.1.2
The Edges of 2:1 and 1:1 Layers

The size of most clay mineral crystallites ranges between 10 nm and 10 µm. These very small dimensions greatly increase the contribution of edges to the overall surface. That is where the specificity of clays lies compared to the other families of minerals. In other words, per volume unit, the number of Si–O or R^{2+}–OH or R^{3+}–OH bonds interrupted by edges is very high. These interface sites are electrically charged. Neutrality is obtained only by adsorption of ions from surrounding solutions (Fig. 5.3). Thus, in the case of kaolinite, two chemical functions appear: silanol (Si–OH) and aluminol (Al–OH) groups. The properties of these groups change according to the pH of solutions:

– at low pH, aluminol groups fix H^+ protons thus yielding Al(III)–H_2O groups that are Lewis acid sites;

– at higher pH, the water molecule is replaced by a $(OH)^-$ group.

The edge of the tetrahedral layers is marked by O^{2-} ions whose available valency in a bond with Si^{4+} is compensated for by the bonding of a H^+ proton. Owing to the high valency of the Si^{4+} ion, the OH group thus formed is strongly bonded to the crystal structure and can only complex hydroxide anions; it cannot fix H^+ protons.

5.1.1.3
Particles, Aggregates, Granules

Particles are considered here as the association of crystallites by surface forces. Most of the time, associations are performed by superposition of (001) faces thus yielding more or less coherent and thicker stacking sequences of layers than those of separate crystallites. These particles exhibit a totally irregular shape in the a-b plane as well as "reentrant" angles (Fig. 1.18). These particles are a few microns in size.

An aggregate of particles only exists if organic (humin, humic or fulvic acids, etc) or inorganic (Fe–Mn oxides or hydroxides, etc) substances are used as a "ligand". The size of aggregates varies from a few microns to a few tens of microns. Inner voids are a few nanometers in size.

Aggregates are themselves fused together into bigger units that constitute the structural elements of soils, alterites or sediments: clods, macro-aggregates, granules. Depending on their size, inner pore dimensions vary from the micron to a few hundreds of microns. Note that granules are scarce in diagenetic or hydrothermal environments. Most of the time, the aggregate is the maximum organisation level.

Each structural organisation level of clays is characterised by different binding energies that give it special physical properties: mechanical strength, field of electrical forces, etc. In the same manner, the varying size of voids/pores (ranging from a few nanometers to a few hundreds of microns) imposes a variable energetical state to the fluids that fill them. Brickmakers know, for instance,

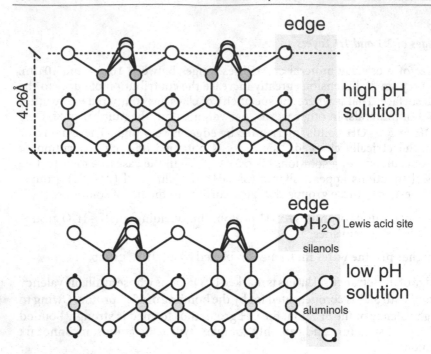

Fig. 5.3a,b. Basal and edge surfaces of kaolinite crystals. **a)** At basic pH, the interrupted bonds of the crystal lattice are neutralised by OH groups. **b)** At acidic pH, aluminol groups fix the H^+ protons thus yielding $Al(III)-H_2O$ groups that are Lewis acid sites

that removing the last "percent" of water from the clay paste requires much more energy (hence more money) than removing the first "percent". Consequently, the study of the physical properties of the solid or fluids amounts to a thorough investigation of the structure of clay organisations at any level.

5.1.2
The Different States of Water in Clay Materials

5.1.2.1
Energetical State of Water in Clays

Thermal Analyses
The water molecules impregnating clay materials can be extracted in two different ways: they can be heated until evaporation (agitation breaks the chemical bonds) or they can be squeezed out of their site through the application of a pressure. The thermogravimetric analysis of Na-montmorillonite crystallites or particles with a size contained between 0.10 and 0.35 μm reveals several water losses at different temperatures (Fig. 5.4a). In an open system (molecules escape or are dragged by an inert gas flow), evaporation of the molecular water weakly bonded to outer and inner surfaces of clay particles occurs at

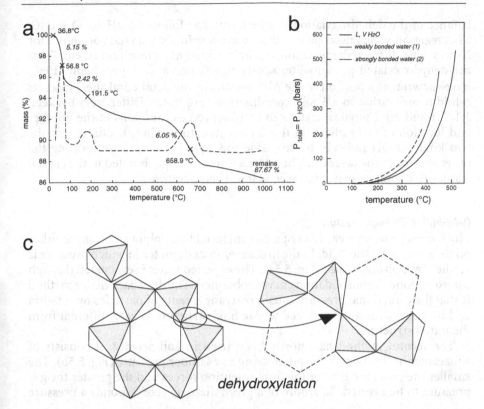

Fig. 5.4a–c Dehydration and dehydroxylation reactions. **a)** Thermogravimetric analysis (direct curve and derivative) of a saturated Na-montmorillonite (Wyoming type) showing two low-temperature dehydration stages and the dehydroxylation stage (after Bish and Duffy 1990). **b)** The two dehydration reactions in a closed system are represented in the pressure-temperature space (after Koster van Groos and Guggenheim 1990). **c)** Loss of the 6-fold coordination of octahedral Al^{3+} cations to a 5-fold coordination

56.8 °C. The release of the water molecules strongly bonded to the interlayer cations (hydration inner sphere) requires a temperature of 191.5 °C. These two reactions can be observed by differential thermal analysis in the form of two partially overlapping peaks (Koster van Groos and Guggenheim 1990). Consequently, the thermodynamic status of the strongly bonded water is not that of free water, but is rather related to that of ice (Mercury et al. 2001). In a closed system, the partial water pressure equal to the total pressure is not an intensive variable independent from temperature; it varies with it. The temperatures of the two dehydration reactions increase up to values contained between 400 and 500 °C (Fig. 5.4b). This is of interest in some geological phenomena (see Sect. 10.2.1).

At higher temperatures (685.9 °C) in an open system (Fig. 5.4a), the very structure of crystallites is reached. The H^+ protons are released; the chemical balance of the dehydroxylation can be written as follows: $2\,OH \rightarrow O + H_2O$. This reaction yields news phases that retain a sufficient crystal continuity for X-rays to be diffracted: metakaolinite, dehydrated montmorillonite, etc. From the dehydroxylated pyrophyllite, Koster van Groos and Guggenheim (1990) have shown that a portion of the Al^{3+} cations in the octahedral sheet loses its 6-fold coordination to a 5-fold coordination (Fig. 5.4c). Differences between 1:1, 2:1 and 2:1:1 phyllosilicates can be observed as functions of the number and location of OH radicals in the crystal structure. Thus, kaolinite exhibits two kinds of OH radicals: those inside the layer (2) and those forming the outer surface of the layer (6). The latter are less strongly bonded to the crystal framework, and are hence released at lower temperatures.

Dehydration Through Pressure

The water can be squeezed out of a clay material by applying a pressure either on the solid or on the fluid. In the first case, an oedometer in which pressure is applied by a piston is used (Fig. 5.5a). The expelled water is evacuated through a porous stone. A consolidation curve is obtained. The drawback of this method is that the directional pressure causes rearrangements of particles by rotation and by shear within the material whose final structure is very different from the initial structure.

The second method has mostly been used in soil science. It consists of squeezing the fluid out of the pores using a gas under pressure (Fig. 5.5b). The smaller the pore, the greater the fluid retention forces and the greater the gas pressure to be exerted. To a pore of a given diameter corresponds a pressure

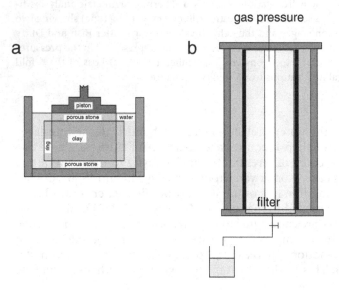

Fig. 5.5a,b. Schematic representation of the two methods exerting a pressure: oedometer (a) and filtration under pressure (b)

Table 5.1. Equivalence between gas pressure exerted, ionic strength of solutions, water activity and pore size

pF	a_{H_2O}	Pressure (bar)	D_{max} pores (μm)	Ionic force
1	0.999993	0.01	150	$2,2 \times 10^{-4}$
2	0.999927	0.1	15	$2,2 \times 10^{-3}$
3	0.99927	1	1.5	$2,2 \times 10^{-2}$
4	0.9927	10	0.15	$2,2 \times 10^{-1}$
4.2	0.9888	15.8	0.095	$3,38 \times 10^{-1}$
4.67	0.9669	46.4	0.032	1
5	0.927	100	0.015	2.2
5.48	0.8	305	0.005	6
5.7	0.695	500	0.003	
6	0.484	1,000	0.0015	

that causes expulsion of the fluid and its replacement by gas: this is the entry of air point. The diameter can then be determined by applying the Jurin-Laplace law:

$$P = \frac{2A \cos \alpha}{D} \tag{5.1}$$

P gas pressure applied

A surface tension

α contact angle at the solid-liquid-gas interface

D maximum pore diameter

For hydrophilic solids like clays, the cosine of the angle formed by the connection of the meniscus to the pore walls ($\cos \alpha$) is equal to 1.

Considering a pore with a given diameter, the pressure exerted at the entry point of air decreases with fluid salinity (ionic strength). An equivalence chart between gas pressure, ionic strength, water activity and pore size can then be drawn up (Table 5.1).

Amounts of water contained in a known mass of smectite can be calculated. Let's consider the Wyoming montmorillonite whose half-cell formula is as follows: $Si_4 O_{10} Al_{1,7} Mg_{0,3} (OH)_2 Na_{0,3}$. The unit cell mass m is 734 g; the values of the a and b parameters are 5.21 and 9.02 Å, respectively. When relative humidity (p/p_0) passes from 79 to 52%, the interlayer spacing decreases from 14.8 to 12.5 Å (Brindley and Brown 1980).

The value of the surface S of a unit cell is: $S = 2ab$ or $5.21 \times 9.02 \times 2 = 93.99 \, Å^2$ or $93.99 \times 10^{-20} \, m^2$. The specific surface S_0 in $m^2 \, g^{-1}$ is given by SN/m where N is the Avogadro number ($N = 6.022 \, 10^{23}$) or: $(93.99 \times 10^{-20} \times 6.022 \times 10^{23})/734 = 771.1 \, m^2 \, g^{-1}$.

The volume of lost water in $m^2 g^{-1}$ when the interlayer spacing is reduced by 2.3 Å can be calculated as follows: $771.1 \times 2.3 \times 10^{-10} = 1{,}773.5 \times 10^{-10}\, m^3$ or $1{,}773.5 \times 10^{-4}\, cm^3$.

Assuming that water density in the interlayer space is close to 1, the mass of lost water will then be 177.35 mg per gram of clay.

Under laboratory conditions (25°C, 1 atm), the pressure $P_{2\to1}$ required to extract a water layer from montmorillonite whose interlayer spacing decreases from 14.8 to 12.5 Å can be calculated by the thermodynamic relationship of gases: $PV = nRT$. One shows that $P_{n\,H_2O} = -(RT/V_{M\,H_2O})\ln p/p_0$ where $V_{M\,H_2O}$ is the molar volume of water under laboratory conditions ($0.01802\,L\,mol^{-1}$), R is the perfect gas constant ($8.31 \times 10^{-2}\,L\,bar\,K^{-1}\,mol^{-1}$), and T is the absolute temperature (298 K). For a relative humidity of 79% (2 water layers), the equilibrium pressure is:

$$P_{2H_2O} = (-0.0831 \times 298 \times 2.303 \times -0.10)/0.01802 = 316.49\,\text{bar}$$

For one water layer (12.5 Å), the equilibrium pressure P_{1H_2O} is given by:

$$P_{1H_2O} = (-0.0831 \times 298 \times 2.303 \times -0.28)/0.01802 = 886.16\,\text{bar}$$

The pressure $P_{2\to1}$ required to remove a water layer is then $P_{2\to1} = 569.67$ bar. Van Olphen (1977) has shown that the pressure required to remove the last water layer is even higher: $P_{1\to0} \cong 4{,}000$ bar.

Shrinkage after Drying

In soil mechanics, the shrinkage curve of a clay material is determined by measuring the variation of its volume as a function of the amount of water left in the solid. Considering that the volume of the involved material is equal to the sum of the volumes of solids (V_s) and voids (V_v), the variation of the ratio $e = V_v/V_s$ (volume of voids) as a function of the water index $\vartheta = V_{water}/V_{solid}$ reveals the behaviour of both the pore water and the water adsorbed on the surface of clays (Fig. 5.6).

Tessier (1984) explains the shrinkage curve by a three-stage process:

- the first stage (linear variation of e as a function of ϑ) means that the volume of voids decreases proportionally to the volume of water (A). The sample remains saturated,

- the second stage is characterised by the loss of linearity (B): the shrinkage is smaller than the extracted volume of water. This means that air enters the pores (entry of air point),

- the third stage C corresponds to the loss of water without variation in the volume of the sample (shrinkage limit).

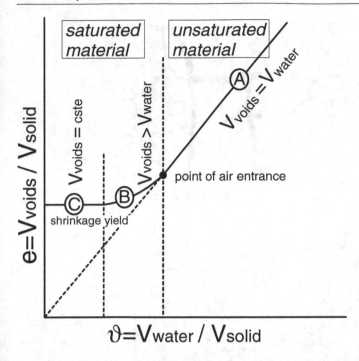

Fig. 5.6. Shrinkage curve of a clay mineral. Volume of voids $e = V_v/V_s$ and water index $\vartheta = V_{water}/V_{solid}$ where V_s, V_v and V_{water} represent the volumes of solids, voids and water, respectively. Details in the text

The diameter of those pores which lose their fluid at the entry of air point can be calculated by applying the Jurin-Laplace law (Table 5.1) This diameter depends on the nature of clay; the greater the gas pressure, the smaller the diameter: 1 bar for kaolinite; 10 bar for illite; 1 kilobar for smectite.

Fracturation of Clay Materials
The effects of the drying of clay materials on the scale of the square metre are expressed by the aperture of fracture networks (Fig. 5.7a). The fractures and their intersections are spatially distributed according to the nature, and particularly the porosity, of the soil (Velde 1999). The regularity of fracture networks can be measured by fractal analysis (the box method is suited to 2-D spaces; the measured fractal dimension then varies between 1 and 2). A relationship exists between the porosity and the distribution of pores (Fig. 5.7b). The aperture of the connected networks plays an essential part in the flow of the solutions in the soils subjected to drying during summertime. The geochemical behaviour and the microstructure of vertisols and saline soils in arid climate are conditioned by this aperture (see Sect. 6.2.4.2).

Modelling the hydrous behaviour of soils and clays is delicate because it depends on the distribution laws of pores, of solids-solution interfaces or of

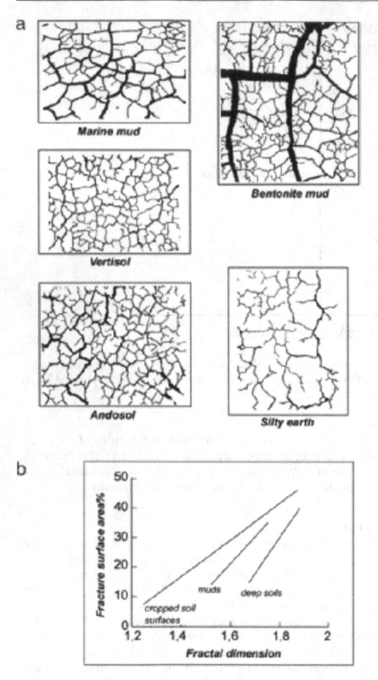

Fig. 5.7a,b. Macroscopic effects of the shrinkage on clay materials and soils. a) Examples of fracturations of soils and muds under the effect of drying-induced shrinkage (Velde 1999). b) Geometrical relationships between distribution and size of fractures

solids themselves. Van Damme (1995) emphasises the difference between the undersaturated state and the saturated state:

1. the organisation of soils can be modelled by a fractal distribution between upper and lower bounds separated by several decades. The hydrous behaviour in the undersaturated state is then properly described by static structural scaling laws, namely independent from the water content;

2. in the saturated state, the hydrous behaviour is described by dynamic scaling laws (dependent on the water content).

Even if these details go well beyond the scope of this book dedicated to beginners, the approaches connecting the microstructure with the laws of physical behaviour must be acknowledged as those carrying the great progresses that will be performed in the understanding of clays. On this account, Van Damme's article is an inescapable starting point.

5.1.2.2
Water Adsorption Isotherms

Water is more or less heterogeneously bound on the accessible surfaces of clays: the layer in direct contact with the solid is not necessarily complete before additional layers start forming (Fig. 5.8a). Despite this drawback, the affinity of clays for water can be globally measured by its adsorption isotherms. Since the binding energy of water on clays is of type $\Delta G = RT \log a_w$ (a_w: water activity), isotherms measure the variation of the amount of water fixed by clay as a function of a_w. They can be established experimentally at constant temperature using two complementary methods:

1. measurement of the amounts of extracted water (desorption) by weighing after filtration under pressure. This method is used for high values of water activity ($0.89 < a_w < 1$).

2. measurement of adsorption or desorption by the drier method (the relative humidity of air is controlled by $H_2SO_4 + H_2O$ mixtures) for low values ($0 < a_w < 0.9$).

The resulting curves can be plotted in various ways according to whether the amount of extracted water is expressed in absolute coordinates (mg or g g^{-1}) or in reduced coordinates (Fig. 5.8b,c) on the one hand, and according to whether the relative humidity is expressed by the water activity a_w or by the partial pressure of water (p/p_0) on the other hand. The number of adsorbed layers is then determined by dividing the amount of extracted water (mass x or volume v) by the water amount corresponding to a monolayer (mass x_m or volume v_m), assuming that the first layer is continuous. The adsorption is then described by the BET function (Brunnauer, Emmett and Teller) where c is a constant:

$$\frac{v}{v_m} = \frac{c(p/p_0)}{(1 - p/p_0)\left[1 + (c-1)p/p_0\right]} \tag{5.2}$$

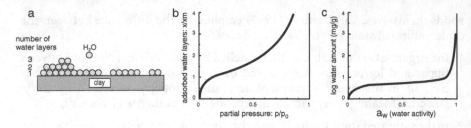

Fig. 5.8a–c. Adsorption of water by clays. **a)** Diagram showing that layers 2 and 3 can "nucleate" before completion of layer 1. This explains the shape of the adsorption isotherm in the reduced coordinates. **b)** Water adsorption isotherm of a kaolinite expressed in reduced coordinates, after Newman (1987). **c)** Desorption isotherm of a kaolinite, after Prot (1990)

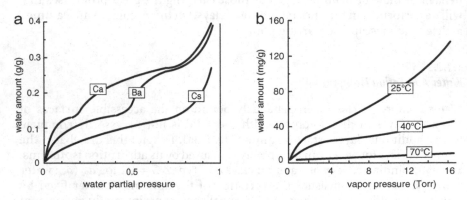

Fig. 5.9a,b. Adsorption isotherms of a Wyoming montmorillonite. **a)** Influence of saturation by different interlayer cations, after Newman (1987). **b)** Influence of the measuring temperature (Ca-saturated smectite)

The shape of water adsorption isotherms varies for a given clay as a function of the cations saturating the interlayer spaces (Fig. 5.9a). The amounts of adsorbed water, and particularly the inflexion points of curves, depend on the affinity of interlayer cations for water molecules. Thus, the Ca^{2+}-saturated Wyoming montmorillonite is more "hydrated" than if it was saturated by Ba^{2+} or Cs^+: the water content of $0.1\,g\,g^{-1}$ is reached for a partial pressure of 0.1 against 0.2 and 0.5, respectively. For a given cationic saturation, the adsorbed water content decreases with the temperature at which it is measured (Fig. 5.9b).

5.1.2.3
"Swelling" of Interlayer Spaces

Influence of Interlayer Cations
The concept of water adsorption on surfaces does not correspond to the same physical reality for all types of clays. Indeed, in neutral minerals (kaolinite, pyrophyllite, talc, serpentines), or in minerals with a high interlayer charge

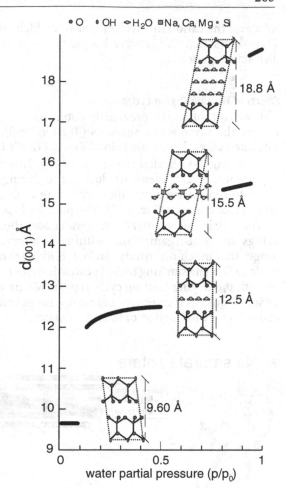

Fig. 5.10. Variation in d-spacings of a Wyoming montmorillonite as a function of the water partial pressure (p/p_0)

(greater than 0.8 per Si_4O_{10}) such as illites and micas, as well as in 2:1:1 minerals (chlorites), only the external crystal surfaces exhibit an adsorption capacity. Whatever the water partial pressure in their environment, the $d_{(001)}$ spacing does not vary. This is not the case of smectites or vermiculites, which can adsorb water molecules in their interlayer spaces. These molecules, whose number increases with the water partial pressure, are organised into layers and increase the $d_{(001)}$ spacing. Thus, for a Na^+-saturated Wyoming montmorillonite, the $d_{(001)}$ spacing varies from 9.60 Å (0 water layer) to 12.5 Å (1 water layer), 15.5 Å (2 water layers) and 18.8 Å (3 water layers) when p/p_0 increases (Fig. 5.10).

The maximum hydration state of di- or trioctahedral smectites depends on the nature of the cation saturating the interlayer spaces. It becomes infinite for Na^+ and Li^+; it corresponds to 3 water layers for Ca^{2+}, Mg^{2+} and Ba^{2+}. Potassium allows for 2 water layers in montmorillonites (octahedral charge) and only 1 water layer in other smectites. Vermiculites receive 2 water layers whatever the cation involved except K^+, which allows for only one water layer. This shows that the relative humidity must be known before interpreting diffraction patterns.

Effects of Wetting-Drying Cycles

The wetting of a clay previously subjected to an intensive drying does not restore the same water amounts (Tessier 1990). In the case of Na^+- or K^+-saturated clays, hysteresis is low (Fig. 5.11a); it is much higher in the case of Ca^{2+}- or Mg^{2+}-saturated clays (Fig. 5.11b). This means that clay particles have experienced rearrangements during the drying period. Extraction of water tends to reorganise crystallites according to face – face contacts rather than edge – face contacts (Fig. 5.22). The pore walls become thicker.

The very intensive extraction of water adsorbed by smectites or vermiculites brings about reorganisations within those crystallites whose layers are no longer that much disorderly stacked. Slides are due to shearing forces (De La Calle 1977). The stacking order (c direction) increases with the drying intensity or with the drying-wetting cycle repeat (Mamy and Gautier 1975; Eberl et al. 1986). It looks as if smectites and vermiculites "remember" the reorganisations caused by the extraction of interlayer water.

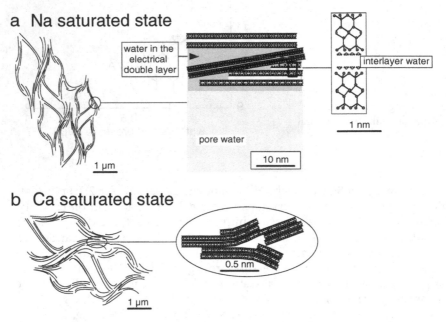

Fig. 5.11a,b. Effect of ionic saturation on the arrangement of smectite particles. **a)** Water-saturated Na-smectite is composed of layers with two water layers. Crystallites are more or less disorderly stacked (*tactoids*), forming the relatively thin walls of a polygonal network that traps the pore water. **b)** Ca-smectite forms a polygonal network of greater size, *quasi-crystals* are thicker and are more coherently stacked

5.1.3
Cation Exchange Capacity (CEC)

5.1.3.1
External CEC and Internal CEC

Clays have the property of fixing reversibly some cations contained in the surrounding solutions. The cation exchange capacity (CEC) corresponds to the number of negative charges likely to fix cations in this manner. It is expressed in centimols per kg ($cmol\,kg^{-1}$), which is a translation in the international system of units of the milliequivalents per 100 g (meq), which have been traditionally used for decades. Cations can only be exchanged if they are weakly bonded to the external or internal surfaces (interlayer spaces) of crystals.

The external CEC depends on the number of bonding sites of cations on the external surfaces. These negatively charged sites correspond to charges resulting from the tetrahedral or octahedral substitutions of those sheets forming the (001) faces, or to defects emerging on these faces. To this can be added the interrupted bonds of the $hk0$ faces. Therefore, the external CEC is a direct function of the crystal size: for a given volume or mass, the bigger the external surfaces, the smaller the crystal size. Consequently, the measurement of the external CEC gives information on the mean crystal sizes. The properties of external sites depend on pH (Fig. 5.3); this is why they are called *variable charges* of the clay material.

The internal CEC reflects the charge deficiency of 2:1 layers in the case of vermiculites and smectites. Consequently, the internal CEC depends on the *permanent charges* of clay species. One might think that the higher the structural charges, the greater the CEC. This would mean that the CEC of micas should be greater than that of smectites or vermiculites. In reality, it is the opposite because when structural charges are too high, cations are irreversibly fixed in the interlayer spaces (Fig. 5.15).

Example: theoretical calculation of the CEC of a montmorillonite

The exchange capacity is defined as the amount of cations retained by all the negative charges (permanent charges) in 100 g of clay at pH 7. It is expressed in milliequivalents (meq) per 100 g of clay. The milliequivalent is equal to (charge/mass) $\times 1,000$; it is equal to one centimole of unit charge per kilogram of dry matter ($cmol\,kg^{-1}$). The exchange capacity is calculated following the relation:

$$CEC = (charge/mass) \times 1000 \times 100 \tag{5.3}$$

Let's consider a montmorillonite half unit cell whose formula is: $Si_4O_{10}\,Al_{1.7}$ $Mg_{0.3}\,(OH)_2\,Na_{0,3}$. The mass of the half unit cell is 367 g; the charge is 0.33 so

$$CEC = (0.33/367) \times 105 = 89.9\,meq/100\,g \tag{5.4}$$

This value of 89.9 meq/100 g corresponds to the exchange capacity of the interlayer sites. The CEC related to the external surfaces of crystals (or quasi-crystals) must be added. The overall value for smectites varies from 100 to 120 meq/100 g.

5.1.3.2
Cation Exchange

Selectivity coefficient K_s

A clay, whose negatively charged exchangeable sites (X_2) are saturated by Mg^{2+}, is dispersed in a $CaCl_2$ solution and energetically stirred. Once in the neighbourhood of these sites (diffusion), the Ca^{2+} ions replace the Mg^{2+} ions, which enter the solution. Although the exchange is not a classical chemical reaction, because only low-energy bonds are involved, it can be written following the same formalism (McBride 1994):

$$CaCl_{2\,(sol)} + MgX_{2\,(arg)} \Leftrightarrow MgCl_{2\,(sol)} + CaX_{2\,(arg)} \qquad (5.5)$$

The exchange equilibrium constant K_{eq} is considered as the equilibrium constant of a classical reaction and is expressed from the activities of reactants and products:

$$K_{eq} = \frac{(MgCl_2)\,(CaX_2)}{(CaCl_2)\,(MgX_2)} \qquad (5.6)$$

The activity of solids is equal to 1 by convention,

$$K_{eq} = \frac{(MgCl_2)}{(CaCl_2)} \qquad (5.7)$$

So written, the constant K_{eq} imposes the following rule: as long as the solid can exchange Mg^{2+} ions, the activities of those ions in solution are constant. This is not the case in real exchanges because the activity of solids is not equal to 1. Indeed, MgX_2 and CaX_2 values vary as functions of the proportion of exchangeable sites occupied by Mg^{2+} and Ca^{2+}. The true variable is the ratio of the concentrations of adsorbed ions $[CaX_2]$ and $[MgX_2]$:

$$M_{Ca} = \frac{[CaX_2]}{[CaX_2] + [MgX_2]} \qquad (5.8)$$

$$M_{Mg} = \frac{[MgX_2]}{[CaX_2] + [MgX_2]} \qquad (5.9)$$

Therefore, since the equilibrium constant K_{eq} does not describe the exchange phenomenon, it is replaced by a constant K_s-or selectivity coefficient–which is expressed as follows:

$$K_s = \frac{(MgCl_2)\,M_{Ca}}{(CaCl_2)\,M_{Mg}} \qquad (5.10)$$

Deviation from Ideality

Ion exchange modifies the composition of clay which varies between calcic and magnesian end members. If this variation is considered similar to a solid solution, the selectivity coefficient is likely to deviate from ideality by the addition of an energy of mixing. The more similar the charges and diameters of ions, the lower this energy; the more different, the higher this energy. This means that the concentrations [CaX₂] and [MgX₂] need to be "corrected" by selectivity factors: $(CaX_2) = f_{Ca} M_{Ca}$ and $(MgX_2) = f_{Mg} M_{Mg}$. The exchange equilibrium constant then becomes:

$$K_E = \frac{(MgCl_2) \times f_{Ca}M_{Ca}}{(CaCl_2) \times f_{Mg}M_{Mg}} = K_S \frac{f_{Ca}}{f_{Mg}} \tag{5.11}$$

where K_s is the selectivity coefficient.

The example of the exchange of Rb^+ cations in a Na-saturated montmoril-lonite shows that K_s is greater when fewer exchangeable sites have fixed this cation (Fig. 5.12a). For monovalent cations, the selectivity order is as follows: $Cs^+ > K^+ > Na^+ > Li^+$; for bivalent cations: $Ba^{2+} > Sr^{2+} > Ca^{2+} > Mg^{2+}$. This order is defined by the size (diameter) of the association cation – hydration sphere: the cation with the smaller diameter displaces the one with the greater diameter. Selectivity varies with electrolyte concentration: the more diluted the solution, the greater the selectivity (Fig. 5.12b).

If the exchange is performed at constant temperature, the selectivity coefficient (K_s) determines a cation exchange isotherm. Exchange isotherms are modified when pH conditions vary. Indeed, the behaviour of the H^+ ion is similar to that of the other cations with which it competes to bond to the vari-

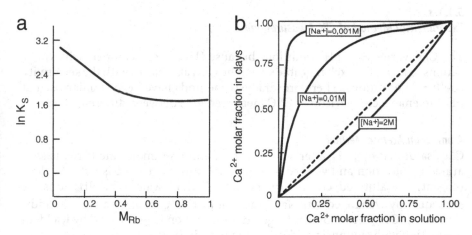

Fig. 5.12a,b. Variation of the selectivity coefficient K_s. **a)** Rb^+ - Na-montmorillonite exchange. K_s varies as a function of the respective proportions of the exchangeable sites occupied by Rb^+ and Na^+ ions (M_{Rb}). **b)** Relationship between Ca^{2+} contents in the exchanger (montmorillonite) and the solution. The more diluted the electrolyte, the greater the selectivity (K_s) for Ca^{2+}

able or permanent charges. This imposes that the measurement of isotherms be performed under controlled pH conditions. What does the selectivity co-efficient mean at the crystallite level? Obviously, it corresponds to exchange energies that differ according to sites. If one considers smectites, for which variable charges are considered negligible, the exchange sites are located in the interlayer zone. Talibudeen and Goulding (1983) have shown through microcalorimetric analyses the occurrence of six groups of sites whose enthalpy of exchange (exothermal reaction) varies from 5.7 to 10.9 kJ/eq. For low-charge smectites, the enthalpy of most of the exchange sites varies from 5.7 to 7.5 kJ/eq. They have a few sites of higher enthalpy. These results have been confirmed by the study of Cs for Ca exchanges (Maes et al. 1985). Although not proved yet, it seems obvious that the variation in the exchange energy is related to the charge heterogeneities at the surface of 2:1 layers. These heterogeneities could result from the way in which ionic substitutions are distributed between tetrahedral and octahedral sheets (annex 3).

Molecular dynamics

The properties of cation exchange in interlayer zones are now addressed from the angle of molecular dynamics. The distribution of the electric charges at the surface of clay layers is calculated from the crystal structure, as defined by X-ray diffraction and by infrared spectrometry techniques (absorption and Raman). Although well beyond the scope of this book, it is important to keep up with the latest advances, the first reports of which can be found in the following publications: Fang-Ru Chou Chang et al. (1998), Smith (1998), Young and Smith (2000), Pruissen et al. (2000).

5.1.3.3
Experimental Method of CEC Measurement

The CEC varies with pH conditions because H^+ protons compete with other cations to fix on the exchangeable sites. The CEC also varies with the selectivity coefficient of cations. Therefore, analysis methods have been standardised at pH 7 to enable comparison of CEC measurements between different clays.

Ammonium Acetate Method
Clay, saturated by NH_4^+ after suspension in ammonium acetate (1 N), is separated by filtration and washed with ethanol until no acetate is left (Nessler's reagent). So saturated, clay is suspended in distilled water. The NH_4^+ ions are subsequently exchanged with Na^+ by addition of Na(OH) or with Mg^{2+} by addition of Mg(OH)$_2$. Ammonia nitrogen determination is performed by Kjeldahl distillation with sulphuric acid using Tashiro's indicator.

Strontium Chloride Method
Clays are saturated in a $SrCl_2$ solution, then filtered and washed with ethanol until chlorides have disappeared (absence of white precipitate with $AgNO_3$).

The Sr^{2+} cations bonded in clays are displaced by a normal HCl solution. The suspension is filtered and determination of Sr^{2+} cations in solution is performed by atomic absorption.

Variable Charges and Permanent Charges

The method advocated by Anderson and Sposito (1991) permits measurement of both the CEC related to variable charges (external surfaces of crystals) and the CEC related to permanent charges (interlayer spaces). The principle rests on successive cation exchanges between the clay and various solutions under constant pH conditions.

A fixed amount of clay (100 mg) is dispersed in 10 ml CsCl (0.1 N) at pH 7. The variable and permanent charges of clay crystals are saturated by Cs^+ cations. After washing with ethanol and drying, clays are dispersed in 10 ml LiCl (0.01 N) at pH 7. Under these imposed conditions, the Li^+ cations can only replace those Cs^+ cations bonded to the surfaces of the crystal $hk0$ faces, or edges if they are xenomorphic. The Cs^+ cations passed into solution are determined by atomic absorption; the measurement permits calculation of the CEC related to variable charges.

Clays are subsequently dispersed in an acetate solution of $C_2H_7NO_2$ (1 N) at pH 7. The NH_4^+ cations replace those Li^+ cations bonded to the edge surfaces and those Cs^+ cations bonded to interlayer spaces and $00l$ interfaces. To measure the Cs^+ amounts in solution permits calculation of the CEC related to the accessible permanent charges, namely not irreversibly compensated for by the bonding of K^+ ions.

5.1.3.4
Values of the Total CEC for the Main Clay Species

The total CEC is equal to the sum $CEC_{variable\ charges} + CEC_{permanent\ charges}$. The variable charge contribution is weak for smectites or vermiculites and strong for kaolinites and illites. Measurements are performed at pH 7. Table 5.2 shows the values measured for the main clay minerals. The CEC of mixed-layer minerals is intermediate between those of pure end members. It appears that relatively good correlation exists between the CEC of I/S mixed layers (Hower and Mowatt 1966) and the relative proportions of both components (Fig. 5.13).

Table 5.2. CEC values of the main clay mineral species

Mineral	CEC $(cmol\,kg^{-1})$
Kaolinite	5–15
Illite	25–40
Vermiculite	100–150
Montmorillonite	80–120
Chlorite	5–15

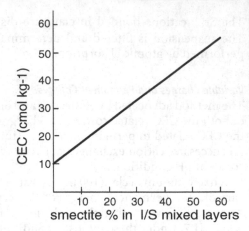

Fig. 5.13. The exchange capacity in a series of illite-smectite mixed layers varies like a simple mixture between the end members smectite and illite (Hower and Mowatt 1966)

5.1.4
Anion Exchange Capacity (AEC)

The anion exchange capacity of clays involves only those sites on the edges of crystals where OH groups cannot totally compensate for their valency. Figure 5.3 shows that these sites are located on interrupted bonds between the structural cations (Cat^+) and the oxygens or OH groups of the tetrahedral and octahedral sheets. The AEC is favoured by the low pH values that permit the bonding of a proton to these OH groups, thus forming a water molecule. This water molecule is much easier to displace because it is very weakly bonded to the structural cation. The AEC is obviously greater for allophanes because of their great number of interrupted bonds, or for hydroxides owing to the great number of OH groups. The reactions leading to the adsorption of anions in

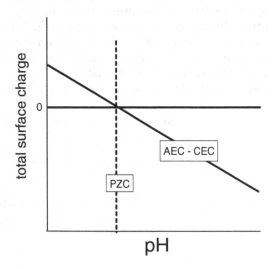

Fig. 5.14. Anion and cation exchange capacity (AEC and CEC) balance as a function of pH. The point of zero charge (PZC) corresponds to the equality AEC = CEC

phyllosilicates can be written as follows:

$$> Cat^+ - OH + A^- \rightarrow > Cat^+ - A^- + OH^- \tag{5.12}$$

or

$$> Cat^+ - H_2O + A^- \rightarrow > Cat^+ - A^- + H_2O \tag{5.13}$$

Theoretically, the sum AEC+CEC cancels out for a pH value corresponding to the "point of zero charge" (PZC). Clays behave similarly to amphoteric compounds (see Sect. 5.2.2). At this point, the exchange capacity of the phyllosilicate internal and external surfaces is minimised. This implies that the electrostatic repulsive forces are minimised too (Fig. 5.14). Under such conditions, clay particles get closer and flocculate.

5.1.5
Layer Charge and CEC

5.1.5.1
High Charge and Low Charge

The charge (tetrahedral+octahedral) of the layers making up the crystal structures of smectites can vary approximately from 0.30 to 0.65 per Si_4O_{10}. Such a difference (from simple to double) entails changes in the chemical and physical properties of crystallites. Indeed, cations are weakly bonded in the interlayer space of low-charge layers; they are totally exchangeable and polar molecules such as water, glycol or glycerol can enter this space. By contrast, the much more strongly bonded cations in the interlayer space of high-charge layers are not all exchangeable. The K^+ ions in particular adopt a configuration similar to the one they have in the structure of micas or celadonites. In that case, the layers lose their expansion capacity by absorption of polar molecules.

The presence of high-charge or low-charge layers can be readily identified using a method based upon the differences in chemical properties. Suspended clays are saturated with potassium by agitation in a 1 M KCl solution (pH 7). After rinsing, they are deposited on a glass slide (Malla and Douglas 1987). The resulting oriented sample is heated at 110 °C for 12 h so that all the molecular water is removed from the interlayer space. The K^+ ions are then irreversibly bonded in the high-charge sites and prevent polar molecules from entering. The structure of smectite behaves like that of a mica (10 Å). Since low-charge sites remain accessible to polar molecules, the expansion takes place after glycerol saturation (18 Å). Therefore, a calcium-saturated smectite exhibiting a homogeneous swelling of all the interlayer spaces – whether of high or low charge – behaves like a complex mixed-layer mineral consisting of layers with 2, 1 or 0 glycol layers when saturated with potassium.

This simple method can be improved by saturating anew the K-smectite with Ca^{2+} ions. This permits identification of the various types of expandable layers:

– low-charge smectite: Ca- or K-saturated $\rightarrow$ 2 layers of ethylene glycol (EG),

– intermediate-charge smectite: Ca-saturated $\rightarrow$ 2 EG; K-saturated $\rightarrow$ 1 EG,

– "vermiculite"-like: Ca-saturated $\rightarrow$ 1 EG; K-saturated $\rightarrow$ 0 EG.

Nevertheless, the method is not accurate enough to measure the value of these charges. This measurement is performed using a more complex method of difficult implementation: the absorption of alkylammonium ions. This method, devised by Lagaly and Weiss (1969), is thoroughly described by Mermut (1994). A simplified empirical method has been proposed by Olis et al. (1990).

5.1.5.2
CEC and Layer Charge

The theoretical calculation as defined above (section 5.1.2.1) gives an approximate value of the CEC provided that it is strictly proportional to the interlayer charge. This proportionality is represented by a straight line of simple mixture between phyllosilicates without interlayer charge (pyrophyllite) and those with the maximum charge (micas). In reality, this proportionality is not verified by a linear relationship for two reasons (Fig. 5.15). First, the CEC is not strictly

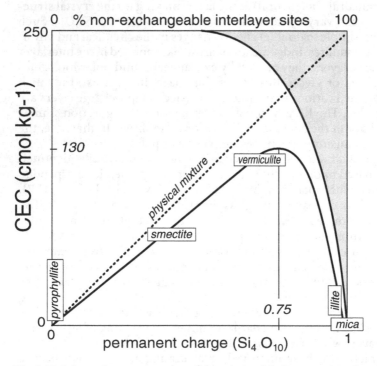

Fig. 5.15. Schematic representation of the relationship between CEC, interlayer charge and rate of sites irreversibly fixing K^+ or NH_4^+

dependent on the number of exchangeable sites that are negatively charged through ionic substitutions in octahedra or tetrahedra, but rather on their activity coefficient (section 5.1.2.2). The second reason is related to the bonding irreversibility of some cations such as K^+ or NH_4^+.

The number of sites losing their exchange property through such bonding increases with the layer charge. When this number is significant, the interlayer spaces decrease in thickness and become totally inaccessible to cation exchange. The 2:1 layers "collapse", like those of illites or micas. These collapsed layers widely prevail over expandable layers if the charge exceeds 0.75 per $Si_4 O_{10}$. The stacking sequence of collapsed layers and layers remaining expandable are diagnosed by diffraction patterns that are similar to those of mixed-layer minerals of type illite-smectite.

5.2
Physical Properties

Introduction

Clay rocks – containing more than 60% clay – are abundant in the Earth's outer skin. Their mechanical properties are mostly controlled by the characteristics of the internal and external surfaces of the small-sized phyllosilicates. This means that the stability of rock structures or human constructions rely on their behaviour rules. The impact on economy is great. It is even greater when these clay materials transform into muds able to flow down slopes. The study of rheological properties is crucial both in the northern countries where "quick clays" cover large surface areas and in the equatorial countries where natural or human-made slopes in huge weathered rock mantles collapse. The purpose of this section is to show how the mechanical and rheological properties of clay materials depend on the surface properties of their constituent crystallites.

5.2.1
Specific Surface

5.2.1.1
Theoretical Calculation of the Specific Surface

The specific surface of a clay (S_0) corresponds to the sum of the surfaces of all the exchangeable sites accessible to a given ion or molecule. These sites stretch along basal faces and edges of crystals in proportions that vary according to the type of mineral and to the local pH conditions. The maximum value of the specific surface of a phyllosilicate is equal to the sum of the surfaces of all the faces of each elementary layer. Faces have a different significance depending on the mineral species considered. Thus, the surfaces of the *hkl* faces are negligible in comparison with those of the (001) faces for smectites whose crystallites have a very reduced thickness, whereas they are much larger for kaolinite, chlorite, illite or mica.

The theoretical calculation is based on the "unit formula" of the mineral involved and on the parameters of its crystal unit cell (remember that the unit cell dimensions depends on the polytype; it is composed of 1 layer in turbostratic stackings, 1 M and 1 Md polytypes). Let's consider the example of a montmorillonite studied by McEwan (1961), whose composition is as follows:

$$[Si_{7.76}Al_{0.24}] \, O_{20} \, (Al_{3.10}Fe^{3+}_{0.38}Fe^{2+}_{0.04}Mg_{0.52}) \, (OH)_4 \, Na_{0.68} \tag{5.14}$$

$a = 0.521$ nm and $b = 0.902$ nm.

The mass of the unit cell is 750.22 g; if edges are considered negligible, the surface of layers is approximately equal to $2ab = 0.9399$ nm^2. S_0 is calculated as follows:

$$S_0 = \frac{2ab \times N_A}{M} = \frac{(0.902 \times 10^{-18}) \, (6.022 \times 10^{23})}{750.22} \approx 754.4 \text{m}^2 \, \text{g}^{-1} \tag{5.15}$$

N_A Avogadro number

The calculated value of about 750 m^2 g^{-1} can be considered as representing the maximum specific surface for a phyllosilicate whose interlayer spaces are all accessible to the exchange of ions or molecules. If, for structural reasons, some interlayer spaces are not accessible (irreversibly bonded K$^+$ or NH^{4+} cations in phyllosilicates with high interlayer charges), the specific surface decreases (Fig. 5.16).

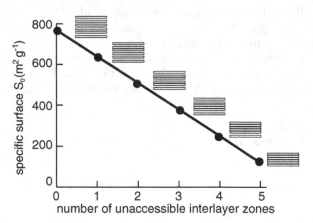

Fig. 5.16. Decrease in the theoretical specific surface as the number of inter-layer spaces inaccessible to exchanges of ions or molecules increases

5.2.1.2
Experimental Measurements of the Specific Surface

Principle
The specific surface S_0 of a clay sample is equal to the total area of the particles accessible to those molecules used for the measurement. In other words, there is not a unique value of S_0 but as many values as there are measurement

Table 5.3. Measurement of the specific surfaces of common clay species by adsorption of nitrogen + water or nitrogen + cetylpyridinium-N bromide (CPB) mixtures

Mineral	Nitrogen + water ($m^2 g^{-1}$)	Nitrogen + CPB ($m^2 g^{-1}$)
Kaolinite	20	15
Illite	100	100
Vermiculite	710	720
Montmorillonite	850	800

techniques. The techniques used involve components able to be adsorbed on various accessible exchange sites. Once the exchange has been completed, the total mass of the adsorbed component is measured. Knowing the mass of a monolayer of the adsorbed component, the total accessible surface can be calculated. Consequently, it is necessary to determine the mass of a complete monolayer, which is difficult considering that the 2nd and the 3rd layers "nucleate" before the first one has been completed (Fig. 5.8). The values obtained for common clay species are given in Table 5.3.

Specific Surface and Size of Tactoids (Na^+) or Quasi-Crystals (Ca^{2+})

The adsorption-desorption isotherms are characterised by a hysteresis due to the retention of N_2 molecules in slit-shaped pores (Fig. 5.17a). The amounts of nitrogen molecules adsorbed depend on the interlayer cation (Table 5.4). Thus, since Cs^+ is the cation with the larger ionic diameter, layers are sufficiently separated for N_2 molecules to enter into or escape from the interlayer spaces. This explains why the hysteresis is hardly marked. The hysteresis pressures are a bit greater for Ca^{2+} and Na^+ cations, but remain low owing to re-arrangements of the layers when Na^+ tactoids are transformed into Ca^{2+} quasi-crystals. Assuming that layers are perfectly superposed squares with sides of 3,000 Å, their number in the tactoids or quasi-crystals can be evaluated for every cation saturating the interlayer spaces (Table 5.4). The thickest quasi-crystals are obtained with Ca^{2+}. This is consistent with the observations of the honeycombed structures (Fig. 5.11).

The recent advances in the measurement of specific surfaces are due to the refinement of adsorption techniques and data processing (Cases et al. 2000):

– argon (at 77 K) is preferred to nitrogen because its molecules are little polarisable as opposed to N_2 which possesses an inductive quadripolar moment that makes it react strongly with the OH groups of the basal and lateral faces. Volumetric analysis at quasi-equilibrium is used to reveal the energetic heterogeneities;

– the DIS (derivative isotherm summation) method permits decomposition of the experimental derivative isotherm into derivative isotherms corresponding to homogeneous surfaces.

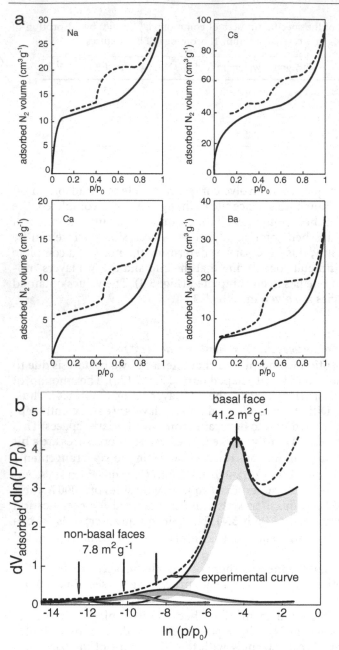

Fig. 5.17a,b. Measurement of the specific surfaces by gas adsorption. **a)** Adsorption-desorption isotherms of nitrogen at 77 °K of a Na-, Cs-, Ca- or Ba-saturated montmorillonite. The hysteresis phenomenon is general whatever the cation. The amounts of adsorbed nitrogen molecules depend on the cation. **b)** Highlighting of the differences between the crystal faces of a kaolin sample – by DIS processing of the derivative isotherm – in the volumetric analysis of argon adsorption in quasi-equilibrium (modified from Cases 2000)

Table 5.4. Specific surfaces and number of layers in tactoids inferred from the measurements of nitrogen adsorption in a dry state (BET method) or in a humid state (Harkin and Jura), after Villiéras et al. (1996)

Cation	Specific surface ($m^2 g^{-1}$)		Number of layers per tactoïd	
	N_2 dry state (BET)	Water sat. state (Harkins and Jura)	Dry state	Water sat. state
Na	42	84	33	8
Cs	130	86	11	8
Ca	17	63	69	12
Ba	25	56	32	14

The results obtained with kaolinite are noteworthy. Cases et al. (2000) have shown how the energetic characteristics and the specific surface of each type of faces can be accurately defined (Fig. 5.17b). Since the CEC is related to the specific surface of the non-basal faces, these authors have also shown that the contribution of silanol sites is 64% whereas that of aluminol sites is 36%.

5.2.2
Surface Electric Charge Density

5.2.2.1
Charge Density

Theoretical Calculation of the Structural Charge Density (Permanent)
The permanent charges of phyllosilicates are due to cation substitutions in tetrahedral sheets (Al^{3+} for Si^{4+}) or in octahedral sheets (R^{2+} for R^{3+}). Assuming that the layers are strictly homogeneous, the charge density σ_0 then corresponds to the number of negative charges per surface unit. It is expressed in Coulomb per square meter ($C m^{-2}$) and can be calculated in a very simple manner: $\sigma_0 = e \times$ interlayer charge$/2(a \times b)$ where e is the elementary electric charge ($1.6022 \times 10^{-19} C$) and a and b are the unit cell parameters along the xy plane. For instance, the charge density for phyllosilicates with tetrahedral charge varies from 0.15 to $0.70 C m^{-2}$ (Table 5.5).

Table 5.5. Theoretical values of the structural charge density of phyllosilicates with a tetrahedral charge

Mineral	Beidellite	Muscovite	Margarite
Structural formulae	$Si_{7,1} Al_{0,9} O_{20}$ $Al_4 (OH)_4 Na_{0,9}$	$Si_{7,1} Al_{0,9} O_{20}$ $Al_4 (OH)_4 Na_{0,9}$	$Si_{7,1} Al_{0,9} O_{20}$ $Al_4 (OH)_4 Na_{0,9}$
a dimension (nm)	0.514	0.519	0.512
b dimension (nm)	0.893	0.901	0.889
Charge density ($C m^{-2}$)	0.157	0.342	0.704

Theoretical Calculation of the CEC-Related Charge Density

The theoretical calculation of the "structural" charge density as defined by Sposito (1984) does not characterise the capacity of clay to exchange its interlayer cations with the surrounding environment (CEC). The phyllosilicates that exhibit this property (smectites and vermiculites) have low-energy bonds connecting the compensating cations of the interlayer spaces to the framework of the 2:1 layer (Sect. 5.1.2). The CEC-related charge density measures the number of these low charges per surface unit. Since the CEC ($cmol\,kg^{-1}$) as well as the specific surface S ($m^2\,kg^{-1}$) are measurable quantities, the CEC-related charge density is readily calculated: $\sigma_{CEC} = e \times CEC \times 10^{-2}/2\,ab$. Thus, a low-charge smectite with a CEC equal to $120\,cmol\,kg^{-1}$ will have a charge density σ_{CEC} equal to $0.209\,C\,m^{-2}$.

The low-charge smectite can be either a beidellite or a montmorillonite. The unit formulae, and hence the molar mass and the unit cell parameters, are different:

- beidellite $[Si_{7.4}Al_{0.8}]O_{20}Al_4(OH)_4\,Na_{0.6}$, molar mass $739.16\,g\,mol^{-1}$; unit cell parameters $a = 0.518\,nm$, and $b = 0.899\,nm$,

- montmorillonite $[Si_8]O_{20}Al_{3.4}Mg_{0.6}(OH)_4Na_{0.6}$, molar mass $732.83\,g\,mol^{-1}$; unit cell parameters $a = 0.521\,nm$ and $b = 0.902\,nm$.

Despite these differences, the theoretical CEC of beidellite and montmorillonite are identical (see Sect. 5.1.3.1; Example: CEC calculation of a montmorillonite). Indeed, the CEC calculation through the interlayer charge/molar mass relationship yields the following values: beidellite $0.6/739.16 \times 10^5 = 81.2\,cmol\,kg^{-1}$; montmorillonite $06/732.83 \times 10^5 = 81.9\,cmol\,kg^{-1}$. By contrast, the corresponding charge density values are slightly different: 0.138 and $0.140\,C\,m^{-2}$, respectively.

5.2.2.2
The Diffuse Double Layer (Gouy-Chapman and Stern)

Structure of the Double Layer: the Stern – Gouy – Chapman Model

Electrically, phyllosilicate crystals behave like plane capacitors. The planes, namely the (001) crystal faces, are considered to be uniformly negatively charged over their surface. An electric field is locally developed; as a result, the bonding energy of the cations decreases with the distance to the electrically charged surface. The electric field also makes polar molecules orient like water (Fig. 5.18a).

If ions are considered as punctual charges (Gouy-Chapman model modified by Stern 1924), the number of cations (n^+) exponentially decreases with the distance to the charged surface (x), whereas the number of anions (n^-) increases inversely (Fig. 5.18b):

$$n^- = n_0^-\exp^{\left(\nu^- e\frac{E}{kT}\right)} \tag{5.16}$$

$$n^+ = n_0^+\exp^{\left(\nu^+ e\frac{E}{kT}\right)} \tag{5.17}$$

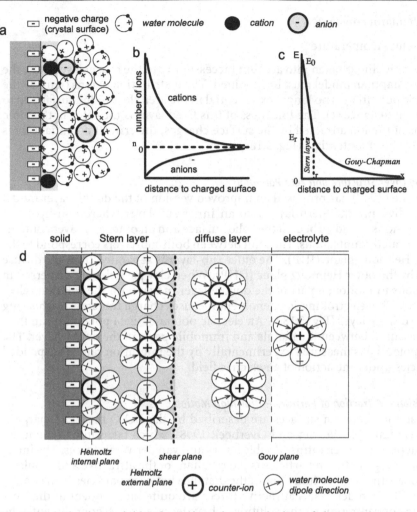

Fig. 5.18a–d. The electric field at the surface of negatively charged clay particles. **a)** Bonding of cations and orientation of polar molecules (water). The *shaded area* is the equivalent of the Stern layer. **b)** The number of ions decreases exponentially with the distance to the crystal surface. If ion crowding (volume) is not taken into account, their number can be considered to vary similarly (Gouy-Chapman model). **c)** By contrast, if their size is taken into account, the law is no longer perfectly exponential; it becomes linear in the area close to the surface (Stern model). **d)** Representation of the Stern-Grahame triple layer model

n_0^+ and n_0^- number of anions and cations per volume unit

v^+ and v^- valency of cations and anions

e electron charge

E electric potential at a distance x from the charged surface

k Boltzmann constant

T absolute temperature

If ion crowding is taken into account (accessibility to the charged surface), the Gouy-Chapman model must be modified. The distribution of ions (namely the electric potential) is no longer exponential in the area close to the surface up to a critical distance (r). The thickness of this fixed layer is the value of the ionic radius of the ion attracted by the surface charges; the critical potential (Er) is named "Stern potential" (Fig. 5.18c).

Double Layer Structure and Zeta Potential
Grahame (1947) has proposed an improved version of the double layer model by subdividing the Stern layer into an inner sub-layer where non-hydrated counter-ions are adsorbed on the solid surface and an outer sub-layer containing hydrated counter-ions. The limit between both sub-layers correspond to the inner Helmoltz plane (IHP). The outer sub-layer is separated from the diffuse layer by the outer Helmoltz plane (OHP). When clay particles are dispersed in solutions in motion, a part of the double layer moves with the solution (electrolyte). This electrokinetic phenomenon causes charge transfers by shearing of the diffuse layer (Fig. 5.18d). An electric potential (zeta potential) can then be measured between the mobile and immobile parts of the double layer. The zeta potential is measured experimentally by the migration rate of suspended particles under the action of an electric field.

Repulsion – Attraction of Particles – Brownian Motion
Interactions between surfaces are described by the DLVO theory (Dejarguin and Landau 1941; Verwey and Overbeek 1948), which takes into account the interactions between diffuse double layers and van der Waals forces. The interaction energy between particles is the resultant of the attractive and repulsive potentials that vary as functions of the distance between particles (van Olphen 1977). The van der Waals attractive forces are quite independent of the electrolyte concentration in the solution. They decrease hyperbolically with the distance. The repulsive forces produced by the electric field of the double layer decrease exponentially with the distance. The greater the electrolyte concentration, the weaker the repulsive forces (Fig. 5.19). For a given electrolyte concentration in the solution, the resultant potential shows a "well" that determines the attraction distance of particles. In other words, flocculation will take place when the energy barrier preceding the "well" is passed. The higher the electrolyte concentration, the lower this energy barrier.

The stability of suspended clay particles is not only related to the electrostatic repulsion between particles. The Brownian motion contributes to the fight against gravity effects. The particles are put in perpetual motion by the thermal energy through incessant shocks of the solvent molecules. The effects of the Brownian motion decrease with the size of particles but increase with temperature. Einstein in 1905 was the one who defined the quantitative laws of this thermal agitation.

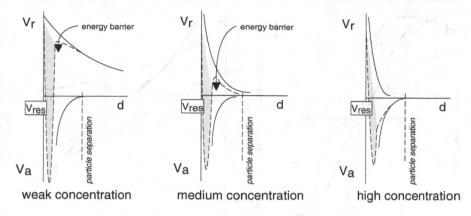

Fig. 5.19. Variation in the interaction energy as a function of the distance between particles for three values of the electrolyte concentration. V_a: attractive potential; V_r: repulsive potential; V_{res}: resultant potential. The potential well corresponds to the *shaded area* (from van Olphen 1977)

Point of Zero Charge PZC

Kaolinite exhibits three different types of sites: gibbsitic planes, siloxane planes and crystal edge planes where bonds are interrupted (Fig. 5.3). In the absence of substitutions of Al for Si in kaolinite tetrahedra, the siloxane plane has no electric charge and therefore can be ignored here. Neutralisation of surface charges through acid-base titration then enables the number of protons bonded to the gibbsitic and edge sites to be measured (Fig. 5.20). At pH 7.5, the amount of bonded protons is zero: it is the point of zero charge (PZC). The total density of bonded protons increases as the pH value decreases from 7.5. The edge sites are saturated first in an environment close to neutrality. The gibbsitic sites start saturating as soon as the pH value is lower than 5.

Fig. 5.20. Variation in the number of H^+ protons bonded to the kaolinite surface as a function of pH (after Stumm 1992). Measurement of the excess density of H^+ due to the bonding of protons to the $(OH)^-$ groups of the gibbsitic surfaces and the crystal edges (interrupted bonds). The point of zero charge (PZC) of edges is reached at pH= 7.5

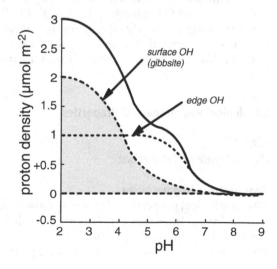

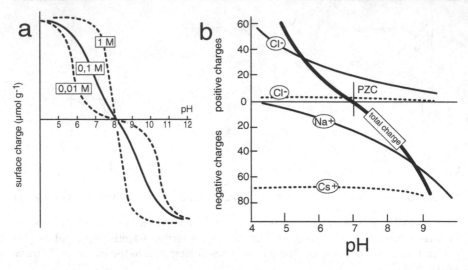

Fig. 5.21a,b. Determination of the point of zero charge (PZC). **a)** Theoretical behaviour of an amphoteric substance. **b)** Variation in the amounts of positive and negative charges of an allophane and a smectite as a function of pH

Clays behave like amphoteric substances, namely like acids or bases according to the pH conditions of the environment. From a theoretical standpoint, the titration curves should have a symmetrical shape in relation to the PZC (Fig. 5.21a). The shape of the curves obtained in the pH range below the PZC for the three concentrations of some electrolyte (not adsorbed) shows that for an equivalent charge, the greater the concentration, the higher the pH value. Conversely, in the pH range over the PZC, a given charge will be reached for pH values that are greater as the electrolyte concentration is low.

The amounts of bonded negative and positive charges vary as a function of pH. In the case of allophanes (Fig. 5.21b), the PZC can be determined because these amounts are nearly equivalent; it is situated at pH 7. By contrast, these amounts are very different in the case of a smectite brought into contact with CsCl. In this case, the PZC cannot be determined.

5.2.3
Rheological and Mechanical Properties

5.2.3.1
Physical State of Suspensions

Aggregation – Fragmentation
The rheological properties of clay suspensions depend on the size of aggregates and the interactions they exert upon each other. Indeed, the crystallites or particles of clay minerals (see definitions, Chap. 1) can agglomerate by contact in three ways: (1) face-face, (2) edge-face, (3) edge-edge (Fig. 5.22a). The stability

Fig. 5.22a,b. Aggregates in suspensions. **a)** The three types of bonding between particles. **b)** Smectite suspensions show an alveolar structure due to the flexible deformation of crystallites

of these aggregates depends on the clay concentration of the suspension, the ions in solution and the energy of the environment released upon mechanical or thermal agitation. The apparent viscosity (η) increases with the aggregate size. It decreases when these aggregates are fragmented.

Smectite suspensions have a special organisation owing to the small thickness of the crystallites that are able to deform by flexion. Their aggregates form "honeycombed" structures in which water can be found in various states (Fig. 5.22b). If the clay concentration is high, the suspensions become true gels trapping great amounts of water in the pores of the structure.

Apparent Viscosity – Shear Rate/Viscoelasticity Relationship

In the simple shear, the apparent viscosity η (Pa s) is related to the shear rate $\dot{\gamma}$ (s^{-1}) and to the shear stress τ (Pa) by the following relationship:

$$\eta = \tau/\dot{\gamma} \tag{5.18}$$

The rheological behaviour of suspensions can follow several different laws (Fig. 5.23a):

- Newtonian model: $\tau = \mu\dot{\gamma}$; no threshold and constant viscosity,

- Binghamian model $\tau = \tau_0 + \eta_{pl}\dot{\gamma}^n$; threshold stress, constant viscosity;

- Ostwald model: $\tau = k\dot{\gamma}^n$; no threshold, viscosity varies with $\dot{\gamma}$;

- Herschel–Buckley model: $\tau = \tau_0 + k\dot{\gamma}^n$ where k and n represent the consistency (permeability) and the suspension index, respectively.

The power rule (Ostwald model) gives the apparent viscosity: $\eta = k\dot{\gamma}^{n-1}$. If $n < 1$, viscosity decreases with $\dot{\gamma}$ (shear rate thinning); if $n > 1$, viscosity increases with $\dot{\gamma}$ (shear rate thickening). Those fluids behaving in accordance with the Binghamian or Herschel–Buckley models exhibit a yield point τ_s. This means that they flow like fluids only when the value of the applied stress excesses τ_s. The threshold stress corresponds to the minimum stress for which a finite number of aggregates move and produce flow. It corresponds to the crossing of an energy barrier that depends on the edge-edge, edge-face or face-face interactions that bind particles together.

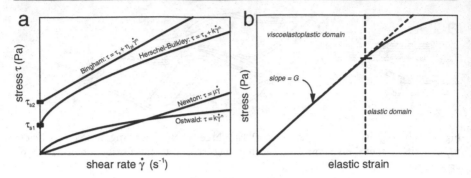

Fig. 5.23. The four types of behaviour of fluids in the stress – shear rate relationship: The Herschel-Bulkley and Binghamian threshold behaviours each feature a minimal stress value (τ_{s1} or τ_{s2}) below which the suspension behaves like a viscoelastic solid

When $\tau < \tau_s$, the suspension behaves like a solid whose framework is rigid over the whole sample. This solid has viscoelastic behaviour. The mechanical equivalent is a parallel combination of springs (elasticity) and shock absorbers (viscosity). The elasticity modulus G appears in the relationship between shear rate and stress (Fig. 5.23b):

$$\frac{1}{G}\frac{d\tau}{dt} + \frac{1}{\eta}\tau = \dot{\gamma} \tag{5.19}$$

Shear Rate Thinning-Thixotropy

Under the influence of shear, aggregates break up until the suspension behaves like a Newtonian fluid (shear rate thinning). When shear stops, the suspension at rest forms aggregates again. The suspension can become thixotropic if it returns to its original state after rest. The apparent viscosity is a decreasing function of the flow duration for a given value of stress kept constant over time. Thixotropy for clay suspensions is equivalent to a soil/gel transition due to the destructuration of aggregates (Besq 2000).

5.2.3.2
Rheological Behaviour of Suspensions

Rheometrical tests enable the relationship between stress and shear rate to be established. Their combination with observations of the deformation range have allowed Pignon et al. (1996) to define four states of a clay suspension flow (Fig. 5.24):

- *state 1:* the gel has an elastic behaviour in the range of small deformations above a threshold (critical deformation), the suspension flows and the state of flow depends on the velocity gradient applied;

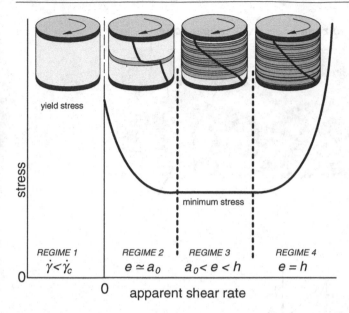

yield stress

stress

minimum stress

REGIME 1   REGIME 2   ¦   REGIME 3   ¦   REGIME 4
$\dot{\gamma} < \dot{\gamma}_c$ $e \simeq a_0$ $a_0 < e < h$ $e = h$

0

0   apparent shear rate

Fig. 5.24. Flow curve of a laponite suspension (after Pignon et al. 1996). The four states are distinguished by the deformation range and the stress variation

– *state 2:* in the decreasing part of the flow curve (small apparent shear rates), the shear is localised in a thin layer (e) whose size is close to that of aggregates (a_0). Forces bonding these aggregates prevail here;

– *state 3:* the thickness of the sheared layer increases with the apparent shear rate. The stress remains constant until the whole sample is sheared. The inter-aggregate forces and the force applied by the stress compete on time scales of similar order of magnitude;

– *state 4:* the whole volume of the sample is sheared homogeneously. An increase in the shear rate causes an increase in the stress (shear rate thinning).

5.2.3.3
Basic Mechanical Data of Unconsolidated Clay Materials

The mechanical properties, the porosity and the permeability of clay soils and materials are controlled by the space organisation of minerals or of their aggregates (Arch and Maltman 1990; Luo et al. 1993; Belanteur et al. 1997). Any change in the microstructure brings about variations in these properties. The compaction affecting sediments during their burial leads to the reduction of porosity through the rearrangement of clay particles (Skempton 1970). To date, few studies have been dedicated to the relationships between the physical properties and the centimetre-to-micron microstructure of clay materials, owing to the difficulty of multi-scale observation. The recent works on the de-

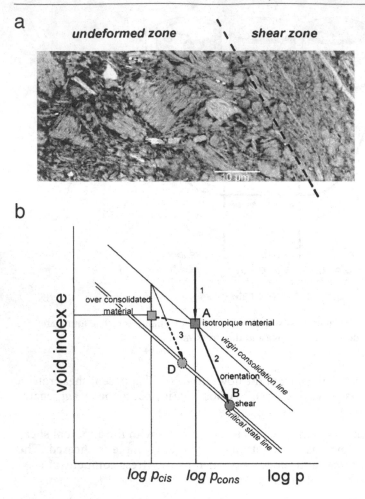

Fig. 5.25a,b. Experimental deformation of kaolinite samples (water/clay= 1.2; drained tests) after Dudoignon et al. (2001). a) Rearrangement of particles in the shearing zone. b) Relationship between the void index and the mean pressure stress p. Detailed explanations are given in the text

formation of kaolinite samples (water/clay=1.2; drained tests) have shown how particles reorient themselves (Dudoignon et al. 2001). The starting material is almost isotropic. In its normally consolidated state, reorientation under shearing effect is very intense (Fig. 5.25a). This effect decreases when the sample is in an overconsolidation state. The void index (e) is modified by reorientation following a rule that depends on the mean stress $p = \frac{(\sigma_1 + 2\sigma_3)}{3}$ (Fig. 5.25b).

The application of a pressure on a waterlogged sample reduces the void index until the virgin consolidation line has been reached (point A). The material is gradually isotropic, meaning that kaolinite particles exhibit all types of contact:

EE, EF and FE. The behaviour of this isotropic material can follow two paths according to whether the consolidation pressure is maintained or released:

– *maintained consolidation pressure:* the increase in the mean pressure (increase in the deviator) produces structural modifications through reorientation of particles (path 2). The void index decreases until shearing takes place when the critical state line has been reached.

– *released consolidation pressure:* the void index remains constant or slightly increases (from A to C). If the mean pressure increases (increase in the deviator), the flow path follows path 3 until the critical state line has been reached in D. Everything takes place as if the sample "remembers" the consolidation pressure initially applied in its microstructure: it is said to be "overconsolidated".

5.2.3.4
Applications to Natural Clay Materials

Clay Gouges
Natural clay materials obey the laws of rheology or mechanics presented above. Their microstructure is the result of consolidation, reorientation and shearing phenomena that can repeat themselves cyclically over their history. To "read" these microstructures amounts to deciphering the mechanical history of these materials. These studies are doomed to a great future because they will allow for a better understanding of the behaviour or faults (hence earthquakes) or ground movements.

Two examples illustrate the complexity of microstructures: clay gouges in faults and vertisol structure. Gouges are complex materials in which rock debris is dispersed in a clay matrix. The latter is blatantly optically anisotropic, emphasizing the sheared zones (Fig. 5.26a). In the Saint-Julien fault of the sedimentary basin of Lodève (France), the clay gouge mainly consists of illite whose crystals or particles are remarkably oriented in the sheared zones (Zellagui 2001; Prêt et al. 2004).

Clay Soils
Clay soils are known for their ability to "work", i. e. to produce forces causing shears able to cut plant roots or fissure human constructions. The soil is in constant reorganisation under the effect of drying and wetting cycles. Clay particles are oriented along more or less plane surfaces, yielding microstructural patterns that are very recognisable by petrographic observation. This is typical of vertisols (Fig. 5.26b) or any other clay-rich material at the Earth's surface. In reality, every mechanical event (expansion, shrinkage, shearing) is recorded by the soil microstructure. This "memory", as for fault gouges, must be read in order to understand the mechanical behaviour of these soils or clay-rich materials.

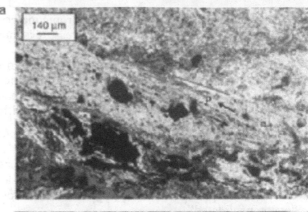

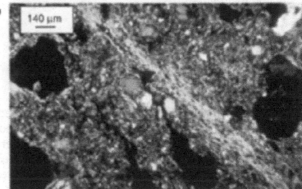

Fig. 5.26a,b. Microstructure of natural clay materials. **a)** Clay gouge of the St Julien fault, Lodève, France (after Zellagui 2001). **b)** Vertisol from Uruguay

Suggested Reading

Van Olphen H (1977) An introduction to clay colloid chemistry, 2nd edn. Wiley, New York, 317 pp

Sposito G (1984) The surface chemistry of soils. Oxford University Press, New York, 234 pp

Stucki JW, Mumpton FA (1990) Thermal analysis in clay science. CMS Workshop Lectures 3, 192 pp

Güven N, Pollastro MN (1992) Clay-water interface and its rheological implications. CMS Workshop Lectures 4, 228 pp

Stumm W (1992) Chemistry of the solid-water interface. Processes at the Mineral-Water and Particle-Water interface in natural systems. Wiley, New York, 428 pp

McBride MB (1994) Environmental chemistry of soils. Oxford University Press, New York, 406 pp

Wilson MJ (1996) Clay mineralogy: spectroscopic and chemical determinative methods. Chapman and Hall, London, 367 pp

CHAPTER 6

Clays in Soils and Weathered Rocks

6.1
Atmospheric and Seawater Weathering

6.1.1
Introduction

Weathering is defined here as all the interactions occurring between fluids and rocks at low temperature. It takes place at the interfaces between continents and atmosphere ($0 < T < 25\,°C$) and between oceanic basalts and seawater ($-5 < T < 4\,°C$). The penetration of meteoric or sea fluids into rocks causes destabilisation of their initial mineral components and favours the formation of new hydrated species. The chemical balance of these reactions is expressed as the loss of some elements that pass into solution whereas the most insoluble ones concentrate in solids.

Complex mineral or organo-mineral reactions take place at the interface between rock substrate and atmosphere: the meteoric alteration of rocks (weathering) and the formation of soils (pedogenesis). These are the two processes transforming the coherent rocks into a loose material partly composed of clay minerals. Like any interface, the weathered rock (alterite) and the soil form a polarised domain:

- *polarisation from bottom upwards:* the weathering patterns are rather governed by the progressive fading of the petrographic features of the parent rock,

- *polarisation downwards:* soils invade the underlying rock over increasing thicknesses as functions of time. They depend on climate (rain, heat) and biological activity.

The mechanical stability of these two types of loose rocks depends on topography. Significant mass movements take place in surface horizons, even on low slopes, whereas deep horizons mostly remain in structural continuity with their parent rock. Therefore, the weathering profiles are often separated from the soil by a discontinuity marked by pebble beds. Most of the time, soils do not form a geochemical continuum with the underlying alterations. For this reason, these two loose formations will be discussed separately here.

Clay minerals are formed in surface environments by weathering of pre-existing minerals that have crystallised under very different temperature-pressure-fluid composition conditions either in magmatic and metamorphic environments such as feldspars, pyroxenes, amphiboles, micas, chlorites and so on ... or in sedimentary environments such as glauconites, sepiolites, zeolites etc ... Any weathering process comprises three stages:

1. dissolution of the primary minerals that are out of their stability field,

2. transfer of the chemical elements so released from the dissolution zones to the precipitation zones (fluid circulation or/and diffusion),

3. precipitation of secondary minerals: clays, Fe–Mn oxyhydroxides ... (nucleation, crystal growth). Oxides, hydroxides and oxyhydroxides of iron and manganese form a complex mineral family hereafter referred to as oxyhydroxides (Bigham et al. 2002; Dixon and White 2002).

Each stage is controlled by local physicochemical conditions. The weathering rate is determined by the slowest of the three stages.

When clays replace a primary mineral whose crystal structure is not that of phyllosilicates, it is called *neogenesis*. When they replace a phyllosilicate (biotite, muscovite, chlorite, serpentine), the alteration process may be either a neogenesis or a layer-by-layer *transformation*. The replacement can take place without any change in the volume or form of the parent minerals (pseudomorphosis); in such a case, the original structure of the rock is conserved. If not, by contrast, the original structure of the rock can be totally erased. The purpose of this chapter is first to survey as simply as possible the fundamental processes governing the formation of clay minerals in weathering conditions. Then, the main natural environments in which these processes take place will be presented. Subsequently, the crystallization sequences of clay minerals will be connected to the major types of alterites. The solutions recovered from springs and rivers retain the geochemical signature of the alteration profiles they have crossed. Only those data that help to understand the mineral reactions will be used here. The bases of the analysis of natural waters can be found in Drever's book (1982).

6.1.2
Mechanisms of Formation of Clay Minerals

6.1.2.1
Dissolution of Primary Phases

Definitions
The literature contains a number of ambiguities regarding the concepts of congruent and incongruent dissolution. Indeed, the incongruent dissolution is generally considered to be typical of primary minerals yielding secondary minerals through weathering: plagioclase + solution A → kaolinite + solution B. In fact, this reaction consists of the congruent dissolution of plagioclase followed by the crystallisation of kaolinite. These terms will be used here according to the definitions suited to the petrographical observations:

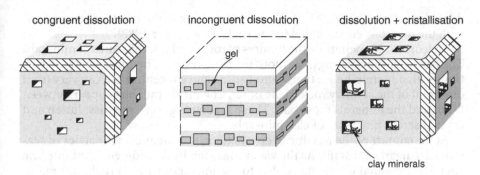

Fig. 6.1a–c. Definition of the terms used in the text. **a)** Congruent dissolution. **b)** Incongruent dissolution. **c)** Dissolution and precipitation

- *congruent dissolution:* the crystal is destroyed in such a way that all the elements enter the solution with the same stoichiometric coefficients as those of the solid. The dissolution first takes place at the most energetical sites (vertices, edges, crystal defects on faces), then spreads in the form of corrosion zones formed by coalescence of the dissolution sites (Fig. 6.1a);

- *incongruent dissolution:* the crystal is partially destroyed so that the elements entering the solution do not have the same stoichiometric coefficients as those of the solid. Generally an alkali depleted amorphous material (gel) remains in the dissolved sites. Besides, another mineral particularity may influence the composition of dissolving solutions on a larger scale. Indeed, minerals whose composition may vary in large solid solution domains exhibit frequently a heterogeneous crystal structure: preferred zones are affected by the dissolution while other ones are spared (Holdren and Spyers 1986). The dissolution then spreads through preferred corrosion of some sectors of the parent mineral such as parts of the twin or of the chemical zonation (Fig. 6.1b);

- *dissolution and precipitation:* the dissolution can be congruent or incongruent; in both cases, if the solutions reach the required oversaturation, the nucleation and growth of secondary minerals are triggered. The corrosion zones are rarely empty of solid matter (porosity); they incorporate most often a mixture of parent mineral debris and newly-formed clays (Fig. 6.1c). The nucleation of secondary phases in altered silicates is made easier by structural connections with the parent mineral crystal lattice (Eggleton 1986).

Chemical Potential Gradients

Meteoric fluids essentially contain H_2O+CO_2 and accordingly are acidic. After more or less long paths through soils and rocks, they reach the open air in springs, and are subsequently drained towards rivers and streams or stored in water tables. Waters from springs, from rivers or ground waters are generally diluted solutions ($> 10 \, \text{mg} \, \text{l}^{-1}$) whose composition is the result of interactions with the minerals encountered in the rocks (see Drever 1982).

Meteoric fluids penetrate the fissures of rocks and microfissures of minerals. The narrower the openings, the more slowly they flow. Therefore, their interactions with minerals, hence their composition and concentration, can vary from one point of the country rock to the other. Chemical gradients appear between fluids and the minerals they soak. These gradients govern the dissolution and crystallisation processes of clay minerals.

At the contact of the basaltic deep-sea floor, the chemical properties of seawater are nearly constant. As the water mass can be considered as infinite, the chemical potential gradients related to the interactions with rocks are maintained over very long periods. This explains that alterations have macroscopic effects, despite the very low temperature.

High gradients (*diluted fluid*): they generally lead to a congruent dissolution of minerals. Edges and vertices of crystals become blunt. Dissolution pits appear at the emergence of the screw or edge dislocations on their faces. The minerals become porous; the intercrystalline joints open (weakening of bonds), bringing about the shedding of crystals. The rock becomes friable and looses its mechanical coherence. Dissolved chemical elements are leached out by waters flowing in the connected pores. At the end of the process, only the less soluble minerals remain: Fe- and Al-oxides and hydroxides. The parent rock structure is faded out. Such gradients occur in zones where fluids are constantly renewed by rapid flows (surface horizons of intertropical soils; network of highly permeable fractures).

Moderate gradients: they bring about the new formation of clays often associated with Fe-hydroxides in the dissolution zones of the minerals. A heterogeneous microcrystalline matrix (plasma) composed of newly-formed minerals and fine debris detached from the parent minerals is formed. These gradients occur in arena-rich domains where the rock or the soil retains its original structure but becomes more and more loose.

Low gradients (*concentrated fluids*): these gradients occur in narrow microfissures where the fluids flow so slowly that they approach equilibrium with the minerals they soak. The dissolution zones are totally sealed by growing secondary minerals. These zones appear as pseudomorphoses, with illite replacing K-feldspars and saponites replacing pyroxenes.

The Mechanism of Dissolution

The dissolution process comprises five stages, whatever the mineral involved (Fig. 6.2a):

1. migration of the solvent through the solution up to the mineral surface;

2. adsorption of the solvent on the surface;

3. migration of the solvent into the crystal lattice up to the bonds it will break;

4. break up of the bonds and release of the ions from the lattice;

5. migration of the ions from the lattice to the solution.

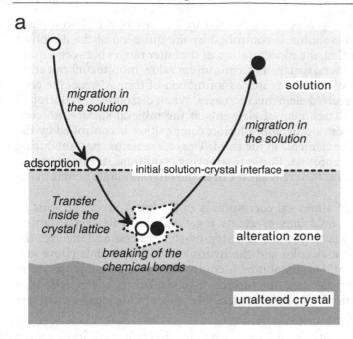

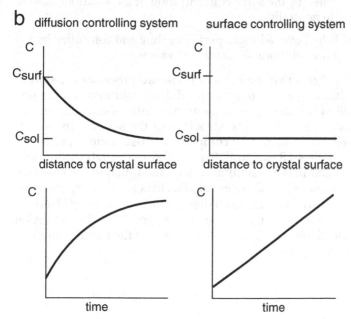

Fig. 6.2a,b. Dissolution. **a**) Schematic representation of the five stages of the dissolution process. **b**) Theoretical states of the chemical gradients and dissolution rates for a diffusion- or surface-controlled system. C: concentration of a chemical element; C_{surf}: concentration of the element at the mineral surface; C_{sol}: concentration of the solution at an infinite distance from the mineral surface

The rate of the whole process is controlled by the rate of the slowest stage (Fig. 6.2b). When dissolution is controlled by the diffusion of the dissolved elements in the solution, the concentration of the latter ranges between a maximum value at the mineral surface and a minimum value (infinite concentration state). The dissolution rate decreases as a function of time whereas the concentration of the dissolved elements increases. When dissolution is controlled by mechanisms of detachment of elements at the mineral surface, no concentration gradient can occur; the solution composition is controlled by the dissolution rate of the elements in the solid. The rate remains constant during the whole dissolution process. Under weathering conditions, the bond breaking stage at the mineral surface is slower than the migration rate of ions in the solutions.

The dissolution of a mineral corresponds to the break up of the ionic or covalent bonds that hold ions or atoms together within its crystal lattice. The dissolution of quartz in water is a step-by-step process and involves the adsorption of water molecules and the hydrolysis of Si–O bonds (Dove and Crerar 1990; Fig. 6.3a). The dissolution is controlled by the coordination states of the chemical components at the surface of solids. Berger et al. (1994) thus show that the alteration of the basaltic glass takes place following two processes with different rates:

1. an initial rate controlled by the silica concentration in the solutions diluted at a pH value close to neutrality;

2. a slower rate established over a longer period of time and controlled by the Al concentration of the solutions for acidic pH values.

Berger (1995) further shows that the rates of these two processes are similar to the dissolution rates of glass in open state (diluted solutions) on the one hand and of crystallised silicates (quartz, albite, anorthite, K-feldspar) on the other hand (Fig. 6.3b). The role of the structure of the surfaces in contact with the solvent has been confirmed by comparative dissolution experiments of smectites and kaolinites (Bauer and Berger 1998). The tetrahedral and octahedral sheets of kaolinites are simultaneously submitted to the chemical attack whereas only the tetrahedral sheets of smectites are first exposed. For kaolinites, the dissolution of the octahedral sheet is faster ($E_a = 33 \pm 8 \, \text{kJ} \, \text{mol}^{-1}$) than that of the tetrahedral sheet ($E_a = 0.51 \pm 8 \, \text{kJ} \, \text{mol}^{-1}$). The activation energy of the smectite dissolution is very close to that of the tetrahedral sheet of kaolinites ($E_a = 0.52 \pm 4 \, \text{kJ} \, \text{mol}^{-1}$).

6.1.2.2
Neogenesis

Neogenesis (new formation) is considered here as the result of two processes:

1. direct precipitation within the pores or fractures from the solutions (replacement of voids). This is the case of kaolinite precipitating in the dissolution pits of feldspars in altered granites.

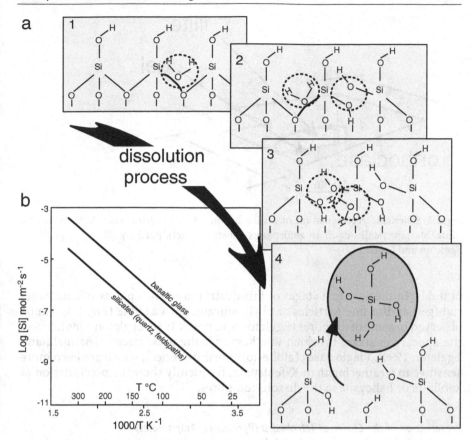

Fig. 6.3a,b. Mechanism of dissolution. **a)** Dissolution of quartz (after Dove and Crerar 1990, modified): 1) Adsorption of water molecules on the crystal surface; 2) Break up of a Si–O bond (hydrolysis) and adsorption of another water molecule; 3) Repetition of the adsorption – hydrolysis process; 4) Release of a Si^{4+} ion from the crystal structure in the form of a $H_4Si_4O_{10}$ molecule into water. **b)** Arrhenius diagram showing the dissolution rates of basaltic glasses and silicates (after Berger 1995)

2. crystallisation of clays by pseudomorphosis of primary minerals other than phyllosilicates. Neogenesis mostly results in the partial or complete replacement (pseudomorphosis) of these minerals. This is the case of saponite replacing olivines.

Direct Precipitation (Weathering Of Feldspars)

Alkaline feldspars (orthoclase, microcline, sanidine, albite, anorthoclase) alter under surface conditions thus becoming more and more porous. As pores grow bigger and more numerous, they cluster into genuine caries partially filled with neogenetic clay minerals mixed with fine debris of the parent mineral. Porous "plasma" is formed, which constitutes a favoured pathway for the penetration

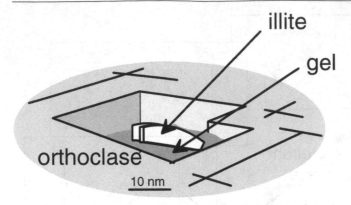

Fig. 6.4. Schematic representation of a dissolution pit in an orthoclase. A phyllosilicate (illite) starts crystallising in an amorphous substance (gel), partially filling the pit (after Eggleton and Busek 1980)

of fluids. In the very first stages of this destruction, amorphous silicate material (gel) and/or fine particles of phyllosilicates such as illite (Fig. 6.4) can be observed in some of the pores (Eggleton and Busek 1980; Eggleton 1986). Smectite appears in stages in which weathering is the most intense (Banfield and Eggleton 1990). Plagioclases (albite-to-anorthite series), which are even more sensitive to weathering than K-feldspars, frequently show the precipitation of kaolinite or halloysite in the dissolution pores.

Preservation of the Chains of Tetrahedra (Pyroxenes, Amphiboles)

Studies of altered pyroxenes by high-resolution electron microscopy (Eggleton and Boland 1982) have revealed the relationships between the crystal lattices of the primary mineral and those of its secondary products (Fig. 6.5). The phyllosilicates produced by weathering (talc and trioctahedral smectite) appear as the result of a topotactic growth because their crystal lattice is strongly oriented by that of pyroxene. This type of growth is very common in the biopyriboles described by Veblen and Buseck (1980). Lattice coherence favours a more rapid growth than that of totally reconstructed secondary phases.

6.1.2.3
Transformation of Pre-Existing Phyllosilicates

Transformation of Micas

Experimental studies of mica weathering have been extensively devised in the 1960s to answer an essential question: how was potassium, a vital element for the growth of plants, released or fixed in soils? These studies focused on the two types of micas most commonly found in rocks: biotites (trioctahedral ferro-magnesian mica) and muscovites (aluminous, dioctahedral mica). These two minerals were subjected to chemical attacks by organic acidic solutions

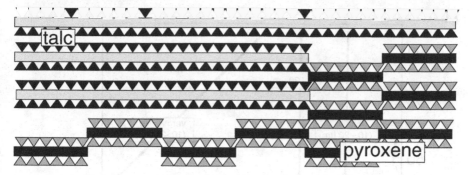

Fig. 6.5. Growth of talc or tetrahedral smectite in the altered zones of an enstatite (after Eggleton and Boland 1982). This type of transformation is said to be topotactic because the crystal lattice of pyroxene is in structural continuity with those of its alteration products

(Robert and Pédro 1972) or inorganic acidic solutions (Scott and Smith 1966). Potassium is much more readily extracted from trioctahedral minerals than from their dioctahedral homologues (Fig. 6.6a). Newman and Brown (1966) have shown that the destruction process of mica occurs in several stages:

- replacement of K^+ by Na^+

- loss of those $(OH)^-$ released after the loss of K^+ ions. The charge of the 2:1 layer decreases, which entails the loss of additional K^+ ions.

- oxidation of the ferrous iron: $4Fe^{2+} + 4$ structural $(OH)^- O_2 \rightarrow 4Fe^{3+} + 4$ structural $O^{2-} + 2H_2O$.

- loss of bivalent cations in octahedral position such as Mg^{2+}.

Generally, the progressive extraction of K^+ ions from the interlayer sheet causes the mica to transform into vermiculite then into smectite. The charge density decreases, bringing about the aperture of the interlayer zones, which are then able to adsorb polar molecules (water or organic molecules). The charge decrease is due to a significant rearrangement of the tetrahedral and octahedral sheets of the 2:1 layers, although the general appearance of the original mica crystal seems unchanged. HERTM observations confirm that vermiculite layers are inserted inside the mica ones (Jeong and Kim 2003; Mukarami et al. 2003). The vermiculitisation progresses from the edges of the biotite crystals (Fig. 6b).

These experiments that simulate the destruction of micas in soils have an additional interest. Indeed, alteration is used as a selective probe of the crystal structure at the beginning of the destruction process of mica. The first stages of the reaction forming regularly ordered mixed-layer minerals show that every second layer in the original mica fixes less strongly the interlayer potassium than its neighbour (Boettcher 1966). These data known since the early sixties have not been fully exploited with this view. They deserve new consideration.

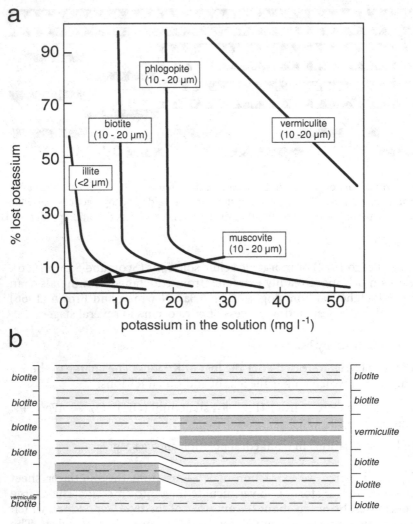

Fig. 6.6. Experimental weathering of micas. **a)** Potassium-retaining capacity of the various mica-like minerals. **b)** Schematic representation of the biotite-to-vermiculite alteration. The propagation of the "vermiculitisation" forms mica-vermiculite regularly ordered mixed layers

The transformation of biotites differ according to the composition of the waters percolating into granitic rocks (Meunier 1977). In highly diluted solutions, biotites are transformed into di- and trioctahedral vermiculites + kaolinite + Fe-hydroxides assemblage (Fig. 6.7). If the solutions become charged with Ca^{2+} ions during their percolation in the sedimentary rocks above granite for instance, they are transformed into a mica-trioctahedral vermiculite regularly ordered mixed layer + kaolinite + Fe-hydroxides.

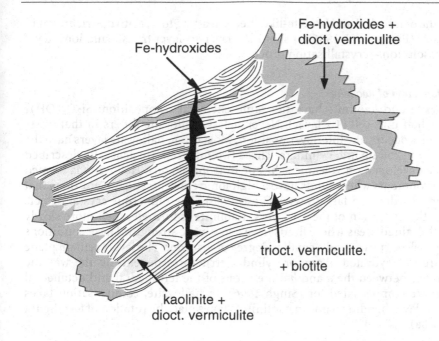

Fe-hydroxides + dioct. vermiculite

Fe-hydroxides

trioct. vermiculite. + biotite

kaolinite + dioct. vermiculite

Fig. 6.7. Weathered biotite from the granitic arena of La Rayrie, Deux-Sèvres, France (Meunier 1977). Phases have been determined using XRD and microprobe analyses

Transformation of Chlorites

Chlorite weathering leads to the formation of vermiculite through a series of mixed-layer minerals. The most noteworthy step of this alteration is the formation of a vermiculite – chlorite mixed layer with a corrensite-type structure. Many attempts at reproducing experimentally these sequences observed in nature have been carried out: deshydroxylation of chlorite at 600 °C then acidic dissolution of the brucite-like sheet (Ross and Kodama 1974) or solubilisation of Fe by dithionite (Makumbi and Herbillon 1972). The formation of corrensite shows that the original structure of chlorite (IIb polytype) is retained in a weathering process where every second layer is transformed. Like for micas, weathering could be used as a probe showing heterogeneities of the crystal structure of chlorites.

It is generally accepted that "vermiculitisation" of chlorites is the result of the oxidation of Fe^{2+} cations. The change of valency has two effects: (1) in the octahedral sheet of the 2:1 unit, $[Fe^{2+}OH]^+ \rightarrow [Fe^{3+}O]^+ + H^+ + e^-$, and (2) loss of Fe^{2+} and Mg in the brucite-like sheet. In nature, it seems that the "vermiculitisation" of chlorite rather starts with the loss of a part of the Fe^{2+} cations of the 2:1 unit because the Fe^{3+} content does not increase significantly from chlorite to the corrensite-type regularly ordered mixed-layer mineral (step 1) then to vermiculite (step 2). The Mg cations are widely leached out from the octahedral and brucite-like sheets, leading to a relative enrichment in Si and Al (Proust et al. 1986). If step 1 is a layer-to-layer transformation, step 2 is a dissolution – crystallisation process.

Transformation of Kaolinite

Halloysite and kaolinite have the same chemical composition: $Si_2O_5(OH)_4$ for the half unit cell (1:1 layer). Nevertheless, halloysite differs in that water molecules occupy the interlayer position. Most often, halloysite layers have the property to roll up into cylinders. These two minerals are generally described in soils developed on volcanic ashes or in tropical soils. Halloysite is formed in environments where fluids have a high ionic concentration, conditions that favour disorder and fast growth (Giese 1988). The numerous crystal defects favour the hydration of the 1:1 layers. Kaolinite is formed much more slowly in well-drained areas where fluids have a low ionic strength. These conditions favour a slow growth and a smaller degree of disorder. The layers remain plane and are not hydrated. Plane or cylinder structures depend on the way the difference between the a and b dimensions of the tetrahedral and octahedral sheets are compensated for (Singh 1996). In halloysite, compensation takes place by layer bending while in kaolinite it takes place by rotation of tetrahedra (Fig. 6.8a).

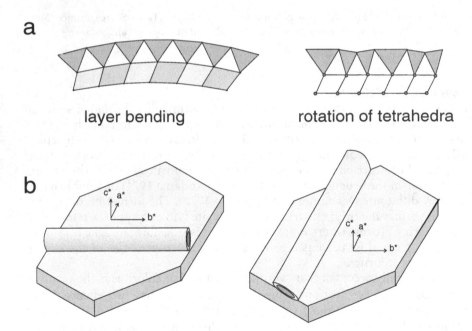

Fig. 6.8a,b. Halloysite and kaolinite crystal structure. **a)** Compensation of a and b dimension differences between the tetrahedral and octahedral sheets by layer bending (halloysite) or by rotation of tetrahedra (kaolinite). **b)** Kaolinite to halloysite conversion by the rolling up of 1:1 layers along b or a direction after hydration (after Singh and Mackinnon 1996)

In nature, kaolinite is considered as the stable mineral phase. The passage from halloysite to kaolinite takes place progressively over time through a dissolution-recrystallisation process. The experimental works by Singh and Mackinnon (1996) show that the reverse path can be travelled when kaolinite undergoes hydration. The kaolinite booklets roll up, thus forming elongated tubes along the *b* direction, more rarely along the *a* direction (Fig. 6.8b).

6.1.2.4
Sequence of Transformations in Profiles

Concept of Microsystem

Soils and alterites are not homogeneous environments. Their physical properties (permeability, porosity) and their chemical and mineralogical composition vary over very short distances. Therefore, the mineral reactions producing the clay minerals take place within microenvironments (from a few square microns to a few square millimetres). Like the rocks from the contact metamorphism zones described by Korzhinskii (1959), soils or alterites consist of a mosaic of reaction microsites.

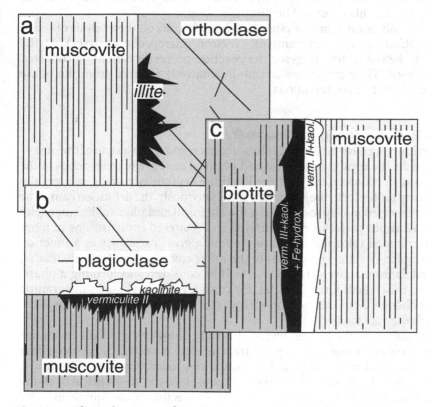

Fig. 6.9a–c. The various types of reaction microenvironments along the open grain joints between muscovite and orthoclase (**a**), biotite (**b**) and plagioclase (**c**)

The example of the incipient weathering of granite provides a better understanding of the diversity of these microsites. At this stage, the meteoric fluids enter the interconnected microfissures that are mainly located at the grain joints. The systematic observation of the contact zones between muscovite crystals on the one hand and their various neighbours (biotite, orthoclase, plagioclase ones) on the other hand shows a clear relationship between the contact type and the forming secondary minerals (Fig. 6.9).

Within each reaction microsite, the dissolution and precipitation reactions (nucleation, growth) are controlled by local chemical conditions established by the interactions between aqueous solutions and the primary minerals at their contact. Each microsite can then be viewed as a microsystem composed of three entities: primary minerals, newly-formed minerals and solutions (Korzhinskii 1959). Then, classical thermodynamics can be applied at this scale.

The microsystem concept is an efficient tool for integrating the mineralogical observations and chemical compositions measured with the electron probe and for interpreting them in terms of phase equilibrium. Indeed, although chemical equilibria between phases are rarely reached, the observed assemblages of secondary minerals comply with the phase rule. The concepts of stability or metastability are not really important at this stage of the analysis of the observed facts. This point will be discussed later.

The identification of microsystems exclusively rests on the analysis of solids (X-ray diffraction on microsamples, electron microprobe analyses). At the moment, the fluids that triggered the reaction processes cannot be collected nor analysed. Their properties can be qualitatively inferred from the phase diagrams built from observations.

Phases, Equilibrium, Stability and Metastability

Theoretically, a phase for a given substance corresponds to a state of the matter such that its chemical composition and physical state are homogeneous in the whole system. Consequently, all the kaolinite crystals occurring in a granite weathering profile form the kaolinite phase. Obviously, the definition cannot be strictly respected in clay rocks where the chemical and physical homogeneity of crystals is not perfect. Crystal defects and scattered compositions in more or less extended solid solutions bring about a great variability as a function of the physico-chemical properties of the microenvironment. Nevertheless, all the crystals from a given clay species will be considered as forming a phase. Thus the trioctahedral vermiculites are considered as a phase in the granitic arena, although their composition varies according to the type of microsystem in which they are formed.

At equilibrium, the phases involved do not show any gradient in the intensive variables (temperature, pressure, chemical potentials ...). At constant temperature and pressure (classical weathering conditions), equilibrium is reached when the chemical potentials of some elements are identical in all the phases involved. For instance, at orthoclase – smectite – illite equilibrium, μ_{Si} is equivalent in the three phases although the SiO_2 concentration in each phase is different (see Sect. 3.1.5.2). A phase is stable in a domain defined by the intensive variables. However a phase may exist outside this domain for reasons of reaction or crystallisation kinetics. Then it is called *metastable*. This does not influence the phase rule, which remains a universal mark in the analysis of weathering profiles. A detailed discussion of the concepts of stability and metastability applied to clay minerals has been proposed by Essene and Peacor (1995).

6.1.3
Weathered Rocks

6.1.3.1
Granitic Rocks

Weathering Profile

Arenas (regolith) form discontinuous mantles at the surface of the granitic or gneissic rock massifs. While, on a large scale, the general organisation presents a fresh rock polarity to the topsoil, in the detail of the outcrop, this vertical logic is rarely obvious. Indeed, this order is disrupted by the action of erosion or by the preferred weathering of the fractured zones, which leads frequently to the observation of little weathered rocks on top of pulverulent arena (Fig. 6.10).

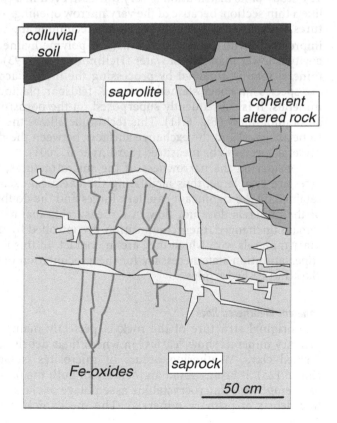

Fig. 6.10. Example of a weathering profile developed on a granite: La Rayrie, Deux Sèvres, France (Meunier 1980)

colluvial soil

saprolite

coherent altered rock

Fe-oxides

saprock

50 cm

This heterogeneity can be found at the centimetre scale where small fragments of apparently fresh coherent rock lie within highly friable levels. Therefore, the use of rock facies is better suited for the description of weathering profiles than the use of horizons, which are only suited to soils. Thus, four great facies can be distinguished: (1) fresh rock, (2) coherent weathered rock, (3) *saprock* or structure-maintained friable arena and (4) *saprolite* or structure-modified friable arena (difference with the definition given by Taylor and Eggleton, 2001). The clay mineral content increases from 1 to 4.

Fresh Rock

All the granitic rocks outcropping at the surface of the continents incorporate weathering minerals, even the most apparently fresh ones. Indeed, the dissolution reactions of primary minerals are triggered by the meteoric water impregnating their microporosity (grain joints, cleavages, microfissures etc.). Plagioclase, biotite and amphibole are the most affected minerals by these early stages of weathering. Clay minerals (kaolinite, ferriferous smectites) precipitate in the voids created by dissolution. Clay crystals, in spite of their small size (> 1 μm), can be observed by transmission electron microscopy.

The connected porosity (permeability) of a granite is low: 10^{-18} to 10^{-20} m^2. The study of its distribution is very difficult, even in a two-dimensional space like a thin section, because of the very narrow openings of pores or microfractures. Only the autoradiography imaging technique applied to sections of rocks impregnated with a radioactive resin (^{14}C-polymethylmethacrylate) whose viscosity is lower than that of water (Hellmuth et al. 1993) proved efficient. The mineral map is obtained by processing the images acquired after the various species have been stained: quartz, K-feldspar, plagioclase, ferromagnesian minerals. It is subsequently superposed on the porosity images obtained by autoradiography (Fig. 6.11). This technique allows the porosity distribution to be analysed and the exchange surfaces between the fluids and the various mineral species to be measured (Sardini et al. 2001).

Considering the narrowness of the microfractures, the solid-fluid interfaces are very large, thus facilitating the chemical exchanges. The solutions retained by the significant capillary forces and the double layer at the surface of the minerals does not flow in a measurable manner. Since the solutions remain unchanged, their composition is controlled by the dissolution of primary minerals with which they are in contact. In this manner they reach the supersaturation state necessary for the precipitation of the hydrated phases: clays and Fe-hydroxides.

Coherent Weathered Rock

The original structure of the rock is perfectly maintained, although some primary minerals show "caries" in which their debris are mixed with newly-formed clays. Weathering occurs in microsites isolated from each other (Fig. 6.12a). It brings about an increase of both the porosity and the amount of *alteroplasma* (microcrystalline assemblage of clays, Fe-oxyhydroxides and fine debris of primary minerals). The mechanical strength of the rock is

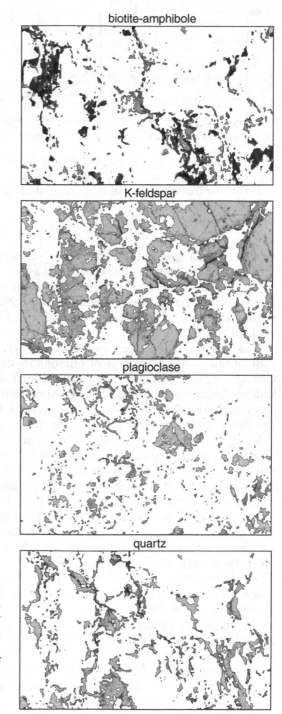

Fig. 6.11. 2-D image of the connected porosity for each mineral species in a granite obtained by superimposition of the mineral map (staining method) on autoradiography (impregnation of a granite with a radioactive resin: ^{14}C-polymethylmethacrylate), from Sardini et al. (2001)

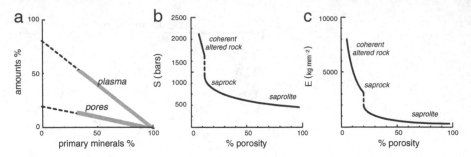

Fig. 6.12a–c. Variation of physical properties of granitic arena with increasing alteration degree. **a)** Amounts of weathering plasma (clays + oxides + fine debris of primary minerals) and voids (measured using point counting on thin sections). **b)** Resistance to a uniaxial stress. **c)** Young's modulus

weakened–decrease in the compressive strength, decrease in Young's modulus (Fig. 6.12b,c)-but yet remains that of a coherent material (Baudracco et al. 1982). The density decreases from 2.6 to 2.1. It is inversely correlated with the increase of the porosity (Righi and Meunier 1995). The chemical composition of the fresh rock does not vary significantly; only the water content increases noticeably.

The new mineral phases, particularly clays, are then formed in a multitude of microsites a few tens of microns in size located in the grain joints, cleavages or microfissures. The weathering process operates like a mosaic of independent microsystems. Although numerous, these microsystems come in three categories only: the contact microsystems, the plasmic microsystems and the fissure microsystems (Meunier 1980). In the contact microsystems, considering the narrowness of the voids, the activity of water is greatly below 1 (see Sect. 5.1.2.2). Flows are slow and low; the chemical transfers are carried out over very short distances only. The system can be considered as closed when its size is that of the microsite. The plasmic microsystems progressively open as the secondary porosity created by the dissolution of primary minerals increases. The activity of some chemical elements is then controlled by the environment outside the system. Finally, developing in the connected lattice, the fissure microsystems are widely open. The activity of silica and alkaline elements in particular are controlled by the external environment

Inside the coherent weathered rock, most mineral reactions occur in the contact microsystems. The assemblages of clay minerals depend on the facing primary minerals. The example of the intercrystalline joints between muscovite and other primary minerals provides a better understanding of how clay minerals are formed at this stage (Fig. 6.9):

– *the neighbouring mineral is orthoclase*: the K^+ content of the fluids is high enough for illite to form locally in replacement of feldspar. Muscovite does not look weathered (solution undersaturated with respect to K-feldspar but close to equilibrium with respect to mica);

- *the neighbouring mineral is plagioclase*: the K^+ concentration in solutions is too low, only kaolinite can form within the plagioclase. Muscovite transforms layer by layer along the joint by loss of the interlayer K into aluminous dioctahedral vermiculite;

- *the neighbouring mineral is biotite*: both micas lose K and their layers are transformed into vermiculite. Two types of polyphase assemblages appear: trioctahedral vermiculite + kaolinite + Fe-oxyhydroxides in biotite, dioctahedral vermiculite + kaolinite in muscovite;

The weathering processes taking place along the other joints (feldspar-quartz or feldspar-feldspar) operate like the internal weathering processes (plasmic microsystems) of feldspars along their internal microfractures:

- plagioclase → kaolinite

- K-feldspars → dioctahedral vermiculite + kaolinite.

The fissure microsystems, at this early stage of rock weathering, already display the mineral assemblage typical of the end of the evolution in temperate climate: kaolinite + Fe-oxyhydroxide.

Saprock

The structure of the rock is maintained although its mechanical strength is very reduced. Porosity ranges from 10 to 20%, resulting from the dissolution of primary minerals and the formation of new fractures generated by local variations in the stress state. The contact microsystems are no longer operative. They disappear in the weathering plasma that develops within each parent mineral (*alteroplasma*). Illite formed at the muscovite-orthoclase joints is no longer identifiable. The plasmic microsystems still retain the signature of the parent minerals (Fig. 6.13):

- K-feldspars: dioctahedral vermiculite+kaolinite ($\pm$ beidellite)

- plagioclase: kaolinite

- biotite: trioctahedral vermiculite + kaolinite + Fe-oxyhydroxides

- muscovite: dioctahedral vermiculite+kaolinite

New fractures open up and are connected with the pre-existing network of fissures. The resulting network drains throughout the whole weathering profile. The solutions flow faster and are more diluted. They transport clay particles and Fe-oxyhydroxides that settle on the fissure walls in red concentric deposits (*cutans*). The only stable clay in these fissure microsystems is essentially kaolinite.

Saprolite

In temperate climate, saprolite forms more or less deep zones within the regolith. In tropical climate, it makes up most of the weathering mantle (Nahon

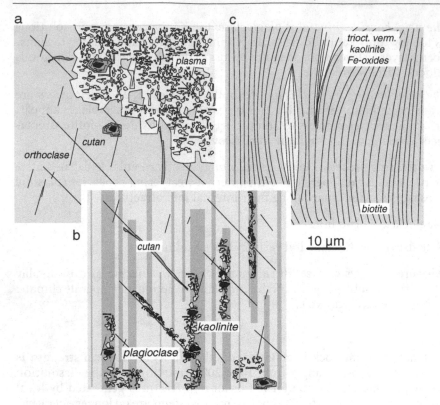

Fig. 6.13a–c. Mineral composition of plasmic microsystems (alteroplasma) in the saprock. **a)** Orthoclase: dioctahedral vermiculite+kaolinite. **b)** Plagioclase: kaolinite. **c)** Biotite: trioctahedral vermiculite+kaolinite+Fe-oxyhydroxides. Cutans (fissure microsystems), whatever the host mineral, are always composed of kaolinite+Fe-oxyhydroxides

1991; Tardy 1993). The initial structure of the rock is no longer recognisable (difference with definition given by Taylor and Eggleton, 2001). The mechanical strength and Young's modulus are very low and correspond to those of unconsolidated rocks. Porosity is high in places (about 40–50% in the remaining small pieces of saprock) and much more reduced in others, owing to the settlement of arena under its own load. The debris of the parent minerals mixed with their weathering products form a new environments (*pedoplasma*), imposing different chemical conditions that bring about a second generation of mineral reactions. Particularly, in the vicinity of dissolving K-feldspar debris, the illite+kaolinite assemblage replaces the aluminous dioctahedral vermiculites (Dudoignon 1983). Fe oxidation grows more intense and the trioctahedral vermiculites disappear to the benefit of ferriferous dioctahedral vermiculites.

Fissures

Fissures – whose opening ranges from the micrometre to the centimetre – form a connected network throughout the whole profile, including the fresh rock. They are coated with cutans composed of the kaolinite + Fe-oxyhydroxide assemblage, whatever their size. This network is used as a drain by the solutions percolating in the granitic massifs and reappearing in springs. The chemical analysis of these "granitic waters" show that they are in equilibrium with only one clay species: kaolinite (Feth et al. 1964). This indicates that the other possibly transported species (vermiculites, smectites, illites) are dissolved.

Geochemical Processes of Granite Weathering

The petrographic observations carried out in the weathering profiles developed in various climates show that particular mineral assemblages are formed in each type of microsystem (Harris and Adams 1966; Bisdom 1967; Wolff 1967; Sikora and Stoch 1972; Gilkes et al. 1973; Rice 1973; Eswaran and Bin 1978; Dudoignon 1983; Boulangé 1984):

Fresh rock: K-feldspars, plagioclase, biotite, muscovite, quartz.
Coherent weathered rock:

– mica-orthoclase contact microsystems: illite

– mica-mica, mica-plagioclases, quartz-feldspars, orthoclase-plagioclase contact microsystems: di- and trioctahedral vermiculites + kaolinite ± Fe-oxyhydroxide ± beidellite, each parent mineral giving its own secondary products as it does in alteroplasma microsystems (see below).

Saprock (alteroplasma):

– plasmic microsystems in micas: di- and trioctahedral vermiculites + kaolinite ± Fe-oxyhydroxide ± beidellite

– plasmic microsystems in K-feldspars: dioctahedral vermiculite (aluminous or ferriferous) or beidellite+kaolinite

– plasmic microsystems in plagioclases: kaolinite

Saprolite (pedoplasma):

– temperate climates: illite+kaolinite

– tropical climates: kaolinite + Fe-oxyhydroxide + gibbsite

Fissure microsystems:

– temperate climates: kaolinite + Fe-oxyhydroxides

– tropical climates: gibbsite + Fe-oxyhydroxides.

To simplify the chemical system representing the weathering processes of granitic rocks, the differences between biotites and muscovites may be ignored because the solubilized Fe is rapidly fixed in the form of minerals that remain inert: oxyhydroxides. Magnesium plays a part in the determination of the secondary phases only in the very first stages of weathering, when triocta-hedral vermiculites form. The latter subsequently transform into dioctahedral vermiculites by oxidation of the Fe^{2+} ions. Let's consider here that biotite and muscovite alter into a vermiculite + kaolinite assemblage. The secondary par-ageneses can then be represented in a phase diagram Na,K–Al–Si (Garrels and Christ 1965).

Plagioclases are always replaced by kaolinite (or halloysite). No other phase with a greater silica content (smectite, micas, zeolites, albite) crystallises. The absence of micas can be explained by the fact that only the potassic species (illite) can form under surface conditions, margarite and paragonite crystallis-ing at higher temperatures. The K^+ contents are too low for micas to form. Likewise, zeolites and albites are forbidden because the activity of the silica in solution and the pH value are not high enough. The activity of the alkaline elements is too high for smectites to crystallise. The too low local pH value reduces the kaolinite + albite + zeolite stability field to the kaolinite – solution tie line (Fig. 6.14a). Quartz, although theoretically stable, cannot crystallise un-der surface conditions. Amorphous silica occurs for higher Si activities only. Therefore, in the absence of these minerals, the Si^{4+} activity is buffered by the newly-formed phase having the greatest silica content: smectites.

In the weathering process, the mineral reactions take place in microsystems that are first closed then more and more open as porosity increases. Some of the most soluble chemical components, Na and K in particular, then be-have like mobile elements (calcium being clearly less abundant; Garrels and Howard 1957). Their chemical potential ($\mu_{Na,K}$) is an intensive variable playing a strategic part in clay crystallisation. Moreover, since hydrolysis reactions can be considered as the introduction of H^+ protons into solid phases, the pH of the solutions too becomes one of the most important intensive variables in the weathering process. The system to be considered should then be Si–Al–$\mu_{Na,K}$, pH. However, another simplification can be brought: the pH being rapidly buffered by minerals, it varies by only two units between the microfractures of the fresh rock and rainwater. The variation of the potential $\mu_{Na,K}$ is greater because the alkaline element concentration becomes infinitely low in highly diluted solutions flowing in fractures. The really active chemical system can then be reduced to Si–Al–$\mu_{Na,K}$ and readily derived from the ternary system by the equipotential method (see Fig. 3.10).

The decrease in the potential $\mu_{Na,K}$ as the weathering intensity increases can be represented by a succession of equipotential lines in order of decreasing values, from the highest values (concentrated solutions, low permeability) to the lowest values (diluted solutions, high permeability). The number and rules of line plotting (see Sect. 3.1.5.2) are determined by the number of three-phase assemblages and their position in the ternary diagram (Korzhinskii 1959; Spear et al. 1982; Meunier and Velde 1986). The plotting of these lines (7 + 2 = 9) is determined by the sequence of secondary minerals observed in the various microsystems (Fig. 6.14b).

The Si–Al–$\mu_{Na,K}$ system (Fig. 6.15) shows how the composition and nature of solids (extensive variables Si, Al) vary as functions of those of solutions ($\mu_{Na,K}$)

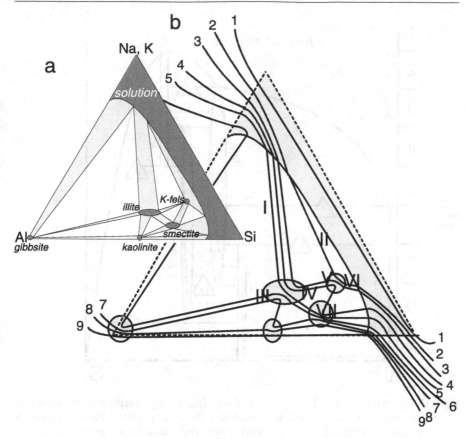

Fig. 6.14a,b. Phase diagram in the system Na,K–Al–Si. **a)** Diagram representing the assemblages of clay minerals identified in the granitic arena, assuming that the three components (alkaline elements, Si and Al) are inert. **b)** The highly soluble alkaline elements are considered as mobile elements. Plotting of the equipotential lines $\mu_{Na,K}$ describing the assemblages identified

represented by dotted paths. The $[K^+]/[H^+]$ ratio of solutions is buffered by the muscovite-K-feldspar contact microsystems at a value such that K-feldspars are no longer stable. Illite (low-temperature equivalent of muscovite) forms and grows within feldspars. No secondary mineral forms along the quartz-K-feldspars contacts because the solutions are not buffered by mica; they reach equilibrium with feldspar. Along the K-feldspar – plagioclase contacts and in plasmic microsystems (internal alteration of minerals in granite), the solutions are buffered for lower values of $\mu_{Na,K}$. The Si and Al contents in the solutions are controlled by illites, vermiculites, smectites and kaolinite. In temperate climates, these solutions flowing freely in fissures are much more diluted, and only kaolinite is stable. The main part of the saprolites developed in tropical climates is also composed of this monomineral assemblage. The latter

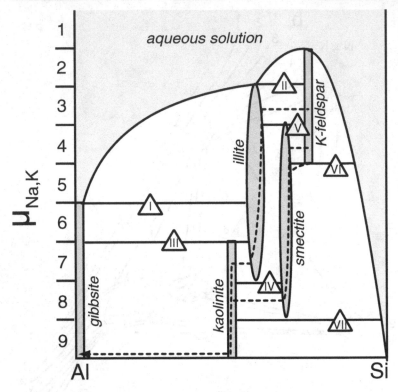

Fig. 6.15. Composition (Al–Si) – chemical potential ($\mu_{Na,K}$) system representing the weathering of granitic rocks. This system is derived from the ternary system (Fig. 6.14) through the equipotential method (Meunier and Velde 1986). The solid solutions are symbolised by *oval chemical composition domains*. The composition of fluids in the various microsystems lines the *dotted paths*

disappears in the most intensively drained zones to be replaced by gibbsite. The illite+kaolinite assemblage is typical of granitic saprolites developed in temperate climates. Indeed, the new friable rock formed by the collapse of the structure-maintained arena is a mixture of clays and debris of primary minerals. The new rock containing less soluble silica (quartz is abundant but little soluble), the alkaline element potential – and notably the K potential – is still high because of the abundance of K-feldspar debris.

6.1.3.2
Ultrabasic Rocks

Grained Rocks (Lherzolite)
The case of lherzolite, whose weathering profile has been described by Fontanaud and Meunier (1983), shall be taken here as an example to facilitate the understanding of the petrographic analysis of the ultrabasic rock weathering. The

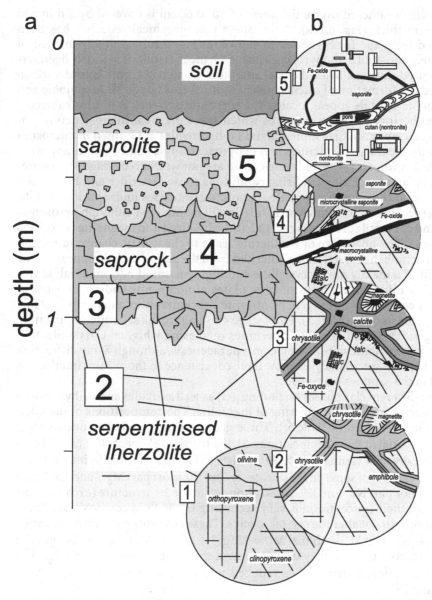

Fig. 6.16a,b. Weathering of a lherzolite (after Fontanaud and Meunier 1983). **a)** Weathering profile. **b)** Schematic representation of the weathering sequence. The fresh lherzolite (stage 1), essentially composed of olivine and pyroxenes (clino- and orthopyroxenes), has been serpentinised (metamorphism): formation of calcite and chrysotile veins; local crystallisations of amphiboles and magnetite (stage 2). In levels where the rock microstructure is maintained (saprock), weathering causes the early crystallisation of talc in pyroxenes (stage 3) then of saponite and Fe-hydroxides (stage 4). Finally, the rock loses its structure (saprolite) in which nontronite and an amorphous substance (gel) precipitate (stage 5)

lherzolite weathered over a thickness of 50 to 60 cm is covered by a thin soil 20–25 cm thick (Fig. 6.16a). A five-stage petrographical sequence has been defined from the freshest rock to saprolite (Fig. 6.16b). Initially formed of olivine, clino- and orthopyroxenes (stage 1), the lherzolite has widely been serpentinised during a hydrothermal alteration or metamorphic episode, which caused the formation of calcite and chrysotile veins (stage 2). Amphibole and magnetite crystals appear locally. The serpentinised rock is in turn subjected to weathering, the early stages of which leave the rock microstructure unchanged (saprock). Clay minerals and Fe-hydroxides are formed in microsites along the cleavages or microfissures of pyroxenes. Talc and Fe-oxides appear in pyroxenes whereas chrysotile and calcite are not weathered (stage 3). Fissures open up and increasing oxidising conditions cause Fe-oxides and hydroxides to settle on the microfissure walls. Saponite crystallises under various habits; the serpentinised rock microstructure is still maintained although pyroxenes, talc and chrysotile are weathered. Calcite is dissolved and magnetite is oxidised (stage 4). The final stage of weathering leads to the drastic change in the microstructure: the crystal frame inherited from the parent rock disappears to the benefit of a mainly microcrystalline and highly fissured new material: saprolite. New minerals are formed: large plates of nontronite and an amorphous substance along fissures (gel). Nontronite concentrates into deposits (cutans) sealing the fissures (stage 5). From stage 4 to stage 5, the saponite composition changes. It's Al and Fe content becomes increasingly higher. Chrysotile, talc, saponite and nontronite do not form a paragenesis, although X-ray diffraction and chemical microanalyses show their coexistence in the < 2 μm fractions in most of the samples collected.

Five different clay phases (including gel) as well as oxides and Fe-hydroxides have been formed by weathering of lherzolite. The compositions of the phyllosilicates are given in Table 6.1. Three groups of chemical components seem to be essential in these compositions: Si, the trivalent elements R^{3+} (Al^{3+}, Fe^{3+}) and the divalent elements R^{2+} (Fe^{2+}, Mg^{2+}). The cations able to integrate the interlayer sheet of some phyllosilicates are for the most part Mg^{2+} and Ca^{2+}, and for a minor part Na^+ and K^+. Weakly bonded to the 2:1 structure (exchangeable cations), their concentrations vary according to the fluid composition. These variations do not alter the crystal species. These elements globally represented by an equivalent M^+ such that $M^+ = 2Mg^{2+} + 2Ca^{2+} + K^+ + Na^+$ can be ignored in a first step because they do not control the mineral phase formation. The active chemical system is then reduced to three groups of components Si^{4+}, R^{2+} and R^{3+}.

Figure 6.17a shows a phase diagram in the system $R^{2+}-R^{3+}-Si$ summarising the observations. The tie lines are consistent with petrographical observations and delineate 6 three-phase assemblages if one considers that quartz or amorphous silica can precipitate. Talc does not co-exist with nontronite. The magnesian gel is a metastable substance appearing when saponites transform into nontronites (stage 5). The saponite-quartz line forbids the talc + Fe-oxides assemblage. These two species can be observed at stage 3 but oxides are not hematite (maghemite?) and are inherited from the serpentinisation

Table 6.1. Chemical composition of the phyllosilicate mineral phases resulting from the weathering of a serpentinised lherzolite (Fontanaud and Meunier 1983). These compositions are recalculated in the system $Si-R^{3+}-R^{2+}$

Mineral	Composition	Si		R^{2+}		R^{3+}	
		Amount	%	Amount	%	Amount	%
Talc	$Si_4 O_{10} Mg_{2.80} Fe^{2+}_{0.20} (OH)_2$	4	57	3	43	0	0
Saponite	$Si_{3.44} Al_{0.56} O_{10} Al_{0.09} Fe^{3+}_{0.53}$ $Mg_{2.80} (OH)_2 M^+_{0.56}$	3.44	46	2.8	38	1.18	16
Nontronite	$Si_{3.66} Al_{0.18} Fe^{3+}_{0.16} O_{10} Fe^{3+}_{1.82}$ $Mg_{0.18} (OH)_2 M^+_{0.52}$	3.66	61	0.18	3	2.16	36
Gel	$Si_2 O_5 Al_{0.09} Fe^{2+}_{0.40} Mg_{2.47} (OH)_4$	2	40	2.87	58	0.09	2

episode; thus talc + Fe-oxides are not a paragenesis. The chemical parameters controlling all the observed mineral reactions are the Mg^{2+} leaching and the Fe^{2+} oxidisation. They operate simultaneously and reduce the concentration of the bivalent elements (R^{2+}), which are consumed in the formation of tri-octahedral clays (Fe^{2+} and Mg^{2+} are assumed to substitute for each other in all proportions). Therefore, the driving force during the weathering process of ultrabasic rocks appears to be the chemical potential of these elements: $\mu_{R^{2+}}$. The sequence of observed mineral reactions can then be represented by a series of equipotential lines plotted in order of decreasing chemical potential values $\mu_{R^{2+}}$ from a to h (Fig. 6.17b).

The $\mu_{R^{2+}}-R^{3+}-Si$ diagram shows that the rock composition changes as the chemical potential of the bivalent elements decreases (dotted paths, Fig. 6.17c). Notably, silica losses lead to the accumulation of Fe-oxyhydroxides, which form compact and thick horizons in tropical climate. Saponite takes part in two different two-phase assemblages, with talc on the one hand and nontronite on the other hand. The petrographic observations show a change in its habit, meaning that saponite recrystallises and experiences also chemical modifications: Mg-rich when stable with talc, it becomes richer in Al and Fe^{3+} when stable with nontronite. Saponite is unstable for low values of $\mu_{R^{2+}}$, yielding nontronite and a metastable magnesian phase: gel. The latter disappears in highly permeable zones.

Knowing the compositions of the minerals formed by weathering of the serpentinised lherzolite (Table 6.1), it is possible to calculate their dissolution equation so as to determine the equilibrium conditions with the solutions in the system $SiO_2-MgO-H_2O$ at 25 °C, 1 atmosphere.

Saponite

$$(Si_{3.44}Al_{0.56}) O_{10}Al_{0.09}Fe^{3+}_{0.53}Mg_{0.27}(OH) 2 Mg_{0.28} + 4.52 H_2O + 5.70 H^+$$
$$\rightarrow 3.44 H_4SiO_4 + 0.65 Al(OH)_3 + 0.27 Fe_2O_3 + 2.35 Mg^{2+} \tag{6.1}$$

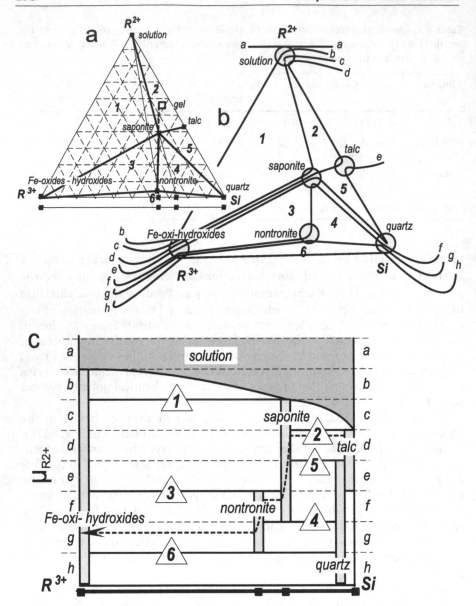

Fig. 6.17a–c. Phase diagram in the system R^{2+}–R^{3+}–Si. **a)** The soluble bivalent elements do not form a mineral phase: Fe^{2+} is oxidised into Fe^{3+} (oxyhydroxide), Mg^{2+} remains in solution and is subjected to leaching. **b)** The assemblages observed from the fresh rock to saprolite are described by the eight *equipotential lines* (see Figure 6.16). **c)** Phase diagram in the system μ_{R112+}–R^{3+}–Si

$$K_{sap} = \frac{a_{H_4SiO_4}^{3.44} \cdot a_{Mg^{2+}}^{2.35}}{a_{H^+}^{4.70}} \tag{6.2}$$

$$\log K_{sap} = 3.44 \log a_{H_4SiO_4} + 2.35 \log \left[\left(a_{Mg^{2+}} \right) / \left(a_{H^+} \right)^2 \right] \tag{6.3}$$

$$\Delta G_r^\circ = \left[3.44 \Delta G_{f \cdot H_4SiO_4}^\circ + 0.65 \Delta G_{f \cdot Al(OH)_3}^\circ + 0.27 \Delta G_{f \cdot Fe_2O_3}^\circ + 2.35 \Delta G_{f \cdot Mg^{2+}}^\circ \right]$$
$$- \left[\Delta G_{f \cdot sap}^\circ + 4.52 \Delta G_{f H_2O}^\circ \right] = +24.03 \, \text{kcal mol}^{-1} \tag{6.4}$$

$$3.44 \log a_{H_4SiO_4} + 2.35 \log \left[\left(a_{Mg^{2+}} \right) / \left(a_{H^+} \right)^2 \right] = +17.62 \tag{6.5}$$

Talc

$$Si_4O_{10}Mg_{2.80}Fe_{0.20}^{2+}(OH)_2 + 4.20\,H_2O + 5.6\,H^+$$
$$\rightarrow 4\,H_4SiO_4 + 0.20\,FeO + 2.8\,Mg^{2+} \tag{6.6}$$

$$K_{talc} = \frac{a_{H_4SiO_4}^{44} \cdot a_{Mg^{2+}}^{2.8}}{a_{H^+}^{5.6}} \tag{6.7}$$

$$\log K_{talc} = 4 \log a_{H_4SiO_4} + 2.8 \log \left[\left(a_{Mg^{2+}} \right) / \left(a_{H^+} \right)^2 \right] = +20.66 \tag{6.8}$$

Gel

$$Si_2O_5Al_{0.09}Mg_{2.47}Fe_{0.40}^{2+}(OH)_4 + 4.94\,H^+$$
$$\rightarrow 2\,H_4SiO_4 + 0.09\,Al(OH)_3 + 0.40\,FeO + 2.47\,Mg^{2+} + 0.33\,H_2O \tag{6.9}$$

$$\log K_{gel} = 2 \log a_{H_4SiO_4} + 2.47 \log \left[\left(a_{Mg^{2+}} \right) / \left(a_{H^+} \right)^2 \right] = +19.31 \tag{6.10}$$

Nontronite

$$\left(Si_{3.66}Al_{0.16}Fe_{0.16}^{3+} \right) O_{10}Fe_{1.82}^{3+}Mg_{0.18}(OH)_2Mg_{0.26} + 6.15\,H_2O + 0.88\,H^+$$
$$\rightarrow 3.66\,H_4SiO_4 + 0.18\,Al(OH)_3 + 0.99\,Fe_2O_3 + 0.44\,Mg^{2+} \tag{6.11}$$

$$\log K_{nont} = 3.66 \log a_{H_4SiO_4} + 0.44 \log \left[\left(a_{Mg^{2+}} \right) / \left(a_{H^+} \right)^2 \right] = -11.03 \tag{6.12}$$

These equations enable the plotting of the different mineral phase solubility curves in the system $\log[(a_{Mg^{2+}})/(a_{H^+})^2]$ as a function of $\log a_{H_4SiO_4}$ (Fig. 6.18). The fluids are enriched in dissolved silica through both the serpentinisation and the weathering processes. The "microsystem" effect expresses itself

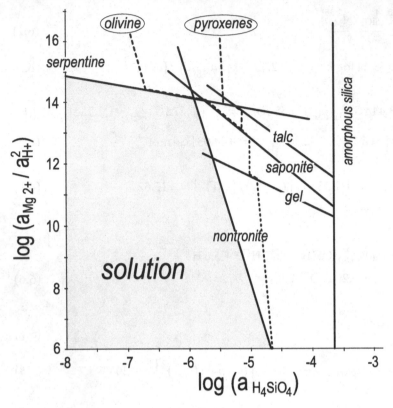

Fig. 6.18. Solubility diagram of weathering phases of a serpentinised lherzolite (Fontanaud and Meunier 1983). The variation in the composition of fluids is represented by *dotted lines*

through differing paths according to whether solutions acquire their composition by weathering of serpentinised olivines or of pyroxenes. Serpentine and talc co-exist in the very first stages of weathering. As porosity increases with weathering, new conditions are imposed and erase the microsystem effects: saponite is the most abundant species in the rock. It becomes unstable in turn along the fissures where more diluted and more oxidising solutions flow, causing gel and nontronite (+ Fe-oxyhydroxide) to precipitate.

Serpentinites

The weathering of serpentinite can be compared to that of the studied lherzolite described above because both rocks are chemically very close ($Al_2O_3\% < 2\%$). By contrast, the mineralogical composition and the microstructure are totally different (Ducloux et al. 1976). Serpentinite is mainly composed of phyllosilicates (antigorite, chrysotile, talc and Mg-chlorite) whose mean size is greatly below 0.1 mm. Contrary to lherzolite, serpentinite is a homogeneous rock of the mm³ scale. Owing to the very high reactivity of the constituent magnesian phyllosilicates under weathering conditions, the rock is extensively altered and shows a sudden transition between the alterite-soil system and the unweathered substrate. The thickness of the weathering profiles depends on topography: thin on the summit rock parts, they grow thicker on the slopes and reach their maximum in the depressions that generally correspond to fractures (Fig. 6.19a). The initial structure suddenly disappears: the saprock is very thin (a few cm). Clay-rich horizons (more than 50% clay) form with thicknesses differing according to the topographical situation of saprolite and exhibit a prismatic structure. They are topped by a thin soil composed of a horizon rich in organic matter (A horizon) and of a clay accumulation horizon (B horizon).

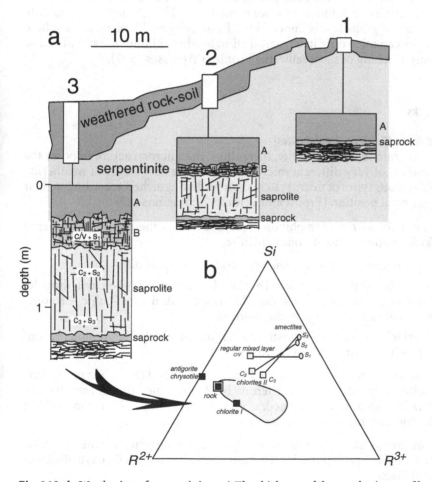

Fig. 6.19a,b. Weathering of serpentinites. **a)** The thickness of the weathering profiles varies with the slope (1: rock summit; 2: slope middle; 3: slope bottom, fractured zone). **b)** Representation of the composition of rock and unweathered minerals (*black squares*) and secondary minerals (*grey squares* and *ovals*) in the system Si–R^{2+}–R^{3+} for profile 3

The saprock becomes porous, thus testifying to the intense dissolution undergone by the serpentinite minerals. A secondary assemblage composed of a Si-rich chlorite and a Mg-saponite appears. This assemblage is destroyed in saprolite (prismatic horizon): saponite disappears whereas nontronite is formed. The composition of the Si-rich secondary chlorite changes with smectites until it has disappeared in saprolite to the benefit of a chlorite-vermiculite regularly ordered mixed layer (Fig. 6.19b). This sequence of mineral transformations is due to the progressive loss of bivalent elements through leaching (Mg^{2+}) or oxidation (Fe^{3+} escaping from silicates to form oxides). The secondary silicates get progressively richer in Al^{3+}, which concentrates in the newly-formed minerals owing to its low solubility. These newly-formed minerals are richer in Si^{4+} than the parent minerals because of the high level of the silica activity in solution (Barnes et al. 1978). The formation of kaolinite – and hence of gibbsite – is impeded by the activity of dissolved silica. These two phases can form only in a tropical climate where abundant rainfalls cause an intense leaching of the weathering profiles (Trescases 1997).

6.1.3.3
Basic Rocks

Grained Rocks (Gabbros, Amphibolites)
Like in all grained rocks (granite, lherzolite), the microstructure imposes the juxtaposition of very different microsystems in the first stages of weathering (Fig. 6.20). Four types of microsystems can be distinguished according to their petrographical position (Proust and Velde 1978; Ildefonse 1980):

- *contact microsystems* (amphibole – plagioclase) in the coherent weathered rock: formation of ferriferous beidellite,

- *internal microsystems of primary minerals* (*alteroplasma*):
  a) Amphiboles: they are replaced by the talc – nontronite – Fe-oxyhydroxides assemblage in the saprock and the trioctahedral vermiculite – Fe-oxyhydroxides assemblage in the saprolite,
  b) Plagioclases: they are transformed into dioctahedral aluminous vermiculite in the saprock and saprolite.

- *Secondary microsystems in the saprolite* (*pedoplasma*): both types of vermiculite recrystallise into ferriferous beidellite. The latter forms the clay matrix in which new trioctahedral vermiculites grow, with a composition differing from that of the previous ones,

- *Fissure microsystems*: cutans coating the fissure walls whatever their position in the profile are formed by the Fe-bearing beidellite+Fe-oxyhydroxides assemblage.

Like for ultrabasic rocks, the mineral reactions are governed by magnesium leaching and Fe oxidisation. The greater amount of aluminium within the rock should favour the formation of Al-bearing secondary minerals such as

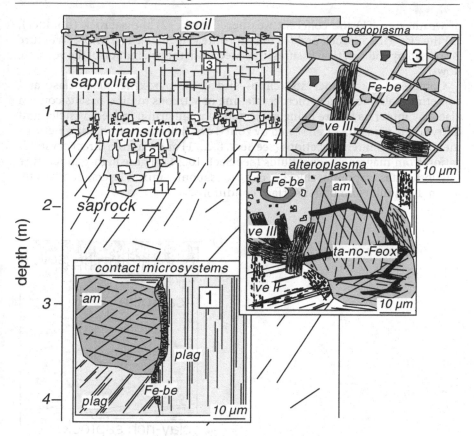

Fig. 6.20. Weathering profile developed on the gabbro from Pallet, Atlantic Loire, France (after Ildefonse 1980). *Insets* represent the microstructure of the saprock (1), the saprock-saprolite transition (2), saprolite (3). *plag*: plagioclase; *am*: amphibole; *ta*: talc; *Fe-be*: ferriferous beidellite; *no*: nontronite; *ve II*: dioctahedral vermiculite; *ve III*: trioctahedral vermiculite; *sm*: mixture of smectites; *Fe-ox*: Fe-oxyhydroxides

chlorites, but this is impeded by the greater activity of dissolved Na^+, K^+ and Ca^{2+}. Vermiculites then ferriferous beidellites are formed. The occurrence of nontronite is only transient in the very first stages of the internal weathering of amphiboles.

Basalts

Basalts form vast surfaces on the continents (trapps from Deccan, Parana, Karoo, Siberia, Columbia River ...). Generally, they are highly weathered and correspond to intensively cultivated areas in tropical climates owing to their fertility. Soils and alterites form a red mantle on the black rocks. Between the fresh rock and the organic soil, the material structure, porosity and chemical composition change according to four facies, which are in increasing order of weathering: (1) the coherent weathered rock, (2) the saprock (boulder), (3) the clay-rich saprock and (4) the saprolite (Fig. 6.21). The whole structure is crossed by wide vertical fissures inherited from the prismation of the volcanic flows.

Composed of phenocrysts (clino- and ortho-pyroxene, plagioclase and sometimes olivine) and a microcrystalline or vitreous matrix, basalts contain also clay minerals before being subjected to weathering. Indeed, vesicles and diktytaxitic voids are often sealed by zoned crystallisations from diagenetic or metamorphic mineral reactions (see Sect. 8.2.2.1) or even from direct crystallisation from magma residual fluids (see Sect. 10.2.1). These high-temperature phyllosilicates (saponite, chlorite, chlorite/saponite mixed layers, nontronite) are in disequilibrium under surface conditions.

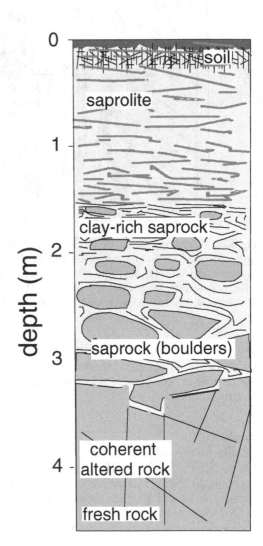

Fig. 6.21. Typical weathering profile on basaltic rocks. The thicknesses of the various horizons vary according to climates

Table 6.2. Mineralogical composition of the different parts of weathering profiles on basalts

Altered level	Olivine	Glass or argillaceous mesostasis	Plagioclase	Pyroxenes	Fractures
Fresh rock	Iddingsite	Celadonite + saponite	–	–	Celadonite + zeolites
Altered coherent rock	Fe-saponite + nontronite	Halloysite + Fe-beidellite	–	–	Celadonite + calcite
Saprock (boulders)	Fe-beidellite + halloysite	Halloysite + Fe-beidellite	Halloysite	Fe-beidellite + goethite	Fe-beidellite + halloysite + goethite
Argillaceous saprock	Goethite + halloysite	Secondary porosity (dissolution)	Halloysite	Fe-beidellite + goethite	halloysite + goethite
Saprolite		Halloysite + gibbsite + Fe–Mn oxyhydroxides			

In the early stages of weathering, the basaltic rocks form a mosaic of microsystems which are more or less independent. Thus, the forming secondary minerals are strongly influenced by the local chemical conditions (Table 6.2). Nevertheless, short-distance transfers of elements probably occur under the effect of local chemical potential gradients. This is particularly well observable in altered olivines: in spite of the fact that they do not contain aluminium, they are replaced by aluminium-bearing clay minerals (saponite, nontronite). Several sequences of mineral transformations can be followed as long as the rock initial structure persists (clay-rich saprock). The tendency to homogenisation of the secondary products whatever their origin (olivine, pyroxene, plagioclase) is related to long-distance transfers of the chemical elements which are evacuated by the fluid flow within an increasingly porous rock. The minerals making up the most weathered horizons (clay-rich saprock, saprolite) are composed of the less soluble elements: Al and Fe^{3+} (Table 6.2).

Glass is rare in basaltic rocks except in their chilled margins that have undergone a sudden quenching. The vitreous matrix is particularly unstable under weathering conditions. It rapidly dissolves (pore) or transforms first into hydrated glass then into very small-sized clay minerals (cryptocrystalline) to finally yield halloysite (Ildefonse 1987). However, massive basaltic rocks do not contain glass. The mesostasis more or less opaque under the polarising microscope generally accepted as being glass is in fact composed of a multiphase assemblage of clay minerals + K-feldspar ± quartz, which was formed early in the last stages of flow cooling (see Sect. 10.2.1). Zeolites formed during diagenetic transformations (see Sect. 8.2.2.1) or hydrothermal alteration stages (see Sect. 9.1.4.2). The meteoric fluids bring about the dissolution and recrystallisation of these early clays.

Weathering causes the Si and Mg depletion and the Al and Fe^{3+} enrichment of basaltic rocks. Saponite and nontronite formed in the very first stages are replaced by a ferriferous beidellite + Fe-oxyhydroxides assemblage. The collapse of the rock structure makes the weathered material homogeneous by mixing the various clay products and the debris of parent minerals. This new environment imposes chemical conditions that bring about the disappearance of some species (feldspars) and the change in beidellite composition. These conditions favour the formation of the ferriferous beidellite + halloysite + Fe–Mn-oxyhydroxide assemblage.

The common point in the weathering of basic and ultrabasic rocks is the Al and Fe^{3+} progressive enrichment of alterites whereas Si and Mg are evacuated by fluids, as indicated by the composition of spring or river waters (Drever 1982). Thus, the chemical potentials of Si and Mg (mobile components) will be the driving forces for the crystallization of secondary phases while the amounts of Al and Fe^{3+} will control their respective quantities. Therefore, the phase diagrams $\mu_{Si,Mg}$–Al–Fe^{3+} enable the major mineral reactions of the weathering process to be represented (Fig. 6.22). Ultrabasic rocks differ essentially from basic ones by their low aluminium, alkaline and alkaline-earth elements contents. Nevertheless, these elements do not influence identically the formation of secondary phases. Indeed, the low aluminium contents of ultrabasic rocks are not an obstacle for the precipitation of Al-bearing phases since chlorite and chlorite/smectite or chlorite/vermiculite regularly ordered mixed layers (structure similar to that of corrensites) are formed. It only controls their relative quantities. On the contrary, these phyllosilicates do not appear in weathering profiles of basic rocks because the chemical potentials of Na and Ca are too high and favour the crystallisation of trioctahedral vermiculites

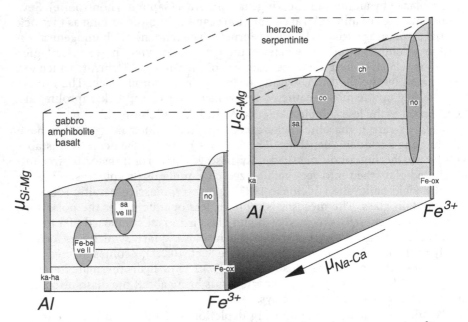

Fig. 6.22. Tentative phase diagram in three-dimensional space: $\mu_{Na,Ca}$–$\mu_{Si,Mg}$–Al–Fe^{3+} corresponding to the weathering of basic and ultrabasic rocks. *ch*: chlorite; *co*: corrensite; *no*: nontronite, *Fe-ox*: Fe-oxyhydroxides; *sa*: saponite; *ve III* and *ve II*: tri- and dioctahedral vermiculites; *Fe-be*: ferriferous beidellite; *ka*: kaolinite; *ha*: halloysite

or high-charge saponites. This difference is increased in microsystems where weathering is more intense. The trioctahedral vermiculites disappear while the ferriferous beidellite – dioctahedral vermiculites – Fe-oxyhydroxides assemblage forms in saprolite. Saprolites in ultrabasic rocks are characterised by the nontronite+Fe-oxyhydroxides assemblage because aluminium is too scarce to produce ferriferous beidellites or dioctahedral vermiculites.

6.1.3.4
Sedimentary and Metamorphic Rocks with Phyllosilicates

Glauconite (Limestones and Sandstones)
The glauconitic rocks are mainly limestones or sandstones. Weathering produces the same effects in both cases; only the thickness of the profile varies (Courbe et al. 1981; Lovel 1981). The greater the drainage of the environment (high porosity, slopes or hill tops), the deeper the profiles. Glauconites are potassium dioctahedral phyllosilicates that generally come in the form of very well-calibrated rounded grains, 75% of which have a diameter contained between 100 and 200 µm (see Sect. 7.2.3). In the first stages of weathering, the external boundaries of these grains become fuzzy and a greenish alteroplasma forms. The latter becomes increasingly red as weathering gets more intense. Electron microprobe analyses show that Fe and K contents gradually decrease in clay minerals (white arrow, Fig. 6.23). Potassium is leached out from the rock by percolating waters (long distance transfer) while the Fe goes out of phyllosilicates (short distance transfer) to precipitate as an independent phase (oxyhydroxides).

Glauconite progressively transforms into iron-rich illite/smectite mixed layers then into smectites. Smectites close to montmorillonites seem to co-exist with nontronites. In upper horizons, cutans bordering fissures are made up by a kaolinite + Fe-oxyhydroxides assemblage. All mineral reactions can be represented in the system μ_K–Al–Fe. The more intense the weathering, the lower the μ_K and the less available Fe for phyllosilicates. The final stage in temperate climates is then the kaolinite + Fe oxyhydroxides assemblage. The comparison of the phase diagram inferred from weathering observations with the one representing glauconitisation (Fig. 7.15) is interesting. The weathering sequence is opposite to that of the sediment glauconitisation (Hower 1961; Velde 1976) indicating that the formation of glauconite is a reversible process at Earth surface conditions.

Illite (Marl)
The weathering of a low-permeability Toarcian marl has made possible the observation of the early stages of the diagenetic illite transformation under surface conditions (Lafon and Meunier 1982). The fresh rock mainly composed of the illite + chlorite + pyrite + dolomite assemblage is black. Small amounts of large-sized quartz and detrital micas are observed. The rock becomes ochreous (pyrite oxidation) without disturbance of the sedimentary structures. Carbonates are dissolved and porosity increases. Chlorite and il-

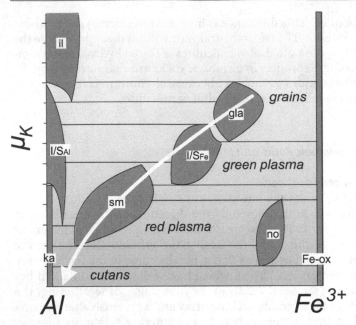

Fig. 6.23. Weathering of glauconites in the system μ_K–Al–Fe^{3+}. The *arrow* is plotted on the electron microprobe point compositions of the grains, the green and red weathering plasmas and the cutans coating fissures. *gla*: glauconite; *il*: illite; I/S_{Al}, I/S_{Fe}: aluminous and ferriferous illite/smectite mixed layers, respectively; *sm* and *no*: smectites close to montmorillonite and nontronite respectively; *ka*: kaolinite; *Fe-ox*: Fe-oxyhydroxides

lite disappear while a new assemblage is formed: illite/smectite mixed layers + kaolinite + Fe-oxides. Owing to the very low permeability of the fresh rock, the weathered zones are thin.

The invasion of the rock by weathering fluids brings about pyrite oxidisation and pH depression in microsystems. Carbonates and chlorites are dissolved under the imposed acidic conditions. Since illites show a greater specific surface than detrital micas, they react more rapidly by losing potassium. The K- and Al-depleted layers become expandable. Illite transforms into illite/smectite mixed layers and the released aluminium is consumed in the kaolinite crystallisation. Fe-oxidisation is accordingly the driving force of these chain reactions in acidic environments (Fig. 6.24).

Chlorite and Micas in Gossan (Micaschists)

The particular interest of the gossan system lies in the fact that it imposes chemical conditions that are totally different from those of common alterites or soils. The selected example is the Rouez gossan (Sarthe, France) that has developed on a huge lens of sulphides (mainly pyrite) and has been exploited for its precious metal concentrations of high economic value (Bouchet 1987). The sulphides are embanked in a quartz-rich brecciated gangue intersecting

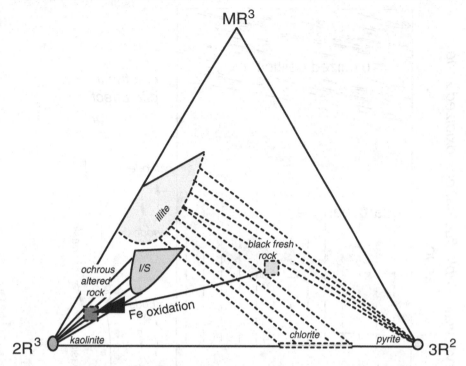

Fig. 6.24. Representation in the system MR^3 µm–µm $2R^3$ µm–µm $3R^2$ of the weathering of a Toarcian marl (after Laffon and Meunier 1982). The passage from the diagenetic assemblage to the weathering assemblage is controlled by the Fe-oxidation and the potassium loss

pelitic rocks (or schists) at the anchizone-epizone boundary. They are composed of chlorites and micas intermediate between illite and phengite. The chemical compositions of these phases change within a metasomatic aureole developed about the sulphide body. Notably, the Fe content of chlorites increases significantly (Fig. 6.25).

The gossan structure comprises from top to bottom:

– an oxidised zone: kaolinite + Fe-rich chlorite/saponite mixed layers;

– an intermediate zone: several types of talc/smectite mixed layers are formed according to the composition of smectite: ferrous smectite, stevensite, saponite.

– a cemented zone: sepiolite + talc/smectite + chlorite/saponite + kaolinite.

– a cemented pelitic zone: kaolinite + sudoite/smectite mixed layers.

The fissure walls in the whole weathering profile are coated with Fe-rich amorphous silicates.

The weathering of sulphides is controlled by oxidation-reduction cells (Thornber 1975, 1983). Electron exchanges take place between a conductor (sulphides) and an electrolyte (weathering solutions). Cathodes are located in zones where

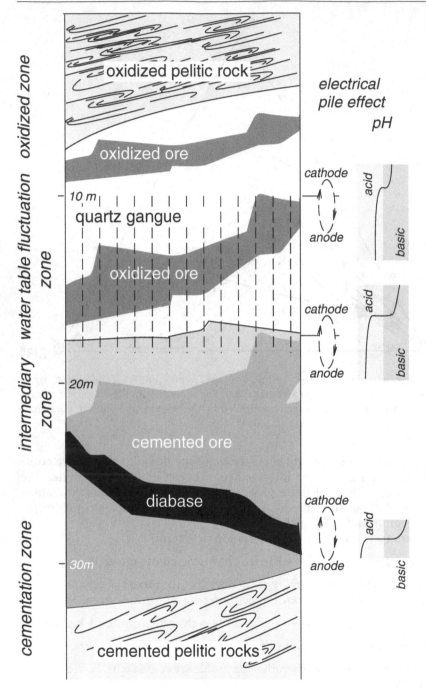

Fig. 6.25. Gossan developed on the Rouez sulphide body, Sarthe, France (after Bouchet 1987). Oxidation causes the formation of electric cells through the exchange of electrons: $Fe^{2+} \rightarrow Fe^{3+} + e^-$ and produces local pH gradients (Thornber et al. 1981)

the oxygen dissolved in water is reduced (generally above the hydrostatic level) according to the following reaction:

$$O_2 + 2H_2O + 4e^- \rightarrow 4(OH)^- \tag{6.13}$$

This reaction is the starting point of oxidation. It shows the relationships between $(OH)^-$ concentration and electron exchanges. These relationships are readily calculated and classically presented in the form of Eh-pH diagrams (Fig. 6.26). Anodes are located under the water table. Sulphides are transformed into sulphates and locally change the pH value:

– metal/sulphur ratio < 1: pH becomes acidic

$$FeS_2 + 3.5O_2 + H_2O \rightarrow Fe^{2+} + 2SO_4^{2-} + 2H^+ \tag{6.14}$$

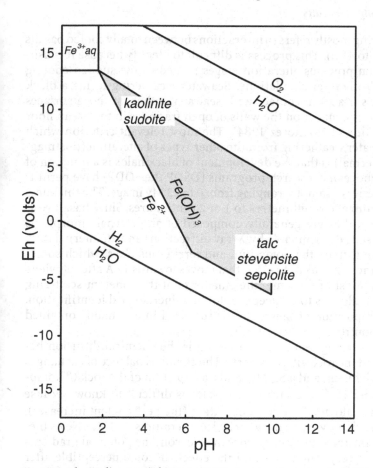

Fig. 6.26. Eh-pH diagram (after Sato 1960) showing the natural condition domain (*shaded area*). The main clay species making up the Rouez gossan are indicated in the zones corresponding to their formation conditions

– metal/sulphur ratio > 1: pH becomes basic

$$Cu_2S + 2.5O_2 + H_2O \rightarrow 2Cu^{2+} + SO_4^{2-} + 2(OH)^- \tag{6.15}$$

Hydroxides are formed: $Fe^{2+} + 2\,H_2O \rightarrow FeOOH + 3H^+ + e^-$. The effects on pH stop when oxidation is complete.

Several cell systems are observed in the Rouez gossan. Some of them are now fossil systems (oxidised zone) while others are active in the zone of water table fluctuation or even in the cemented zone at the contact between diabase and ore. Local variations in pH explain the juxtaposition of clay minerals of differing compositions such as kaolinite and sudoite on the one hand and talc, stevensite, sepiolite on the other hand.

6.1.3.5
Seawater Weathering of Basalts

Seawater weathering mostly refers to interactions between totally cooled basalts and seawater (−5 to 4 °C). This process is difficult to identify because it is usually superposed on previous alteration stages related to the sudden cooling of lava or to hydrothermal circulations. Seawater weathering forms a black halo on the rock surfaces in contact with seawater, either on those surfaces not coated with sediments or on the walls of open fractures in the ocean floor (Andrews 1977; Alt and Honnorez 1984). The most relevant criterion which differentiate seawater weathering from the other types of alteration (postmagmatic or hydrothermal) is that the development of black halos is a function of the basalt age. Suboceanic research programs (DSDP then ODP) have permitted the study of series of samples ranging from 0 to 23 My in age. The thickness of halos ranges from a few millimetres to 1 or 2 centimetres. Since basalt vesicles remain empty, they are generally composed of an external amorphous zone (no clay crystals determinable by X-ray diffraction) and an internal dark zone in which are formed the potassium and ferriferous clays, which sometimes yield a diffraction peak at 14–15 Å that moves towards 17 Å after ethylene glycol saturation. Most of the time, the small size of the coherent scattering domains does not allow us to go deeper into the mineralogical identification. The chemical composition of these clays is scattered in a domain contained between that of nontronite and celadonite.

The process governing seawater weathering is the chemical diffusion between seawater and the rock vitreous parts. The chemical balance of exchanges is K and H_2O enrichment and a Si, Mg, and Ca depletion of the rock (Thompson 1973). Whether Fe^{2+} migrates to the water is difficult to know because its decrease in the halo may be due to oxidation (the Fe^{3+} content increases). Diffusion rates are slow owing to the very low temperatures (< 4 °C). Nevertheless, considering the immensity of the ocean reservoir, the potential gradients are maintained for very long times and the effects become perceptible after 3 to 6 million years (Hart 1970; Staudigel et al. 1981). The formation of these K-nontronites buffers the K^+ content of seawater (Bloch and Bischoff 1979) by consuming the inputs of those streams draining the continents.

6.2
Soils

Introduction

Soils are loose rocks in which biological (plants, animals) and microbiological activities trigger two types of reactions altering the organic as well as the mineral matter. The dissolution of the minerals of the constituent rocks of the substratum is influenced by the conditions imposed by the wetting processes (Eh-pH-P_{CO_2}). Each one of the dissolved minerals forms a microsystem in which clays occur either through genuine neogenesis or through the transformation of pre-existing phyllosilicates (micas, chlorites). An overall presentation of the relationships between the microstructure, clay minerals and the geochemical behaviour of soils can be found in some general papers (Pédro 1989–1993; Lynn et al. 2002) or books (Brady and Weil 1999 among others). The purpose of this section is first to present the major mechanisms involved, then to give the mineralogical characteristics of the main types of soils. The soil nomenclature used here is that recommended by the Soil Survey Staff (1998–1999).

6.2.1
Clays in Soils

6.2.1.1
Soil Forming Processes

Biological Factors
In a soil, vegetation adds organic matter to the mineral components thus forming a litter whose mass varies with latitude (Fig. 6.27a). The amounts of organic matter in the soil then depends in the first place on the abundance of litters but

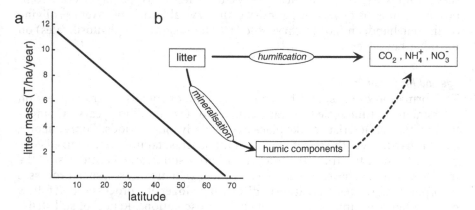

Fig. 6.27a,b. Organic matter in soils. **a)** Variation in the amount of litter produced by vegetation as a function of latitude. **b)** Transformation process of organic matter in soils

also on the resistance of plant debris to biodegradation. Indeed, microorganisms, through their own biological activity, transform the litters into CO_2, NH_4^+ and NO_3^- (mineralisation). The organic residues become humic components (humification). Humification is a slow and complex process in which both soil microorganisms and non-biological physico-chemical reactions take part. The amount of organic matter in the soil is influenced by the difference in velocity between mineralisation and humification processes.

Humification is a complex process leading to the degradation of lignin and to the synthesis of aromatic compounds (Fig. 6.27b). The action of organic acids (humic and fulvic) is double: they provide protons and they are complexing agents of metals, Al and Fe notably. Both actions are very important in the weathering of primary minerals and in the transport of dissolved elements.

Mineralogical Composition of the Parent Rock

In temperate climates, the crystallochemical nature of rock has an obvious effect on the soil nature, as shown by the rather accurate superimposition of the large-scale soil distribution maps ($1/100,000^{th}$) upon geological maps. The soils developed over ancient massifs composed of magmatic and metamorphic rocks stand out against those formed over sedimentary rocks. The soil map of Fontenay-le-Comte for instance (Ducloux 1989) is framed by the metamorphic basement of the Armorican Massif to the North, the Quaternary sediments of the Marais Poitevin to the south and the Jurassic calcareous plain to the centre. In tropical or equatorial climates, soils are very thick and mainly formed of kaolinite or halloysite in combination with Fe-oxides. The initial composition of the parent rock is not as determining as in temperate areas except for trace element distribution. The rock is of decisive importance in thin, relatively recent soils only; this influence decreases in thick and ancient soils, where weathering is intense.

The rock microstructure is a factor as significant as its chemical composition: the soils developed on lherzolites are different from those developed on serpentinites. Although chemically very close, these two types of rocks are composed of minerals of very different size and crystal lattice: pyroxenes+olivine on the one hand, antigorite-chrysotile (1:1 ferromagnesian phyllosilicates) on the other hand (see Sect. 6.1.3.2).

Age and History of Soil

The formation of soils, and hence of clay components, is a slow process on the human life time scale (several centuries to several million years). Changes in climatic characteristics take place over equally long periods. Consequently, a soil is hardly ever the result of a constant process, taking place in fixed conditions. Most of the time, the interpretation of soil characteristics using the influence of actual processes alone is erroneous. In most cases, such processes incorporate the superimposition of different pedogenetic processes that follow one another over time at the rate of climatic oscillations. Relics of soil structures typical of hot climates are frequently found reworked by a pedogenetic process in cold climates.

The soil structure itself is not the result of a single process taking place at a constant rate. Accordingly, incorporation of organic residues into the surface horizon requires a few tens of years whereas the effects produced by the leaching of clays into deep horizons are visible only after several millennia. The formation of a kaolinitic horizon several meters thick, like in Brazil or Africa, requires at least one million years (Tardy and Roquin 1992). The most ancient geomorphological surfaces in West Africa date from the middle of the Tertiary Period. The soils developed on these surfaces are very ancient and are deeply eroded.

6.2.1.2
Kinetics of the Mineral Reactions in Soils

A few studies have attempted to measure durations for the transformation of clays in soils. Such durations not only depend on the kinetics of the mineral reactions involved but also on the rate at which the fluids come in contact with these minerals individually. This rate is controlled by climatic factors, seasonal variations and the microstructural characteristics of the materials determining the soil permeability. Measuring a duration is difficult, if not impossible, except in some favourable cases where temporal reference points permit measurement of time. These soils are then referred to as forming a chronosequence.

Chronosequence of Clay-Rich Soils in the Marais Poitevin (Western France)
The study of a time sequence of clay soils (over 60% clay) developed in similar pedogenetic conditions has been made possible by the conquest of seven polders over the sea from 1665 to 1912 (Sodic Entisols – Aquic Entisols). The most ancient soil has been developing for 330 years, the most recent for 80 years. The < 0.1 μm clay fraction extracted from soil samples collected at a depth of 36 cm is the most reactive. The potassium of the micaceous phases has been transferred to the expandable phases: smectite+mica → illite + mixed-layer minerals. The energy necessary for this reaction is provided by the wetting-drying cycles, the effects of which are assumed to remain the same at constant depth. The progress of the reaction is not a linear function of time (Fig. 6.28). At first rapid, it tends towards equilibrium after 200 years.

Chronosequence of Podzols (Spodosols) on Till Deposits
Two examples are selected in post-glacial environments. In Finland, soil age is easily given by the present day altitude above sea level because the isostatic upwelling rate of the Baltic Shield after the last glaciation is perfectly known. A chronosequence of podzols developed on highly homogeneous tills (ground moraines) has been studied for ages ranging from 6,600 and 10,000 years bp (Gillot et al. 1999). The clay reactions take place essentially in the E horizon (eluvial horizon) of those podzols over 6,600 years in age. The more ancient the soil, the more significant the Al, Fe^{3+} and Mg depletion brought about by leaching (Fig. 6.29a). At the same time, the amount of mixed-layer minerals decreases with time until only smectite remains in the most ancient soil

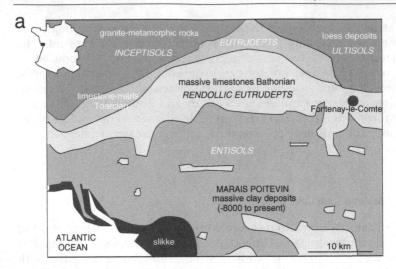

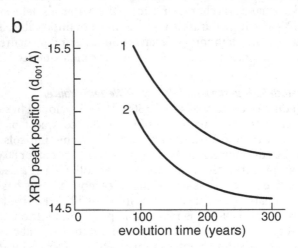

Fig. 6.28a,b. Soil-geological formation dependence. **a)** Simplified pedological soil map from the 1/100,000*th* Fontenay-le-Comte map (Ducloux 1989). **b)** Sodic and aquic entisols from the Marais Poitevin (western France). Migration of the XRD peak position of the Ca-saturated and air-dried < 2 μm fraction as a function of time: smectite (1), illite/smectite mixed layers (2)

(10,000 years). The smectites formed are dioctahedral with a variable tetrahedral charge (beidellites): high charge and low charge coexist in each sample. High-charge layers disappear as a function of time (Fig. 6.29b).

The age of the soils developed on the Mont-Blanc Massif moraines ranges from 80 to 6,600 years (Righi et al. 1999). The podzolic features have progressively developed over this period. While mineral transformations are very

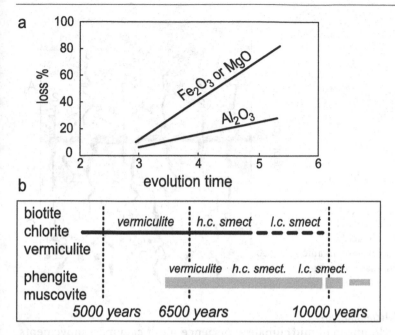

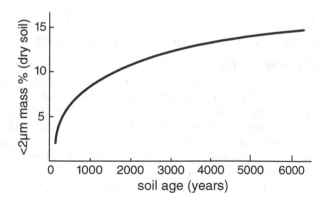

Fig. 6.29a,b. Mineralogical and chemical evolution of E horizons of podzols (spodosols) from a chronosequence in Finland as a function of the evolution duration (10^3 y). **a)** Chemical depletion. **b)** Progressive weathering of the trioctahedral then dioctahedral phases into high-charge smectites (*h. c. smect*) and low-charge smectites (*l. c. smect*)

discrete in accumulation horizons (B horizons), they are by contrast intense in eluvial horizons (E horizons). The trioctahedral primary phases (biotite + chlorite) have disappeared within 1,000 years, yielding mica/smectite dioctahedral mixed layers. The weathering of the dioctahedral primary phases has then taken over, yielding mica/smectite regularly ordered mixed layers. These mineral reactions have caused a logarithmic increase in the soil clay mass as a function of time (Fig. 6.30).

Fig. 6.30. Logarithmic increase of the clay mass as a function of time in soils developed on the Mont-Blanc massif moraines

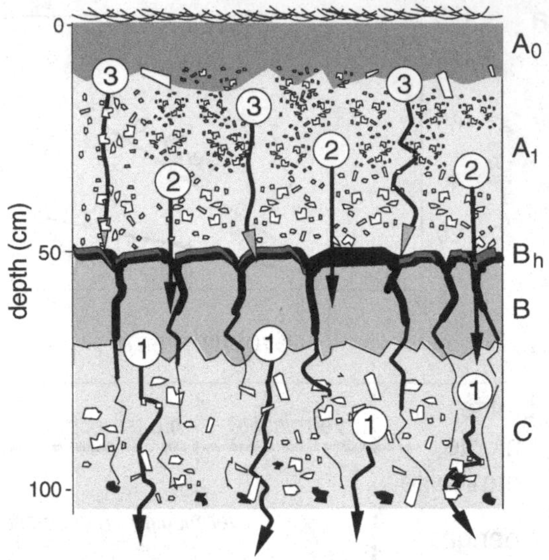

Fig. 6.31. Transfers within a soil.
1) Leaching of soluble ions:
Na^+, Ca^{2+}, Mg^{2+}, NO_3^-, HCO_3^-.
2) Transfer of clay particles. 3)
Transfer of organic acids. *Bh*:
accumulation horizon of organic
matter. *B*: accumulation horizon
of clays

Transfers and Accumulations

The soils developing in humid climate experience significant mass movements that take place at various rates (Fig. 6.31). The most immediate and rapid process is the leaching of cations (Na^+, Ca^{2+}, Mg^{2+}) that brings about the decrease in the soil solution concentration, which in turn causes the desaturation of clay minerals. Clays are dispersed through solutions then dragged down to the deep zones of the profile where they accumulate (B horizon). They may migrate laterally if topography shows slopes steep enough to cause lateral flow. These transfers of clay particles are described under the term illuviation.

The organic matter produced by humification is fixed on the surface of clays. Attached to its support, the organic matter moves only if clays move themselves. Nevertheless, in acid soils such as podzols for instance, some organic compounds may migrate as solutions or as colloids. Generally, these compounds are organic acids complexing metals, notably Al and Fe. They accumulate in depth forming a Bh horizon.

6.2.1.3
Example: Acid Soils in Temperate Climate (Inceptisols)

Mineralogical Evolution

Acid soils develop in temperate climate in areas composed of magmatic (granites) or metamorphic (gneiss) quartzo-feldspathic rocks. The example chosen here is a soil developed on weathered granitic rocks in the area of the Millevaches plateau in France (Righi and Meunier 1991). This soil (pH = 6) comprises four horizons from top to bottom: the A1 horizon rich in organic matter, the partially leached A2 horizon, the clay-rich Bw horizon and the C horizon forming a transition with the weathered granite. The C-horizon is in great part formed of the minerals inherited from the underlying granitic

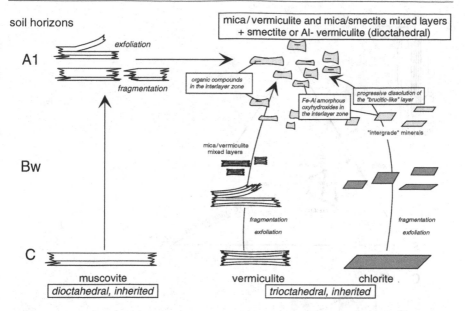

Fig. 6.32. Schematic representation of the transformation of phyllosilicates inherited from the weathered granitic rock in the A1, Bw and C horizons of a brown acid soil (dystic inceptisols). Fragmentation takes place either by chemical weathering or by mechanical deformation. At the same time, the inherited phyllosilicates transform either into mica/smectite or mica/vermiculite mixed layers or into intergrade minerals in which the interlayer sheets are blocked by deposits of organic compounds or of Fe–Al-oxyhydroxides

rock: feldspars, quartz and primary or secondary phyllosilicates (muscovite, chlorite and vermiculite). Feldspars are dissolved and are not found in upper horizons. Particle-size analyses show that the fraction of fine clays ($< 0.2\,\mu m$) has increased by more than 60%. The amount of fine (1 to $0.2\,\mu m$) and very fine ($< 0.2\,\mu m$) particles increase from the C-horizon to the A1 horizon. This attests to the formation of clay minerals in the soil. The inherited coarse-grained minerals are fragmented and can be found in the very fine grain size fractions. At the same time, they are transformed into a complex assemblage of mixed-layer and intergrade minerals. Organic compounds or deposits of Fe-Al-oxyhydroxides settle in the interlayer sheets (Fig. 6.32).

The phyllosilicates inherited from the weathered granite (muscovite, vermiculite and chlorite) are progressively transformed into minerals having increased expandability (smectite and vermiculite). The electron microscope observations show that the peripheral part of the partially chloritised biotite coarse-grained particles is widely transformed (Fig. 6.33a). One portion looks like an amorphous silicate and seems to be the cause of the increase in the intensity of the fluorescence domes observed in the X-ray diffraction patterns of these particles. The smaller these particles, the higher the transformation rate. Transformation occurs by elimination of the bivalent elements of the crystal lattice: oxidation of Fe^{2+} into Fe^{3+} and Mg solubilisation (Fig. 6.33b).

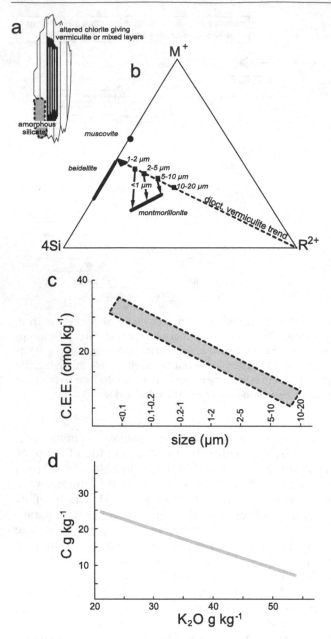

Fig. 6.33a–d. Transformation of the phyllosilicates inherited from granite and new formation of smectite. **a)** Photograph-derived representation of a chloritised biotite particle observed by transmission electron microscopy. **b)** Chemical analyses of the coarse soil fractions extracted from the C horizon (*squares*) and of the fine and ultra fine fractions extracted from the A horizons (*arrows*). **c)** Relationship between the exchange capacity (CEC) and the size of the clay fractions. **d)** Relationship between the fixed organic carbon content and the K$_2$O content

The very fine grain size fractions (0.1–0.2 μm and < 0.1 μm) incorporate swelling minerals whose d-spacing reaches 16.6 Å after glycol saturation. These are smectite-rich mixed-layer minerals that crystallise in the environment of the dissolving large particles. Their crystal lattice cannot form directly from the inherited phyllosilicates by loss of bivalent cations. Their tetrahedral and octahedral sheets, interlayer charge and bulk chemical composition clearly differ from those of the vermiculites resulting from the transformation of the large particles. They are probably the result of a genuine new formation. Their partial expandability (16.6 instead of 17 Å) shall be explained later. Each dissolving large particle form a single microsystem that controls the chemical composition of the "montmorillonitic" clays.

Evolution of the Chemical Properties of Clays
The cation exchange capacity regularly increases from 8.6 to 34.4 cmol kg^{-1} as the size of the analysed clay fractions decreases (Fig. 6.33c). The CEC varies in inverse proportion to the K$_2$O content of these various fractions. This shows that the larger the fractions, the greater their unweathered mica content. These results are confirmed by X-ray diffraction when the intensity of the mica diffraction peaks is compared with that of weathered minerals (vermiculite and chlorite).

Smectite- and vermiculite-like minerals in this acid soil have a strong tendency to show total or partial expandability after ethylene glycol saturation or heating at 300 °C. This is classically interpreted as the characteristic of intergrade minerals, namely those having portions of brucite- or gibbsite-like sheets in their interlayer zone. These discontinuous sheets are dissolved by a sodium citrate treatment. When applied to the soil studied here, this treatment has revealed a noteworthy difference between the minerals with an intergrade tendency from the A horizon and those from the B-horizon. The first incorporate Fe-Al-oxyhydroxides in their interlayer sheet whereas the others contain brucite-like sheet residues from the chlorites they replace. Consequently, the first are the result of a precipitation process of these Fe-Al-oxyhydroxides in the interlayer zone of expandable minerals while the others bear witness to an unachieved weathering process of the structure inherited from the parent chlorite.

The 1–2, 0.2–1, 0.1–0.2 and < 0.1 μm soil clay fractions show noteworthy organic carbon contents after hydrogen peroxide treatment. Figure 6.33d shows that the organic carbon content is inversely proportional to the K$_2$O content, namely to the amount of mica present in the fraction being analysed. The organic carbon content is a direct function of the amount of expandable minerals. The organic carbon is strongly fixed in the interlayer zones of these minerals and hence escapes the classical chemical extractions.

6.2.2
Soils in Cold or Temperate Climates

6.2.2.1
Podzol – Cambisols (Spodosol – Inceptisol)

In temperate climate, podzolization never affects all surfaces exposed to pedo-genesis. This process is only triggered in favourable sites in quartz-rich sandy rocks. Podzols are generally combined with acid cambisols, both being char-acterised by differing associations of clay minerals. These soils are organised in sequences in the landscape. All pedogenetic factors are identical with the exception of the organic matter behaviour. Indeed, podzols are characterised by high complexing organic acid contents.

A sequence has been thoroughly studied in the Medoc area (south-west France). Righi et al. (1988) have shown that for a same parent rock composed of quartz, mica, vermiculite and chlorite/vermiculite mixed layers, clay minerals in cambisols (acid brown soils) are clearly distinct from those in podzols (Fig. 6.34). In the first, micas are transformed into intergrade vermiculite. In the others, micas are transformed into smectite. This difference is due to the fact that Al and Fe are highly complexed with the organic matter in podzols and are not available to precipitate in the form of oxyhydroxides in the interlayer sheets of the expandable minerals.

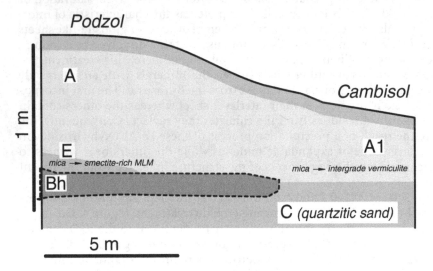

Fig. 6.34. Sequence from podzol (spodosol) to cambisol (acid brown soil; dystic inceptisol) observed in the Médoc area, south-west France (Righi et al. 1988). Smectite accumulates in the podzol E horizon whereas intergrade minerals are formed in cambisols

6.2.2.2
Leached Loess-Soils (Ultisols)

The loess is an accumulation of wind-laid particles predominantly of silt size (2–60 µm). Distributed essentially in the Northern hemisphere, they cover about 10% of the continental surfaces. They were formed during the Quaternary Period in periglacial areas. This explains the weak weathered state of their constituent minerals. Micas and chlorites are often intact. They are combined with kaolinite and smectites arising from ancient soils disintegrated by erosion. Most of the time, loess contains a lot of calcite.

The first stage of pedogenesis is marked by the dissolution of calcium carbonates. The dissolved elements are leached out of the soils. The loess becomes darker and divides into polyhedra. Mica and chlorite crystals experience a beginning of mechanical fragmentation attested by their occurrence in fine grain size fractions. Decarbonatation brings about the suspension of clays by lowering the Ca^{2+} concentration of soil solutions. Suspended clays are moved and redeposited lower, forming typical zoned deposits: cutans (Fig. 6.35a). This process (illuviation) modifies the distribution of clay fractions in the soil by depletion of upper horizons and enrichment of deep horizons (Fig. 6.35b).

The later stages of pedogenesis bring about very significant changes in the soil structure. Thick clay accumulation horizons are formed in depth by the intense illuviation. These horizons form impervious layers retaining a temporary perched water table in the wet season, the level variations of which control the oxidation-reduction conditions. Iron reduced to Fe^{2+} is solubilized and moves with solutions. It precipitates in the form of nodules fixing Fe^{3+} ions when oxidation conditions make it possible in the dry season. Subsequently, those accumulation horizons that have reached a threshold thickness begin to divide into columns or prisms. The outlining fractures make up preferred drains for the solutions. Desaturation causes dispersion of accumulated clays. The eluvial

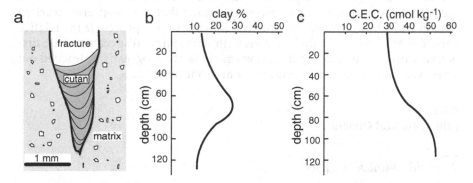

Fig. 6.35a–c. Characteristics of leached loess-soils. a) Cutan: zoned deposit of clay particles suspended in soil solutions and redeposited at the bottom of fractures. b) Distribution curves of the clay fraction (< 2 µm) in leached soils. c) Variations in the CEC as a function of depth

horizon is disintegrating. Spectacular profiles show leaching "tongues" separating clay blocks (glossic structure). The suspended clays accumulate in turn at the bottom of these "tongues".

Clay minerals are slightly transformed during their migration in the soil. The only noteworthy change is the progressive desaturation of swelling species that lose their interlayer adsorbed cations (Ca^{2+}, Mg^{2+}). The latter are replaced by aluminous polycations that polymerise in the form of hydroxides in the interlayer zones. The expandable phyllosilicates are then partially transformed into intergrade minerals. An accurate study of the mineralogy of those smectites contained in the leached soils developed on moraines has shown that beidellite concentrates to the expense of montmorillonite in E eluviation horizons (Spiers et al. 1986). The exchange capacity decreases near the surface (Fig. 6.35c). Using the chemistry of oxygen isotopes, these authors have shown that the observed enrichment was not resulting from the new formation of beidellite but rather from the preferred leaching of montmorillonite.

6.2.2.3
Clay-Rich Soils (Entisols)

Soils developing on clay-rich sedimentary rocks are themselves clay-rich soils and exhibit a particular structure. During the dry season, their upper part is heavily crackled. Gaping shrinkage cracks due to desiccation form a connected network, isolating prisms that are themselves crackled by smaller aperture fissures (Fig. 5.7a). This structure controls water pathways and soil macropermeability. Very fine clay particles ($< 0.1 \, \mu m$) are leached out from the soil. Losses of the order of $400 \, kg \, ha^{-1}$ have been measured (Nguyen Kha et al. 1976). Owing to these selective transfers, the structure changes with seasons. During winter and early spring, rainwaters cannot penetrate in the soil and run off at the surface. By contrast, during the dry season (summer and fall), the wide fractures opening from the surface make up a connected network thanks to which rainwaters penetrate in the upper part of the soil. Fine clays are then dispersed, forming stable suspensions that are themselves leached out towards brooks and rivers. The consequence of this process is the relative enrichment of the upper soil horizons with those mineral species that are dispersed with difficulty. Smectite and vermiculite are progressively leached out whereas kaolinite, mica and fine quartz are concentrated.

6.2.3
Soils in Tropical Climate

6.2.3.1
Worldwide Distribution of Soils

Soils in tropical climates are generally very thick. The mineral reactions (weathering of primary minerals, formation of secondary minerals) are pushed to a more intense degree than in temperate areas, owing to the joint action of two factors:

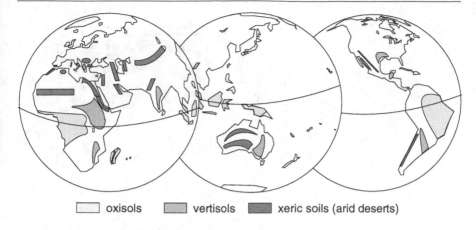

oxisols   vertisols   xeric soils (arid deserts)

Fig. 6.36. Schematic representation of the distribution of oxisols, vertisols and xeric soils on Earth

1. high mean temperatures;

2. very long evolution durations; some surfaces date from the Tertiary. Indeed, these regions have not been planed by glaciers that, after melting, have uncovered surfaces freed from their weathering products in the cold temperate zones (Europe, North America and Northern Asia).

The crystallisation of weathering minerals (clays and Fe–Al-oxyhydroxides) is controlled by climatic parameters, and notably by rainfall distribution and intensity during the year. Abundant and steady rainfall all year round imposes diluted soil solutions and favours the formation of kaolinite (oxisols). Light rainfall followed by periods of intense evaporation lead to soil solutions with a high Si, Ca and Mg content that favour the crystallisation of smectites (vertisols, xeric soils). These different soil types mark the pluvial and arid regions in the intertropical space (Fig. 6.36).

6.2.3.2
Oxisols (Laterites) in Wet Equatorial Zones

The lateritic soil profiles (oxisols) not affected by erosional truncation are organised classically into three zones from the weathered rock at the bottom up to the surface organic horizon, as shown in Fig. 6.37. (Bocquier et al. 1984; Muller and Bocquier 1986; Tardy 1993).

1. *Lower zone (A).* The weathered rock at the bottom of the profile retains the initial structure of the parent rock. The primary minerals are weathered to a greater or lesser extent and replaced by highly crystallised kaolinite, hematite and goethite. This horizon is friable and porous. In its upper part, it gradually passes to a material composed of debris of the weathered rock

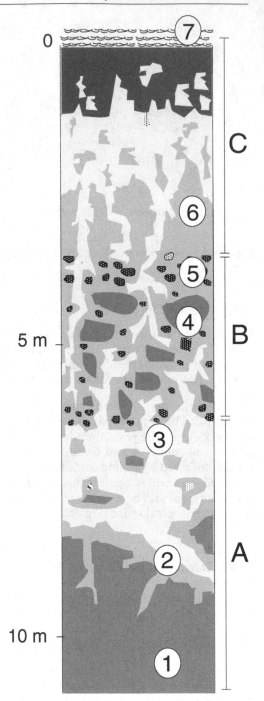

Fig. 6.37. Schematic representation of a lateritic soil profile (oxisol), after Muller and Bocquier (1986). *A*, *B* and *C* indicate the major pedological horizons. 1: saprolite; 2: compact red matrix; 3: friable yellow matrix; 4: ferruginous lithorelict; 5: clay-rich ferruginous nodules; 6: compact yellow matrix; 7: accumulation zone of organic matter

embedded in a red matrix, a mixture of poorly-crystallised kaolinite and Fe-oxyhydroxides. This intensively red zone (prevailing hematite) gradually passes to the yellow intermediate zone (prevailing goethite).

2. *Intermediate zone (B)*. This zone is formed by a nodular horizon. The large-sized nodules (20–80 mm) are ferruginous relicts in which the parent rock structure is retained. They are composed of large-sized kaolinite and hematite. These nodules are numerous in the central part of the interme- diate zone. Their border is not well defined and forms a gradual transition with the enclosing hematite matrix. The small-sized nodules (< 20 mm) are round-shaped and are composed of microcrystalline kaolinite combined with hematite. The initial structure of the parent rock has totally disap- peared. These nodules are the most abundant and are often concentrated in the upper part of the intermediate zone. They are dispersed in a hematite matrix similar to that of the large-sized nodules. Nevertheless, this matrix becomes abundant and highly compact.

3. *Upper zone (C)*. Nodules are scarce and the red hematite matrix is gradually replaced by a goethite yellow one. The replacement is complete 1 m from the surface. The upper part of this zone is porous and impregnated with organic matter.

The major mineral phases forming in oxisols are kaolinite, hematite and goethite, as well as gibbsite in the nodular level of the intermediate zone. This can be explained by the intense leaching of Si^{4+}, Mg^{2+}, Ca^{2+} and Na^+ ions, which reduces the soil-solution interactions to the chemical system Fe_2O_3–Al_2O_3–SiO_2–H_2O (Trolard and Tardy 1989). The range of the Fe–Al solid solutions of hematite and goethite depends on water activity (Fig. 3.8). A part of these oxyhydroxides is fixed on the surface of kaolinite crystals. The aggregates so formed behave practically as a silt. Surface properties are annihilated and the CEC is reduced. Kaolinite properties are partially restored by Fe depletion treatments. In nature, the destruction of oxyhydroxides is triggered either by Fe-complexing organic compounds or by water logging due to the perched water tables that impose reducing conditions.

6.2.3.3
Vertisols

These soils develop in tropical climates with contrasted seasons that are char- acterised by a dry period lasting 4 to 8 months. The small amount of rainfall imposes a very limited soil drainage, and hence a moderate leaching of the soluble Ca^{2+} and Mg^{2+} ions. The latter concentrate in the pore solutions by evaporation of water (solvent elimination; see Fig. 1.19). High concentrations and pH values close to neutrality are conditions favourable to the crystallisa- tion of smectites. Topographically, vertisols cover the surfaces of Ca^{2+}- and Mg^{2+}-rich rocks (basalts, shales, limestones or volcanic ashes) or develop in the lower portions of slopes where ions accumulate by height leaching. They are particularly observed in India, Australia or Sudan (Fig. 6.36).

The chemical composition of smectites in vertisols is variable and is influenced by that of the parent rock, notably as regards Fe contents. Generally, while the compositions of smectitic minerals are scattered between the end members montmorillonite – beidellite – nontronite, the great majority belongs to the domain of ferriferous beidellites (Wilson 1987; Bradaoui and Bloom 1990). The Fe content of clays is proportional to that of the parent rocks.

Smectites in vertisols are poorly ordered. The X-ray diffraction patterns reveal the heterogeneity of their structure. Some of these smectites show an expansion up to 18 to 20 Å after ethylene glycol saturation. These unusually high d-spacings are classically attributed to the presence of organic matter between layers. They could be due also to an interstratification of smectite layers with two glycol sheets (17–17.2 Å) and kaolinite layers (7.15 Å) combined with the effects of low coherent domain size. The latter is perceptible in the diffraction patterns of air-dried samples showing unusual basal spacings for minerals that are not interstratified.

In most cases, clays in vertisols result from the mixture of phyllosilicates inherited from the parent rock (micas, illite/smectite mixed layers) and clays newly formed by direct precipitation or by transformation (beidellites ± rich in Fe). The latter process is a weathering of micas or I/S mixed layers by Fe^{2+} oxidation and K^+ loss. The growth of smectite has been shown by Kounestron et al. (1977). The formation of montmorillonite – rather than beidellite – is made possible by the presence of chlorite releasing much more Mg^{2+} in solutions. When the parent rock itself contains smectites, the latter are subjected to significant compositional changes by the vertisol development. Their global charge increases essentially by the growth of the substitution rate of Al for Si in tetrahedra. High-charge beidellites are then formed (Righi et al. 1998).

The smectites in vertisols undergoing the wetting-drying cycles imposed by the season alternation have their crystalline nature modified. They irreversibly fix the potassium available in soil solutions. Some layers remain collapsed at 10 Å whereas others retain their expandability. The stacking sequences become more ordered. Smectites progressively transform into illite/smectite mixed layers whose chemical composition differs from that of the minerals inherited from the diagenetic parent rock.

The wetting-drying cycles cause significant volume variations in the soil. These variations produce stresses (shear stresses) that result in the clay matter re-organisation (Fig. 5.26b). Cracks appear at the surface and propagate down to 160 cm. The fine granules in the surface horizon disintegrated by drought fall into these cracks that seal in the dry season. The clay and organic matter of the soil is then stirred and homogenised naturally.

6.2.4
Soils in Arid or Semi-Arid Climates

6.2.4.1
Xeric Soils (Aridisols) with Sparse Vegetation

In regions where most rainfall is concentrated during the hottest periods of the year, the water escaping runoff percolates into soils. A significant part is subsequently evaporated during the long dry periods that follow rainfall. The remaining part is stored perpendicular to the sparse clumps of vegetation. In Mexico's arid areas, Delhoume (1996) has shown that a sequence of soils of varying thickness marks the transition between the bare zones (regosols) and the clumps of vegetation, although the parent rock is strictly homogeneous (Fig. 6.38a). The formation of a thicker clay horizon under the clumps of plants is due to mineral reactions and not to a clay enrichment by lixiviation of upper horizons as attested by the absence of cutans (Fig. 6.38b). The parent rock is a lutite composed of beidellite + illite/smectite mixed layers. Clay-rich horizons (B horizons) are mainly composed of montmorillonite. Fibrous clays (palygorskite) occur in thicker soils. Both minerals often co-exist because they develop in similar chemical conditions (Elprince et al. 1979). The solutions extracted from these soils are rich in Mg and Si and happen to be very close to the stability domain of montmorillonite and palygorskite (Fig. 6.38c).

6.2.4.2
Saline Soils Without Vegetation (Aridisols)

In arid or semi-arid climate, intense evaporation causes an alkaline-element overconcentration in soil solutions. Salts precipitate forming more or less temporary crusts at a precise soil level controlled by capillary forces (Fig. 6.39). Evaporation increases the concentration of elements in solution. The precipitated salts are of various natures. Usually, calcium carbonate makes up the main part of the crust, but sulphates and other salts have been observed too. The pH reaches 7.8 (precipitation pH of calcite), sepiolite can form (Van den Heuvel 1966). Palygorskite is not systematically related to the precipitation of calcite. Soils with crusts of salt or gypsum occur in plains or depressions (schot, sebkha). The pH conditions are even higher (pH = 9). The Si and Mg high concentrations reached in solutions are favourable to the precipitation of palygorskite- or sepiolite-like magnesian clay minerals (Paquet 1983).

In these soils, the inherited clay minerals are the Al-bearing species. They become unstable in those horizons where evaporation concentrates Si and Mg. If the reaction takes place in a closed system, Al being the inert element, an aluminous phase is necessarily associated to the magnesian fibrous clays (sepiolite and palygorskite):

– Al-clay minerals (smectite, illite) + $Si_{(aq)}$ + $Mg_{(aq)}$ → palygorskite-sepiolite + kaolinite

In this case, the Al contents of magnesian clays must be in equilibrium with this phase. This is the case of the soils described by Van den Heuvel (1966), in

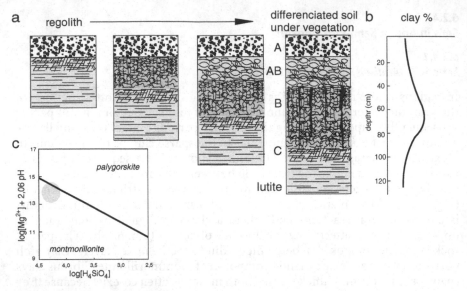

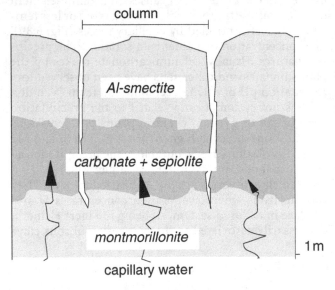

Fig. 6.38a–c. Xeric soils (Aridisols) with sparse vegetation. **a)** Sequence of regosol towards the most differentiated soil under vegetation (after Delhoume 1996); *A*: surface horizon; *AB*: subangular polyhedral structure; *B*: polyhedral-to-prismatic structure; *C*: weathered lutite; lutite: fresh parent rock. **b)** Variation in clay content in the most differentiated soil. The B-horizon is enriched by neogenesis and not by lixiviation of upper horizons. **c)** Phase diagram calculated at 25 °C, 1 atmosphere in the system Si–Mg–Al–H$_2$O (after Leprince et al. 1979)

Fig. 6.39. Schematic representation of the mineralogical zonation and water movements in a xeric soil (Aridisol). The *dark area* corresponds to the calcareous crust

which the Al_2O_3 content of palygorskite is 10.7% and that of sepiolite is 5.4%. In a closed system, the composition of the present phases is controlled by the tie line between sepiolite-palygorskite and kaolinite.

Nevertheless, it is probable that these precipitation horizons by evaporation do not completely constitute a closed system. Transfers by chemical diffusion or by slow flows of solutions are inescapable. Under these conditions, equilibria are modified and differentiated horizons are formed, concentrating the magnesian and aluminous species separately. Under these particular chemical conditions, palygorskite and sepiolite contain less and less aluminium.

6.2.5
Soils on Volcanoclastic Rocks

6.2.5.1
Imogolite and Allophane (Andisols)

Soils developed on volcanoclastic deposits (andisols) cover significant surfaces in highly cultivated areas (and intensively inhabited) owing to their fertility. These deposits are very reactive to the pedogenesis conditions because they are essentially formed of vitreous debris of dacitic-to-rhyolitic composition, typical of explosive volcanoes. Andisols can be found in the Pacific volcanic belt, in the Caribbean, Mediterranean and Canary Islands, in Italy etc. They are characterised by the presence of particular substances that used to be said to be poorly crystallised: imogolite and allophane. In fact, these are minerals whose crystal structure is ordered over short distances (small size of X-ray diffraction coherent domains). Their characteristics have been defined by Cradwick et al. (1972) and Wada (1989):

– imogolite: $(OH)SiO_3Al_2(OH)_2$; it is related to nesosilicates and comes in the form of tubes 20 Å in diameter (Fig. 6.40a);

– allophane: $(SiO_2)_{1-2}Al_2O_3(H_2O)^+_{2.5-3.0}$; amorphous in X-ray diffraction, it comes in the form of hollow spheres 35 to 50 Å in diameter.

Imogolite and allophane are formed in various types of climates (humid, temperate, or tropical). In fact, the presence of glass in the rock is the decisive factor for their occurrence because of its high reactivity under surface conditions. Their stability is maintained in those soils that retain nearly constant moisture content. By contrast, they are unstable under extensive drying conditions and rapidly transform into halloysite. The activity of silica in solution is increased by the drying process and cristobalite can precipitate (Lowe 1986). The greater the SiO_2 content of the glass in the parent rock, the greater the increase in the activity of silica in solution: the Si/Al ratio is higher for rhyolitic glasses than for andesitic glasses. Humus-bearing horizons of andisols are thick. The accumulation of organic substances is facilitated by the formation of allophane-humus complexes. Once in abundance, the organic matter directly forms complexes with the aluminium released by the dissolution of vitreous debris, which inhibits the formation of allophane.

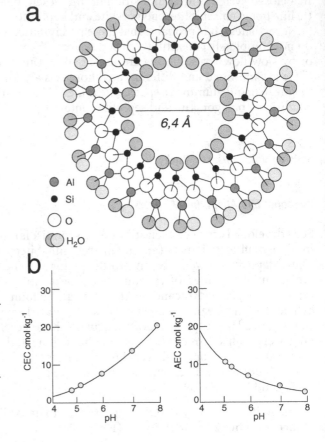

21,4 Å

a

6,4 Å

- ○ Al
- ● Si
- ○ O
- ◐ H₂O

b

Fig. 6.40a,b. Imogolite and allophane. **a)** Crystal structure of imogolite (after Cradwick et al. 1972). **b)** Variation in the cation exchange capacity (CEC) and anion exchange capacity (AEC) of allophane as a function of pH (Wada 1989)

The cation exchange capacity of imogolite and allophane is not constant because the surface electric charges are controlled by the pH and ion concentration of solutions. The CEC (cations) increases whereas the AEC (anions) decreases when pH varies from 4 to 8 (Fig. 6.40b):

- pH = 4: the CEC is almost zero, the AEC is close to 20 cmol kg⁻¹;
  $Al^{VI}(OH)(H_2O) + H^+ \rightarrow [Al^{VI}(H_2O)_2]$;

- pH = 8: the CEC is close to 30 cmol kg⁻¹, the AEC tends towards zero;
  $Si(OH) + (OH)^- \rightarrow SiO^- + H_2O$ (Wada 1989).

6.2.5.2
Kinetics of the Glass-Clay Reactions

The influence of time on the weathering of volcanic glass and on the formation of imogolite, allophane and halloysite in natural environments has been approached thanks to a sequence of palaeosols developed on recent rhyolitic tephras in New Zealand (Hodder et al. 1990). The age ranges between 14,700 and 770 years and the weathering periods (duration separating two successive deposits) vary from 670 to 3,320 years. The authors show that the weathering process comprises two stages with differing kinetics:

1. hydration of glass following a parabolic kinetic law of type $C = k_p t^{1/2}$ where C is the concentration of one element in solution, and k_p is the dissolution rate constant.

2. formation of clay minerals following first-order kinetics $dC/dt = -k_a C$ where $C = C_0 \exp(-k_a t)$, k_a is the reaction rate constant, and C and C_0 are the concentration of one element at times t and t_0 (starting point), respectively.

The activation energies of the various stages have been calculated taking into account the Si/Al ratio of volcanic glasses. The first precipitates are rich in Al (imogolite). The activation energy is close to that required for a diffusion process. Subsequently, the solids become increasingly richer in Si (allophane, halloysite). The activation energies correspond to the dissolution processes of a gel.

Suggested Reading

Colman SM, Dethier DP (1986) Rates of chemical weathering of rocks and minerals. Academic Press, New York, 603 pp

Nahon D (1991) Introduction to the petrology of soils and chemical weathering. Wiley, New York, 313 pp

Tardy Y (1993) Pétrologie des latérites et des sols tropicaux. Masson, Paris, 535 pp

Velde B (1995) Origin and mineralogy of clays. Clays and the environment. Springer, Berlin Heidelberg New York, 334 pp

White AF, Brantley SL (1995) Chemical weathering rates of silicate minerals. Reviews in Mineralogy 31, 583 pp

Ollier C, Pain C (1996) Regolith, soils, and landforms. John Wiley & Sons, LTD, Chichester, 316 pp

Paquet H, Clauer N (1997) Soils and sediments. Mineralogy and geochemistry. Springer, Berlin Heidelberg New York, 369 pp

Taylor G, Eggleton RA (2001) Regolith geology and morphology. John Wiley & Sons, LTD, Chichester, 375 pp

Dixon JB, Schulze DG (2002) Soil mineralogy with environmental applications. Soil Science Society of America book series 7, Madison, 866 pp

CHAPTER 7

Clays in Sedimentary Environments

7.1
Mineral Inheritance

Introduction

Clay rocks make up nearly 70% of the sedimentary rocks (Blatt et al. 1980) and about a third of the rocks outcropping at the surface of continents (Meybeck 1987). Clays in sediments are derived from different sources: (1) erosion of soils and weathered rocks (more than 60% of the Earth's surface), and (2) crystallisation by reaction between saline solutions and silicates. The first form rocks whose granulometric and mineralogical characteristics depend on transport and deposit processes. The second replace minerals or seal pores in rocks that are already formed. The first are inherited from disintegrated rocks; the second are totally neoformed. Between these two end members, intermediate types are possible. Accordingly, clays formed in a continental environment by dissolution of silicates under the influence of diluted solutions undergo chemical modifications – and even recrystallisations – when settled in a saline environment (early diagenesis or eodiagenesis). Several books give a thorough description of the interest of clays as indicators of palaeoclimates, sediment sources and geotectonics (Chamley 1989; Weaver 1989). Only the physico-chemical interactions between sedimentation environments and clays will be presented here, with the exclusion of those controlling weathering (Chap. 6) or diagenesis (Chap. 8). This chapter comprises two separate parts presenting the mineral inheritance and its use as an indicator of sediment sources and climates on the one hand, and the early chemical transformations in the sedimentation environment and the neogenesis of clay minerals on the other hand.

7.1.1
Transport and Deposit

7.1.1.1
Transport Energy and Particle Size

Sediments, and more particularly clays, are transported in the form of suspensions (internal moraines, solid load of streams, aerosols) or fluid beds (flows

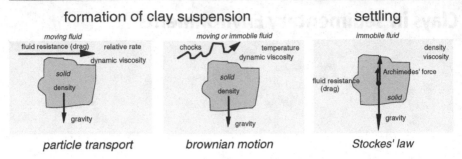

Fig. 7.1. Schematic representation of the main types of forces acting on clay particles in motion

of mud or pyroclastic debris, turbidites etc). The transporting agents, be it ice, water or wind, are fluids with differing intrinsic characteristics (viscosity, dynamic viscosity, temperature, density). Flows (relative velocities) depend upon both these properties and the forces involved (gravity, frictions and impacts). The mass or size of the transported objects as well as the transport distance are complex functions of these parameters (Fig. 7.1). Allen's book (1997) presents the principles of fluid mechanics whose knowledge is indispensable for understanding how fine sediments – and particularly clays – are put into motion. The question raised for clay mineralogy is whether mineral reactions occur at the time of contact with the transporting agent (fresh water, seawater, air) or at the time of changes in the state of clay particles (division, crystal defects, etc). This point will be addressed for each process.

Entrainment

Basically, clays are transported in two forms: isolated minerals and aggregates. Therefore, clay particles range in size between 0.01 and over 20 µm. Fluid (river) or gas (wind) flows can displace the particles within this size range by exerting frictional forces. These forces are proportional to the relative velocity of the particle with respect to that of the flowing fluid. The hydrodynamic force vector is written as follows:

$$\bar{F}_d = \xi \left(\bar{U} - \bar{V}_p \right) \tag{7.1}$$

$\bar{F}_d$ hydrodynamic force vector (N)

ξ friction coefficient $kg\,s^{-2}$ ($\xi = 6\pi\mu_{(T)}a_p$)

$\bar{U}$ fluid velocity vector ($m\,s^{-1}$)

$\bar{V}_p$ particle velocity vector ($m\,s^{-1}$)

$\mu_{(T)}$ dynamic viscosity ($kg\,m^{-1}\,K^{-1}$)

a_p particle radius (m)

The relationship between dynamic viscosity and temperature is given by the Bingham formula: $\mu_{(T)} = 0.6612\,(T-229)^{-1.562}$. The size of transported particles is controlled by the fluid density and viscosity. Accordingly, a glacier can move both finely divided mineral debris – or rock flour – and boulders weighing several tons whereas wind only transports small-sized particles.

Two types of process are involved in the entrainment of clay particles: the suspension process (low clay concentration) and the fluidised beds (high clay concentration). Suspensions can travel great distances. At the mouth of rivers, they form plumes of material that are distributed in the surface water height of the ocean owing to the difference in density between fresh water and seawater. Fluidised beds are moved on continental or marine slopes. They are described under the terms mud flows and turbidites. They behave like Bingham fluids (see 5.2.3.2).

Brownian Motion

The Brownian diffusion coefficient D_{BM} for spherical colloidal particles is given by the Stockes-Einstein equation:

$$D_{BM} = \frac{kT}{6\pi\mu_{(T)}a_p} \tag{7.2}$$

k Boltzmann constant ($1.38 \times 10^{-23}\,\mathrm{J\,K^{-1}}$)

T absolute temperature (K)

$\mu_{(T)}$ dynamic viscosity of the fluid ($\mathrm{kg\,m^{-1}\,K^{-1}}$)

a_p particle radius (m)

Deposition

The velocity of a solid particle falling into water first accelerates (prevailing gravity effect) then decreases under the effect of resistant forces and buoyancy forces. The bigger the particle, the greater this velocity. The balance of the forces present permits calculation of the falling velocity of particles (Stockes' law). If particles are assumed spherical, calculation yields:

$$V = \frac{\frac{2}{9}g\,r^2\,(\varrho - \varrho_0)}{\mu} \tag{7.3}$$

V falling velocity of particles ($\mathrm{m\,s^{-1}}$)

g $9.81\,(\mathrm{m\,s^{-2}})$

ϱ density of particles ($\mathrm{kg\,m^{-3}}$)

ϱ_0 density of liquid ($\mathrm{kg\,m^{-3}}$)

μ viscosity of liquid ($\mathrm{kg\,m^{-1}\,s^{-1}}$)

r particle radius (m)

Since clay minerals are not spherical but plate-shaped, r is referred to as "equivalent" radius and $2r$ is referred to as equivalent diameter (or Stockes' diameter). In reality, suspensions become unstable because the particle size increases by aggregation. Accordingly, in oceanic environment, wind-borne clay particles are aggregated by the organic matter and rapidly fall to the bottom (marine snow). In the same manner, the mineral particles of aerosols aggregate under the effect of impacts in the presence of droplets and fall with the rain.

7.1.1.2
Deposition of River-Borne Sediments

A colloid is a particle whose size is contained between 1 nm and 1 µm. Over 1 µm, it is referred to as a suspended particle. The settling velocity of colloids is greatly slowed down by the Brownian motion (see Chap. 5). Colloids always have a large specific surface. This definition is well suited to smectites whose particle size is most of the time below one micron.

In rivers, sediments are mostly transported through the suspension of detrital particles. This is possible only if the vertical component of the resultant of the forces applied to each particle is not high enough to cause its fall. In fact, gravity is opposed by flow turbulence as regards coarse particles and by the Brownian motion as regards colloidal particles. The transfer towards the mouth of the river takes place through a sequence of falls and resuspension according to the local conditions of flows. The other displacement process (minor) is the flow of fluid beds on bottoms as regards the heaviest particles.

The analysis of the sediments transported by the Amazon River (Gibbs 1967) shows that the great majority of the colloidal particles are smectites whereas the coarsest particles (> 4 µm) are debris of quartz, feldspars, micas and chlorites. Kaolinite forms a clay load intermediate in size (Fig. 7.2). The total sedimentary load is supplied both by the physical disintegration of the rocks outcropping in the watershed and by neoformations of clays in soils and alterites. Thus, Gibbs (1967) has shown that the mineralogical composition and the grain size distribution of the various solid components of the sedimentary load of the Amazon River are identical to those of its tributaries from the Andes high reliefs. This is also the case for the dissolved components. The affluents of the tropical basin carry much smaller amounts of suspended solids (the major contribution is kaolinite) and dissolved elements (the total salinity is 5 to 6 times lower). The persistence of the suspended minerals from the Andes down to the mouth of the river over thousands of kilometres shows that chemical alterations in the river waters are too slow to be measurable. For this reason, the smectite making up the colloidal particles is supplied by the Andean affluents and does not result from the degradation of detrital phyllosilicates during the fluvial journey.

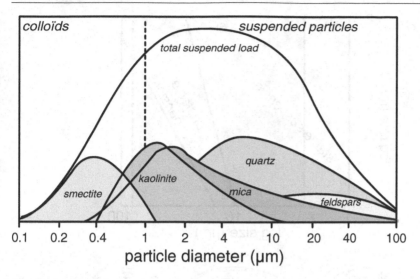

Fig. 7.2. Relationship between grain size and mineralogical composition of the suspended sediments in the Amazon River (after Gibbs 1967)

7.1.1.3
Deposition of Wind-Borne Sediments

Since air density and viscosity are very low as compared to those of water or ice, wind-borne particles are very well screened and belong to the granulometric domains of silts and clays (Fig. 7.3). The suspension of these particles depends on the wind strength (Tsoar and Pye 1987). Silts mostly form temporary suspensions that limit their transit distances. By contrast, clays form stable suspensions that can travel long distances, especially if they reach the altitude of jet streams. Their small size does not permit a direct entrainment by the wind; the exposed clay rock surfaces are subjected to a polishing process and take an aerodynamic shape. The suspension of clays is due to impacts with bigger grains (Nickkling 1994). Dust clouds can be several kilometres high and may transport clays over thousands of kilometres. Air suspensions become unstable when clay particles start forming aggregates. This change of state is not controlled by the Brownian motion but rather by impacts with droplets. The fall of these aggregates far from the coastal zones is the main sediment source in oceans.

Aeolian sediments are derived from bare areas (arid and semi-arid deserts whatever the climate) and from intensively cultivated surfaces. Significant masses are so displaced: soil erosion alone is considered to produce 500×10^6 tons of dust per year in the atmosphere. Dust concentration decreases with distance. No mineral alteration seems to take place during these transits in atmosphere where even the physical fracturation of aggregates is reduced.

On continents, loess are the typical wind deposits. They cover vast regions in the northern hemisphere (Fig. 7.3b). Loess are composite sediments in which

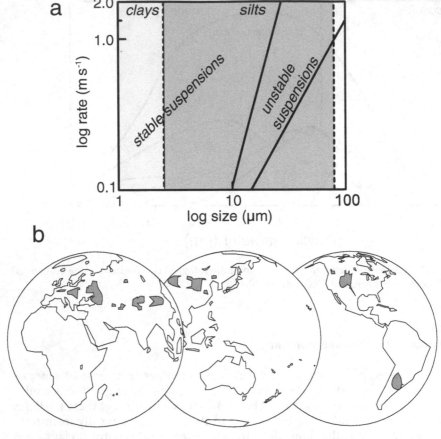

Fig. 7.3a,b. Wind entrainment. **a)** Relationship between wind strength and stability of dust suspensions in air (after Tsar and Pie 1987). **b)** World-wide distribution of loess

fine quartz is the prevailing element. Clays only account for 10 to 15% of the total mass. Their nature varies according to the source areas (Pye 1987).

7.1.1.4
Deposition of Glacier-Borne Sediments

The action of glaciers is noticeable only in high-altitude or high-latitude areas, namely 10% of the Earth's surface today. During glaciations, more than 30% of the Earth's surface was covered with ice, which explains the abundance of tillites (Edwards 1986). The viscosity of ice depends both on temperature and on the state of stress. Therefore, ice behaves like a non-Newtonian fluid (see Sect. 5.2.3.2). Flow rates can reach several tens of meters a year. The motion and pressure of ice exerted on rocks cause their abrasion and yield a rock flour composed of fine debris or disintegrated parent minerals. Since its viscosity is much higher than that of water and air, ice transports debris of any size without screening. This is the main feature of moraine deposits. The finest grains are often composed of isolated primary minerals and show no weathering trace. They are subsequently put back in motion either by subglacial torrents that carry them to great rivers, as is the case with the Amazon River and its Andean tributaries (Fig. 7.2), or by drift ice in the ocean. Glacial clays are then essentially detrital and are formed of minerals occurring in fresh or altered and eroded rocks: illites, chlorites and vermiculites make up most of these tillites or the varved deposits that mark the seasonal cycles.

Wind-borne silts and clay particles can form suspensions congealed in ice when over-freezing conditions are imposed by sudden temperature drops. These "frozen suspensions" form along shallow coasts. Significant amounts of matter subsequently drift with the pack ice (Ehrmann et al. 1992; Kuhlemann et al. 1993). In icebergs, sediments are aggregated in the form of grains up to one centimetre in size (sea-ice pellets). They make up a significant part of the sedimentary load in the Arctic Sea (Goldschmidt et al. 1992).

7.1.1.5
Fluidised Beds

The motion of turbidites is triggered by sliding on slopes. The driving force depends on the difference in density between the sediment and the ambient fluid (see the profile of the sedimentary particle concentration Fig. 7.4a), on the thickness of the fluidised bed and on the slope angle. When this force becomes higher than resistance (frictional forces), the motion is triggered and velocities are distributed according to a vertical gradient (Fig. 7.4a). The feeding of the head, which is lighter, goes more slowly than the rest of the bed. Its volume is increased and flow becomes turbulent (Fig. 7.4b). The ambient fluid flows in the opposite direction at the contact with the fluidised bed. When velocity slows down, the suspension becomes unstable and the deposition process is triggered. The feeding becomes thinner, loses its force and finally stops. The chemical interactions with the ambient fluid are poorly known. They are probably insignificant.

Other phenomena such as subsurface currents and storms cause more or less remote transfers of fine sediments and notably clays. Subsurface currents bring about the suspension and the entrainment of fine particles (12 μm on an average). When the current velocity decreases, these particles (nepheloid layer) re-settle. Storms modify the distribution of coastal sediments. Through the agitation of waves, surface layers of soft sediments are remobilised and transferred from high sea to the coast.

7.1.1.6
Physico-Chemical Transformations of Clays

No clay mineral reaction during suspension, transport and sedimentation could be clearly identified. The interactions with the transporting agent seem

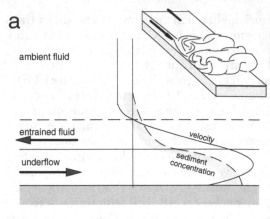

a

ambient fluid

entrained fluid

underflow

velocity

sediment concentration

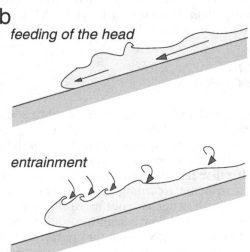

b

feeding of the head

entrainment

deposition

Fig. 7.4a,b. The motion of fluidised beds or turbidites (after Allen 1997). **a)** Schematic representation of the sediment velocity and concentration profiles between the immobile bottom and the ambient fluid. The *inset* shows a three-dimensional view of a moving fluidised bed with the ambient fluid counter-current. **b)** The three stages of the motion: first, the head gets bigger because it goes more slowly than the rest of the bed; the flow becomes turbulent; when velocity decreases, particles settle

to be insignificant and reduced to ion exchanges. The exchange capacity of clays (notably smectites and vermiculites) formed in soils and alterites is saturated by several cations – Ca^{2+}, Mg^{2+}, and K^+ principally. In rivers, saturation is dominated by Ca^{2+} ions. In oceans, saturation is dominated by Na^+ (50%), the remaining part being shared between Mg^{2+} (30–40%), Ca^{2+} (10–20%) and K^+ (5%) (Sayles and Manngelsdorf 1977). These mean data do not take into account the selectivity of the fixation of ions available in waters. Accordingly, potassium is fixed preferentially in the high-charge sites of vermiculites, thus reducing their exchange capacity. Weaver (1989) shows that vermiculites are relatively abundant in rivers of the United States (East Coast and Gulf of Mexico) whereas they are absent from estuarine and marine muds. The preferential fixation of K^+ ions caused the expandable layers to collapse at 10 Å, thus reducing apparently the proportions of vermiculite and increasing accordingly the proportions of illite. This phenomenon explains the inversion in the proportions of smectite (or vermiculite) and illite observed in the estuary of the Guadalquivir River (Melières 1973) or of the James River (Feuillet and Fleischer 1980).

The stability of the suspensions depends on the nature of clay minerals. Smectites form much more durable suspensions whereas kaolinite, illite and chlorite settle more rapidly. This difference is due to the small size ($< 1 \mu m$) of smectite crystallites and to their ability to agglomerate into low-density aggregates. Therefore, a kind of selective screening is performed, smectites being carried off shore and the other species settling on coasts.

7.1.2
Detrital Signature in Marine Sediments

7.1.2.1
Interpretation of the Sedimentary Signal

Sediments are rocks in which detrital elements and neoformed minerals are mixed. These complex assemblages do not constitute parageneses for which formation conditions can be simply defined. The origin of the minerals varies both over space and time. Indeed, at a given instant, rivers and winds carry minerals that are derived from vast surfaces covered by mountains, hills or plains (Fig. 7.5a). Total inputs are received by the final repository. The sedimentary load of the Amazon River (Fig. 7.2) is a good example of the mixture of clays derived both from the Andes disintegration upstream and from the erosion of tropical soils downstream.

There are roughly four great groups of soils whose mineralogical composition exhibits enough contrasts (the most abundant or typical minerals) to be identifiable sources of sediments: glacial soils (illite, chlorite), temperate soils (vermiculites), tropical soils (kaolinite, Fe-oxihydroxides), and arid and semi-arid soils (Al- and Fe-smectites, sepiolite, palygorskite). Variation over time is essentially related to the migration of boundaries between the different types of soils. These boundaries change with climatic evolutions or with tectonics (orogenesis, movement of continents). Any increase in altitude amounts to a displacement towards higher latitudes. Therefore, at a given place on Earth, weathering and pedogenesis conditions change over time owing to the combined effects of tectonics and climatic changes. These changes can be symbolised by a latitude–altitude–time path (Fig. 7.5b). The nature of the dominant clays in sediments change when going back in time in the stratigraphic column.

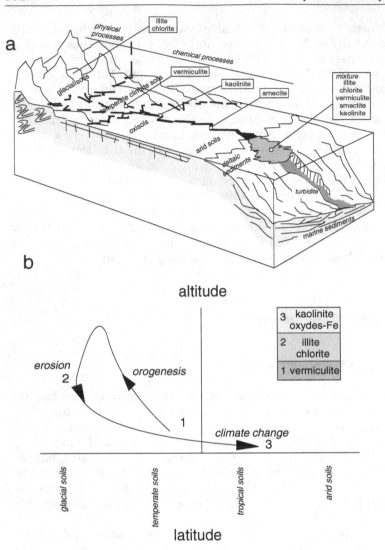

Fig. 7.5a,b. Schematic representation of the diversity of clay minerals making up the marine sediment. **a)** At a given epoch, the genesis conditions vary over space as functions of local climates determined by altitude variations. These variations themselves depend on the latitude of the geological site. **b)** At a given spot on the Earth, the genesis conditions vary over time as functions of the effects of tectonics and global climatic changes. The path represents the modifications of these conditions over time detectable in sediments

7.1.2.2
Sediment Sources

In some favourable cases, the inherited clay fraction in recent or little transformed sediments (shallow burial) can lead to the identification of those continental zones which are input sources. The recent sediments in the Arabian
Sea (Kolla et al. 1981) offer an example of the contribution of various sources
(Fig. 7.6a):

- smectites are derived from the Deccan basaltic plateaus drained by coastal
  rivers and by the Tapali River. These smectites are transported by surface
  currents. At the far south of India, they mix with the smectites carried by
  surface currents of the Gulf of Bengal;
- illite and chlorite are mainly carried by the Indus River that drains the
  Himalayan Mountains. They are dispersed by turbidity currents that remo-
  bilise the delta sediments. These two minerals are also carried to the ocean
  by Northerly winds that sweep across the desert areas of Iran and Makran;
- illite and palygorskite are carried by westerly winds over the deserts of the
  Arabian peninsula and of Somalia;
- kaolinite is concentrated mostly in the equatorial current belt. It is derived
  from soil erosion in southern India, Madagascar and Africa.

The search for sediment sources can have a much finer spatial resolution. Indeed, when different rivers have their mouth on the same coast (Loire and
Garonne in France for example), the sedimentary inputs of the various wa-
tersheds can sometimes be identified. Accordingly, Karlin (1980) has shown
the respective contributions of the Californian rivers and the Columbia River
(Oregon) to the sedimentation along the Pacific coast. The Californian rivers
bring 90% of their sedimentary load – in which chlorite is greatly abundant
– in winter. By contrast, the Columbia River discharges most of its mineral
load – in which smectite is greatly abundant (weathering of basaltic plateaus)
– in summertime. These sediments, although entrained by the littoral currents
along the submarine canyons, clearly mark both types of input (Fig. 7.6b).

7.1.2.3
Palaeoclimatic Signatures

The purely palaeoclimatic interpretation of clay assemblages in sediments is
erroneous if the global history of the genesis conditions is not taken into
account. Error sources are numerous and have been analysed by Hillier (1995)
and Thiry (1999, 2000). In addition to the variability effects over space and
time, several other phenomena equally make the palaeoclimatic interpretation
of sediments complicated:

- the formation of soils and the processes of their erosion are too slow for
  rapid climatic oscillations to be recorded;

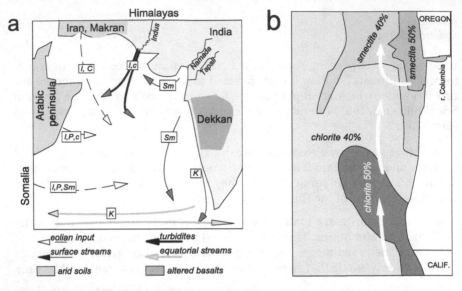

Fig. 7.6a,b. Sediment sources. **a)** Supply zones of the Arabian Sea recent sediments (after Hillier 1995). *I*: illite; *C*: chlorite; *P*: palygorskite; *Sm*: smectites; *K*: kaolinite. **b)** Distribution of chlorite and smectite in the Oregon coast recent sediments, USA (after Karlin 1980)

- a mineralogical screening is performed during transport because clay species come in the form of particles of different sizes (Fig. 7.2);

- mineral reactions occur at the time of deposition (neogeneses at the sediment – seawater interface) or after deposition (transformations during diagenesis).

The sedimentary inheritance progressively fades as the burial depth increases. Some minerals are very reactive and rapidly disappear, such as Al–Fe-smectites and vermiculites. Others recrystallise, like supergene kaolinites made unstable by the presence of Fe^{3+} ions substituted for Al^{3+} and by their numerous crystal defects.

In reality, clay assemblages of sediments record the changes in the conditions of supply, among which climate only accounts for one factor among others (e. g. tectonics, hydrological process of rivers and marine currents and so on). Modifications of ocean currents have at least as much effect on the sediment composition as have climatic changes, but orogenesis remains the dominant factor (Chamley 1989). Only those sites tectonically stable over long periods record climatic changes. In the Mediterranean Sea, Chamley (1971) has shown that the cold and dry periods of the Quaternary glaciations are marked by totally unweathered illites and chlorites while the hotter and rainy interglacial periods are marked by kaolinite- and smectite-rich deposits. Similarly, studying the variations in the proportions of illite and palygorskite in the Arabian Sea sediments, Fagel et al. (1992) have established a correlation between the periodicity of these variations and the Earth's orbital cycles.

7.2
Neogenesis

Introduction

Sedimentary rocks are soft materials comprising mixtures of solid debris (minerals and rocks) with more or less diluted solutions. Typical of atmosphere or seawater interfaces, these rocks are characterised by a high porosity favouring the chemical exchanges between poral solutions and fresh or oceanic water. The presence of organic debris transformed by the microbial activity imposes Eh-pH-P_{CO_2} conditions that trigger dissolution and precipitation reactions of solid phases. Mostly carbonates, hydroxides, chlorides, sulphates and sulphides are formed, but neogenetic clays are formed too. They differ from detrital clays by their chemical composition and sometimes by their morphology. Some of them form species that are typical of sedimentary environments (glauconites). This section is aimed at giving a general presentation of the main neogenesis processes at water-sediment interfaces.

7.2.1
Magnesian Clays: Sepiolite, Palygorskite, Stevensite, Saponite

7.2.1.1
Salt Lakes and Sabkhas

The magnesian clays belong to two families: the 2:1 minerals (stevensite and saponite) and the fibrous minerals (sepiolite and palygorskite). The latter have a higher Si content owing to their structure (see Sect. 1.1.1.2). Since palygorskite and saponite contain more aluminium than sepiolite and stevensite, they are generally considered as resulting from the reaction between detrital minerals and Si- and Mg-rich solutions (Jones and Galan 1988). Sepiolite and stevensite are formed by direct precipitation from solutions.

Salt lakes in desert areas are closed sedimentary basins in which detrital inputs are essentially composed of kaolinite, and to a lesser extent of illite, chlorite and Al-rich smectite. These minerals are derived from the erosion of tropical soils. Their Fe, Ca and Na content is relatively low. The waters supplying these lakes are rich in Mg, Ca, Si and alkaline elements (Millot 1964). A zonation appears in lacustrine sediments between the banks rich in detrital elements (illite, chlorite, kaolinite, dioctahedral smectite) and the centre of the lake where the fibrous clays sepiolite and palygorskite precipitate (Fig. 7.7a). The passage from aluminous clays to magnesian clays has been observed in the Tertiary formations of the Paris Basin (Fontes et al. 1967) or of the south-east part of France at Mormoiron in Provence and at Sommières in the Languedoc region (Trauth 1977). The progressive evaporation of lakes leading to the precipitation of gypsum is accompanied by the formation of saponite, stevensite and sepiolite (Fig. 7.7b). The progressive Mg and Si enrichment of the neoformed phases yields the following compositional sequence (Trauth 1977):

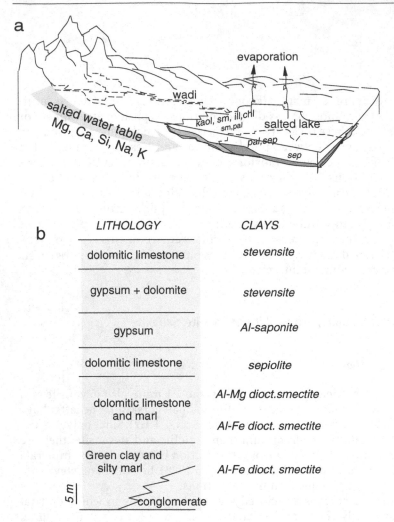

Fig. 7.7a,b. Sedimentation in salt lakes. **a)** Present-day salt lakes in which the detrital inheritance is progressively transformed into a neoformed assemblage (after Millot 1964). **b)**pH Stratigraphic sequence formed by evaporation of the Tertiary lakes in the south-east part of France (after Trauth 1977)

1. detrital smectite: $[Si_{3.92} Al_{0.08}] O_{10} (Al_{1.21} Fe_{0.40} Ti_{0.03} Mg_{0.30}) (OH)_2 Ca_{0.16} K_{0.21}$;

2. transformed smectite: $[Si_4] O_{10} (Al_{1.16} Fe_{0.24} Ti_{0.02} Mg_{0.60}) (OH)_2 Ca_{0.12} K_{0.16}$;

3. neoformed saponite: $[Si_{3.79} Al_{0.21}] O_{10} (Al_{0.78} Fe_{0.24} Ti_{0.02} Mg_{01.35}) (OH)_2 Ca_{0.16} K_{0.21}$;

4. neoformed stevensite: $[Si_{3.97}\ Al_{0.03}]\ O_{10}\ (Al_{0.25}\ Fe_{0.07}\ Li_{0.29}\ Mg_{2.29})\ (OH)_2$ $Ca_{0.08}Na_{0.08}\ K_{0.06}$;

Velde (1985) has proposed an explanation based upon the analysis of the mineral assemblages in a system with three inert components: $Si–R^{2+}(Fe^{2+}, Mg^{2+})–R^{3+}\ (Al^{3+}, Fe^{3+})$. The sequence observed in present-day salt lakes is described by assemblages 1 to 4 (Fig. 7.8a):

1. kaolinite–smectite–amorphous silica: detrital-dominated mineral assemblage. The amorphous silica can yield a flint-like phase (chert);

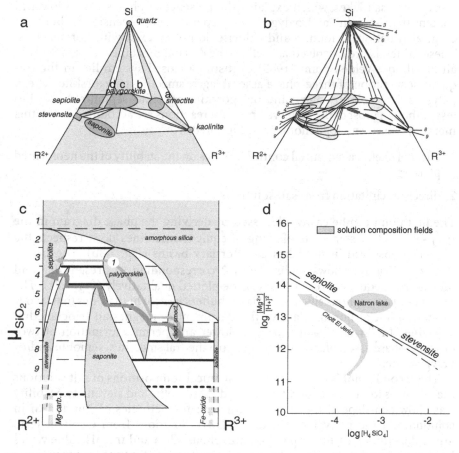

Fig. 7.8a–d. Relationships between detrital aluminous clays and neoformed magnesian clays. **a)** Interpretation of the assemblages observed (Fig. 7.7a) in a closed system (after Velde 1985). **b)** Plotting of the equipotential lines μ_{SiO_2} respecting the sequences observed in salt lakes and French Tertiary basins (Fig. 7.7b). **c)** $\mu_{SiO_2}–R^{2+}\ (Fe^{2+}, Mg^{2+})–R^{3+}\ (Al^{3+}, Fe^{3+})$ diagram inferred from $Si–R^{2+}–R^{3+}$ diagram. Paths 1 and 2 correspond to present-day sebkas and Tertiary basins sequences respectively (see text for details). **d)** Calculation of equilibrium lines between solution, stevensite and sepiolite (after Gueddari 1984)

2. smectite–chert: the most aluminous detrital minerals (kaolinite, sometimes chlorite) disappear whereas dioctahedral smectites become more magnesian as shown by Trauth (1977);

3. smectite–palygorskite–chert: fibrous clays are formed owing to the high Si and Mg activity;

4. palygorskite or sepiolite–chert: only the fibrous clays are stable and nearly reach equilibrium with solutions.

The mineral sequence of the French Tertiary basins is represented by assemblages a, b, c, and d. Silica contents are lower than those observed in present-day salt lakes, which explains the presence of phases with a lower Si content than sepiolite and palygorskite: saponite and stevensite. In both geological situations, kaolinite and chlorite do not co-exist with fibrous clays. These minerals are dissolved as well as the detrital dioctahedral smectites in all likelihood. Velde's point (1985) to justify a closed system lies in the frequent observation of three-phase assemblages: smectite–palygorskite– chert, palygorskite–sepiolite–chert, and palygorskite–saponite–sepiolite. Nevertheless, although most of the observed mineral reactions are accounted for by this method, two essential phenomena are ignored:

1. effect of chemical potential control by silica on the stability of the neoformed phases;

2. direct precipitation from salt solutions.

The first point can be easily addressed by deriving the phase diagram in the μ_{SiO_2}–R^{2+}–R^{3+} system. The plotting of equipotential lines μ_{SiO_2} respects the sequence observed in the lakes and Tertiary basins (Fig. 7.8b). The μ_{SiO_2}–R^{2+}–R^{3+} diagram allows paths 1 and 2 corresponding to present-day and Mormoiron sequences respectively to be plotted qualitatively (Fig. 7.8c). The presence of amorphous silica in salt lakes buffers μ_{SiO_2} at high values that favour the formation of fibrous clays to the detriment of magnesian trioctahedral smectites. In Tertiary basins, μ_{SiO_2} is lower and permits formation of two- or three-phase assemblages involving trioctahedral smectites, sepiolite and/or palygorskite.

The second point is clarified by the chemical compositions of salt solutions that allow us to calculate the stability fields of sepiolite and stevensite. Stability was shown to depend on three parameters: the activities of Mg and Si in solution and the pH value (Gueddari 1984). Solutions from Chott El Jerid are undersaturated with respect to amorphous silica and the pH value varies between 7.5 and 8.3, which allows the formation of stevensite. However, the path determined by the increasing salinity (evaporation) gets close to the stability field of sepiolite. The activity of dissolved silica and pH are much higher in the Natron lake waters (pH = 9–10); these waters are oversaturated with respect to the two phases stevensite and sepiolite.

Sepiolite cannot form below pH 8 (Siffert 1962) and is destroyed at higher pH values. Its stability field corresponds to the one in which carbonates and salts precipitate together (Hardie and Eugster 1970; Gueddari 1984). Indeed, during evaporation, carbonates and gypsum are formed with sepiolite.

7.2.1.2
Ocean Floors

Palygorskite has been frequently observed in deep marine sediments far from any direct continental influence, as is the case for the Arabian Sea (Fig. 7.6). Sepiolite is much rarer. The origin of fibrous clays in this type of environment is still being questioned. Some facts argue for a precipitation at the interface between deep sediments and seawater (Couture 1977; Velde 1985), others for a detrital origin and an aeolian transport (Weaver 1989; Chamley 1989). However, the relative amounts of palygorskite and sepiolite seem to differ according to the type of environment under consideration. Sepiolite prevails in continental environments, palygorskite in deep marine environments. The assumption of a selective screening during transport is difficult to admit knowing that these minerals have very close shape and density. Consequently, the neogenesis assumption shall be seriously considered.

The sediments covering basalts in deep-sea zones form thin, poorly consolidated and very porous layers. Consequently, the mineral reactions taking place there are not due to burial diagenesis. They produce a typical palygorskite-dioctahedral smectite–clinoptilolite phase association (Couture 1997). The formation of these minerals necessitates Si, Mg and Al inputs. Silica is derived from diatom shells and volcanic ashes that have not been dissolved when falling into the ocean (Fig. 7.9). Ashes also provide aluminium, which can be found in interstitial waters despite its low solubility. The aluminium content increases near the interface with seawater (Caschetto and Wollast 1979). Magnesium is derived from seawater weathering of the basaltic surfaces in contact with seawater through sedimentary fluids (Lawrence et al. 1979). This type of alteration, although very slow, releases magnesium in solutions (see Sect. 6.1.3.1). The conditions of formation of palygorskite and of a zeolite (high activities of Si, Mg and alkaline elements, pH above neutrality) are then satisfied at the interface between sediments and seawater. According to Velde

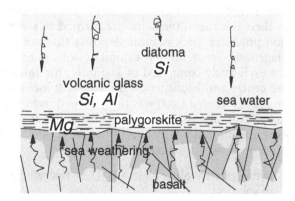

Fig. 7.9. Neogenesis of palygorskite at the interface between deep sediments and seawater (after Velde 1985)

(1985), this paragenesis is practically a by-product of the sea weathering of basalts. High-temperature alterations actually yield other types of zeolites as well as saponite, and even talc (see Sect. 9.1.4.2).

7.2.2
Dioctahedral Smectites

7.2.2.1
Bentonites and Tonsteins

Sedimentary Origin

The formation of bentonites and tonsteins is related to the sedimentation of volcanic ashes in lakes, swamps, lagoons or shallow sea areas (Fig. 7.10a). Bentonites and tonsteins form even beds ranging in thickness from one centimetre to one metre (rarely more), the geographical range of which depends on the energy of the eruption and the power of air currents. These beds can cover huge surfaces and are commonly used as stratigraphic markers (Huff et al. 1991). The volcanic origin of these deposits is attested by two types of petrographical features:

1. presence of minerals inherited from the crystallisation of dacitic or rhyolitic (more rarely basaltic) magmas: β-quartz, biotite, sanidine, plagioclase, apatite, ilmenite, magnetite, zircon, rutile, sphene;

2. vitreous debris or their phantoms (glass shards with bubbles).

Ashes deposited on the surface of continents are subjected to weathering and to andosol-type pedogenesis (see Sect. 6.2.5.1), the eruptive cycles forming a series of buried palaeosols.

Bentonites and tonsteins are monomineral rocks composed of smectite (montmorillonite) and kaolinite, respectively. Their formation process is not fully understood yet. Most specialists think that ash alteration is very rapid and is not due to weathering (Grim 1968). Their chemical composition is not the decisive factor that determines the type of clay mineral formed: whether rhyolitic, dacitic or even basaltic, ashes are transformed into smectite in the presence of water or into kaolinite in the presence of organic acids. According to their thickness, tonsteins are formed in swamps where abundant vegetation produces thick organic deposits that are transformed into coal during diagenesis. Tonsteins may exhibit or not a mineral zonation: the thinnest ones are exclusively composed of kaolinite, the thickest ones contain smectite at the centre and kaolinite on the edges. It looks as if the initial matter (fresh or already altered ash?) was transformed from the interfaces between the volcanoclastic deposit and the organic sediments (Fig. 7.10b). On the basis of experimental studies by Eberl and Hower (1975), Bohor and Triplehorn (1993) propose a two-stage mechanism:

1. hydrolysis: rhyolitic glass + $H_2O \rightarrow$ hydrated aluminosilicated gel + cations in solution,

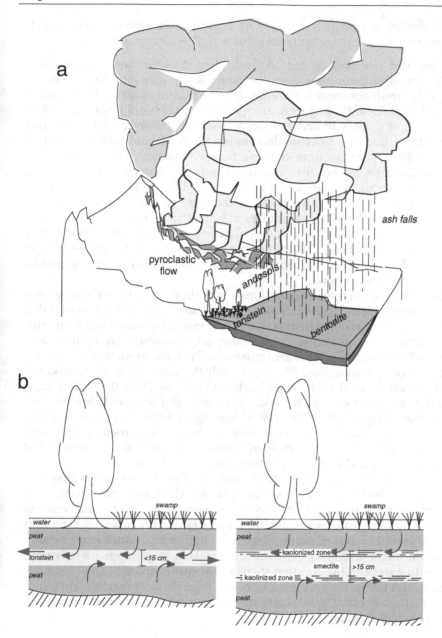

Fig. 7.10a,b. Formation of bentonites and tonsteins. **a)** Schematic representation of the alteration environment of ashes: andosols (continental surfaces), tonsteins (swamps), bentonite (lakes, lagoons, shallow sea areas). **b)** Influence of the thickness of ash deposits on the mineral zonation of tonsteins (after Bohor and Triplehorn 1993)

2. dissolution-precipitation activated by organic acids: hydrated aluminosilicated gel → kaolinite + hydrated silica + H_2O + cations in solution

The proposed reactions do not totally explain the alteration mechanism of ashes. Indeed, volcanic glass does not immediately alter into smectite in the presence of fresh or seawater. Vitreous debris has been observed in sediments several hundreds of thousands to several million years old (Hein and Scholl 1978; Keller et al. 1978; Imbert and Desprairies 1987; Weaver 1989). Their rate of transformation into smectite decreases with depth and remains constant at about 50% at 250 cm from the interface with seawater (Chamley 1971). Besides, the glass to smectite reaction is not stoichiometric. This reaction notably releases silica in amounts that should yield quartz or secondary opal. This is the case of the glass shards of the Otay bentonite from California (Berry 1999). Nevertheless, bentonites are practically monomineral in their body and contain no or very little opal. Their very rapid formation (a few hours to a few days; Berry 1999) is not consistent with an intensive leaching of dissolved elements. Chemical diffusion at the temperature of lake or seawaters is a process much too slow.

There are several alternative solutions which should be taken into consideration: ash alteration may take place either at the time of deposition in waters that are more saline or alkaline than seawater, or before deposition inside the volcano hydrothermal system itself, or after deposition during burial by diagenetic reaction. In the first case, ashes are transformed in lagoons where solutions are concentrated by evaporation. This is probably the case of Uruguay bentonites, which are calcic montmorillonites (Calarge et al. 2001). In the second case, glass is altered very early in its volcanic context and sedimentation involves already transformed ashes. Grim (1968) has suggested that alteration may take place during ash fall, which is improbable. The process proposed by Bohor and Triplehorn (1993) for the formation of tonsteins remains valid if smectite and not glass is assumed to be the starting point of alteration by organic acids. Nevertheless, how the complete transformation of glass into smectite may take place in the volcanic context must be explained. Great amounts of water and very large contact surfaces with magma are necessary. These conditions can be met only in phreato-magmatic systems where highly powerful eruptions produce great amounts of ashes mixed with high-temperature water vapour. Alteration might take place between the emulsion stage where glass is saturated with water and the eruption stage where suspended ashes develop huge surfaces with fluids. High temperatures and huge surfaces may explain a sudden alteration. Eventually, whatever mechanism may be involved partakes of a purely speculative approach to date.

Diagenetic Origin

The second alternative origin of bentonites is a diagenetic-like reaction. The vitreous pyroclastic deposits (not transformed into clay during volcanic processes) undergo mineral transformations during their burial process (Compton et al. 1999). The rhyolitic glass gets hydrated, and then yields smectite by a dissolution-recrystallisation process. The highly porous pyroclastic deposits are gradually compacted; the dissolved elements are leached out along with poral fluids. Several facts support a diagenetic origin:

1. transformation progressivity measurable by the variation in the ratio of glass and clay amounts (mass ratio);

2. isotopic equilibration of hydrated glass and clays with resident waters (seawater for the Monterey sequence, California for instance);

3. presence of CT opal in the adjacent sedimentary formations that could be related to the migration of silica by chemical diffusion from the volcanic glass dissolution zones.

Some problems have yet to be resolved. Indeed, the isotopic equilibrium with seawater ($^{18}O/^{16}O$) is reached at temperatures higher than those found in the sedimentary environment (40 °C). The δD values are not consistent with pure seawater and imply mixtures with fresh water. Despite these uncertainties, the diagenetic process – much longer than the sudden alteration of ashes at the contact with water – seems to be a better explanation of the nearly monomineral character of bentonites.

The diagenetic origin of some tonsteins is confirmed by the formation of pyrophyllite and sudoite along with kaolinite from bentonite beds in the Ardennes coal mine series (Anceau 1992, 1993). These minerals occurred owing to sufficiently acidic and reducing local conditions, as attested by the presence of pyrite. These local conditions have probably been imposed by the presence of cracking organic matter.

7.2.2.2
Sea-Water Hydrothermal Environments

Nontronites of Black Smokers
Seawater hydrothermal activity areas lined by the chimneys of black or white smokers form particular geochemical and biochemical environments characterised by sudden contrasts in physicochemical conditions (Hannington et al. 1995). Indeed, black smokers eject sulfurated acidic solutions at 300 celsius in an oxidising icy seawater (2–4 °C): the mean oxygen content is $4\,ml\,l^{-1}$ (Fig. 7.11a). These contrasts explain the sulphide-sulphate mixture within the chimney walls (Fig. 7.11b). First sulphides then, farther, oxidised metals (mainly Fe and Mn) and nontronites are dispersed by the smoker plume. Purely ferriferous clays are formed by mixture of hydrothermal fluids with ambient seawater under low temperature conditions, namely below 70 celsius (Singer et al. 1984):

$$\left[Si_{3.67}Fe^{3+}_{0.33}\right] O_{10} \left(Fe^{3+}_{1.66}Mg_{0.40}\right) (OH)_2 K_{0.15} \left(NH_4^+\right)_{0.34} \tag{7.4}$$

Crystallisation of these minerals can take place according to two different geochemical paths:

1. nucleation and growth under oxidising conditions. This process is slow (Decarreau et al. 1987). The conditions of formation of these nontronites have been calculated by Zierenberg and Shanks (1983) as functions of the sulphur and oxygen fugacities of solutions. Under oxidising conditions, nontronite and anhydrite precipitate (Fig. 7.11c);

2. initial formation of a polymerised ferrous or ferromagnesian precursor having a brucite-like structure (Harder 1978). This precursor serves as embryo on which siliceous trioctahedral layers polymerise in turn. The Fe valency changes, and the trioctahedral lattice subsequently becomes dioctahedral under oxidising conditions. A Fe^{3+} cation is released from the silicate and forms an independent oxihydroxide phase (hematite, goethite). This path is faster and is consistent with the working process of black smokers. The passage from reducing to oxidising conditions takes place by mixture with seawater as temperature decreases (Fig. 7.11b).

The role of the ferrous precursor has been specified by experimental studies by Decarreau and Bonnin (1986). The ferric smectite is formed according to the following reaction:

$$Si_4O_{10}Fe_3^{2+}(OH)_2 + 2H_2O \rightarrow Si_4O_{10}Fe_2^{3+}(OH)_2 + FeOOH + 3H^+ + 3e^-.$$
$$(7.5)$$

The di-to-trioctahedral transition imposes that 2 out of 6 Fe^{2+} ions in the unit cell be released from the silicated lattice and form a Fe-hydroxide. The charge modifications imposed by the substitutions of Fe^{3+} for Si in tetrahedral sites are not accounted for by the reaction above. The nontronites formed in the ocean bed are exclusively K-nontronites. Their selectivity is noteworthy despite the presence of other cations in solution (notably Mg^{2+} and Na^+). This selectivity is probably related to steric conditions: the presence of Fe^{3+} ions in the octahedral or tetrahedral sheet increases the b cell dimension, hence its capacity to accept interlayer cations with a large ionic diameter (Eggleton 1977; Russell and Clarck 1978).

Ferrous Stevensite and Ferripyrophyllite of the Red Sea

The structure of the Red Sea rift is characterised by the presence of basins full of hot brines at about–2,000 m in depth (Fig. 7.12a), of which three have been explored (Fig. 7.12b). These brines are derived from mixtures of hydrothermal fluids crossing the basaltic basement with solutions flowing in the evaporitic deposits that form the graben flanks. They are stratified, the lower brine having a density of $1.2\,g\,cm^{-3}$ a chlorinity of 156‰ and a present-day temperature of 61 °C and the upper brine having a density of $1.10\,g\,cm^{-3}$, a chlorinity of 82‰ and a temperature of 49–50 °C (Hartmann 1980). The bottom of these basins is covered with metallic ore deposits that have been studied owing to their potential economical value (Bäcker and Richter 1973; Zierenberg and Shanks 1983), and with clay sediments whose descriptions have been re-worked by Badaut (1988).

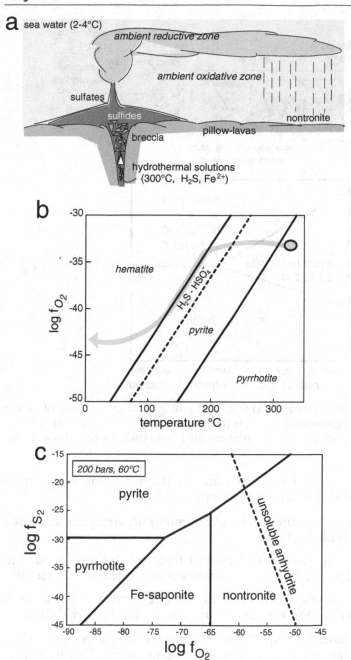

Fig. 7.11a–c. Nontronites in sea-water hydrothermal zones. **a)** Structure of black smokers and mineral deposits. **b)** Variation in the temperature and oxygen fugacity of hydrothermal fluids when mixing with ambient seawater (modified from Hannington et al. 1995). **c)** Phase diagram in the Fe–Al–Na–Si–O–S system at 200 bars, 60 °C showing the stability fields of nontronite and anhydrite (after Zierenberg and Shanks 1983)

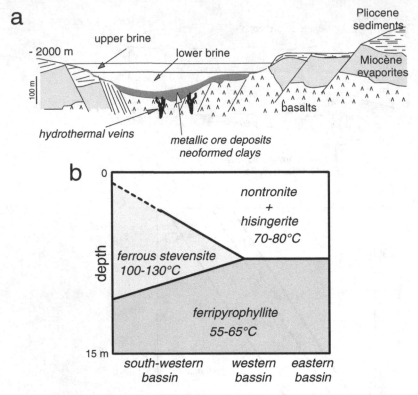

Fig. 7.12a,b. Clays from the Red Sea. **a)** Simplified geological structure of the rift of the Red Sea ridge showing brine stratification in the depression. **b)** Distribution of Fe-rich clay minerals in cores extracted from three basins explored in the rift (after Badaut et al. 1990)

Clays identified in the bore holes from the Ocean Drilling Programmes (ODP) are organised into three main groups:

1. detrital clays (illite, chlorite, kaolinite) commonly occurring in the Red Sea during the Pliocene epoch;

2. magnesian clays (talc, stevensite) derived from the alteration of basaltic rocks more or less rich in glass and sometimes associated with chrysotile;

3. Fe-rich neogenetic clays forming at the contact between sediments and brines and growing mostly in the form of regular laths on detrital clays.

Ferric clays are associated with silicates (hisingerite) and with more or less amorphous Fe-bearing oxyhydroxides (ferrihydrite, ferroxyhite, hematite). Ferrous clays (ferrous stevensite) are associated with sulphides.

The presence of a trioctahedral ferrous clay ($b = 9.32$ Å) has been revealed in sediments protected from the atmospheric oxidation at the time of sampling. They have a low Al content, they may contain a small amount of Zn, and the substitution rate in tetrahedral sites is low. These clays correspond to ferrous stevensites: $(Si_{4-\varepsilon} Al_{\varepsilon}) O_{10} Fe_3^{2+} (OH)_2 M_{\varepsilon}^+$. They are very unstable under oxidising conditions and are transformed into nontronite according to the reaction described by Badaut et al. (1985):

$$Si_4O_{10}Fe_3^{2+}(OH)_2 + 3/2 H_2O$$
$$\rightarrow Si_4O_{10}Fe_2^{3+}(OH)_2 + 1/2 Fe_2O_3 \cdot n H_2O + 3 H^+ + 3 e^- \qquad (7.6)$$

These ferrous trioctahedral clays are close to the precursor devised by Harder (1978). Their isotopic composition ($^{18}O/^{16}O$) shows that they have been formed under temperature conditions that are high for the Red Sea site (100–130 °C).

Nontronites crystallise at the interface between the upper brine and metalliferous sediments, probably by oxidation of ferrous stevensite and partial dissolution of detrital clays under relatively low-temperature conditions (70–80 °C). An example of typical unit formula is given by Bischoff (1972):

$$[Si_{3.15}Al_{0.26}Fe_{0.59}^{3+}] O_{10} (Fe_{1.60}^{3+}Fe_{0.24}^{2+}Mn_{0.02}Zn_{0.15}Cu_{0.07}Mg_{0.19}) (OH)_2$$
$$1/2 Ca_{0.10}Na_{0.55}K_{0.09} \qquad (7.7)$$

The substitutions of Fe^{3+} for Si^{4+} in tetrahedral sites and of Cu^{2+} and Zn^{2+} for Fe^{2+} in octahedral sites are related to the chemical composition of those brines in contact with metallic ore deposits. Note that iron is not completely oxidised in these nontronites.

Ferripyrophyllite forms practically monomineral green clay deposits. Despite the presence of a few expandable layers (charge compensated for by potassium), its chemical composition gets very close to the theoretical end member $Si_4O_{10}Fe_2^{3+}(OH)_2$ (Badaut et al. 1992). It seems to have been formed by early diagenesis in very recent formations (less than 25,000 years) under temperature conditions of about 55–65 °C (Badaut et al. 1990).

7.2.2.3
Al-Fe Dioctahedral Smectites (Early Diagenesis)

Diagenesis begins right after the sediment-seawater interface; it is referred to as "early" when this burial does not lead to a noteworthy temperature rise (Chamley 1989). This mostly involves thicknesses of several hundreds of metres. In the present case, the term "early diagenesis" refers to the first tens of centimetres from the surface of the soft sediments. Their very high porosity (over 50%) allows for easy exchanges with seawater by chemical diffusion from the interface. This phenomenon has also been referred to as "reverse weathering" (Sillén 1961)

Continental clay minerals remain seemingly inert when settled in oceans. Nevertheless, their exchange capacity is no longer saturated by the same ions, namely Ca^{2+} in rivers, Na^+, K^+ and Mg^{2+} in seawater (Sayles and Mangelsdorf 1977). Waters impregnating the first metre of sediment are systematically depleted in K^+ and Mg^{2+}. These seawater-derived elements being "consumed"

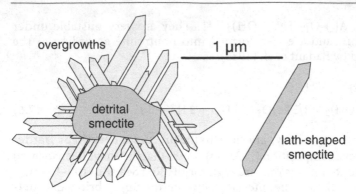

Fig. 7.13. Representation of overgrowths oriented by rotations at 60° on detrital clays and lath-shaped smectites

in solids, they diffuse in the solutions trapped in sediments (Sayles 1979). However, diffusion is unlikely to be maintained only by cation exchange reactions between fluids and clays. Although no unquestionable proof of the neogenesis of magnesian and potassium-bearing clays in the first metre of sediments has been provided to date, it seems that this process cannot be ruled out.

The neogenesis of phyllosilicates is a proved phenomenon in marine sediments. It is expressed by overgrowths on detrital clays (illite, smectite, kaolinite) or by the formation of lath-shaped smectites (Fig. 7.13). Studying black shales from the Albian stage, Steinberg et al. (1987) have shown that overgrowths on illite have an Al–Fe beidellite composition whereas those on smectites have a montmorillonite composition. Typical lath-shaped habits are found in sediments of different ages: Miocene red clays (Kharpoff et al. 1981), Atlantic sediments (Holtzapffel and Chamley 1986). The main factor in the formation of these minerals does not seem to be the burial depth but rather the duration of the exchanges with seawater by diffusion.

7.2.3
Ferric Illite and Glauconite

7.2.3.1
Ferric Illites

Ferric illites (or Fe-bearing illites or glauconite mica) are characterised by aluminium contents that are much higher than those of glauconies (Kossovskaya and Drits 1970; Berg-Madsen 1983). As shown by Velde and Odin (1975), there is no continuous solid solution between these two species (see Fig. 2.17). The crystal structure of these illites is still poorly known. They form in non-marine environments, such as lagoons (Porrenga 1968; Kossovskaya and Drits 1970), xeric soils (Norrish and Pikering 1983), and fluviatile environments (Dasgupta et al. 1990; Backer 1997). They come in the form of green crystals sometimes included in cementing carbonates, thus showing that they crystallise during sedimentation in the vicinity of the water-sediment interface. In xeric soils, they sometimes accompany sepiolite and palygorskite in horizons where salt solutions have been concentrated by evaporation.

7.2.3.2
Glauconies and Glauconite

Formation Environments

According to the Odin and Matter terminology (1981), the green sedimentary grains (glauconies) are mixed-layer minerals whose components are smectite and glauconite mica. The greatest part of the glauconies described in the literature is of marine origin. Nevertheless, some occurrences are reported in lacustrine sediments (see Sect. 7.2.3.1). Glauconies mostly come in the form of rounded grains. Some of them retain the morphology of the detrital elements they have pseudomorphically replaced: sponge spicules, foraminifer shells, volcanoclastic debris, faecal pellets and so on. Under present-day conditions, they form in all oceans except in icy parts. However, they are more abundant in the intertropical domain between -125 and $-250\,m$ (Fig. 7.14a). An open marine environment with a low terrigenous sedimentation rate is necessary (Porrenga 1967). The planktonic and benthic marine organisms provide the organic matters whose presence seems to be a decisive element in determining the chemical properties at the microenvironment scale ($< 1\,mm$). Glaucony is formed by chemical exchanges between solid debris and seawater at the microsystem scale (Fig. 7.14b).

The glauconitisation process takes place at shallow depth (a few decimetres) in the vicinity of the sediment-seawater interface. The greater the size of the pores formed by the grain stacks in the sediment, the more efficient the process. Indeed, this condition facilitates exchanges between poral water and seawater (Odin and Fullagar 1988). The first glauconite glaebules grow in the mineral or in the plankton shell used as substrate, whatever its initial composition: siliceous (sponge spicules, diatoms), silico-aluminous (detrital clays, volcanic glass), or calcareous (bioclasts, oolites). These glaebules give way to flakes or rosettes that finally invade the entire solid substrate. The strontium isotopic ratios ($^{87}Sr/^{86}Sr$) show that dissolution of detrital phases and crystallisation of glauconite take place in a closed microenvironment that progressively opens up (Clauer et al. 1992). In the same time, the potassium content increases progressively. A time scale of these transformations (Fig. 7.14c) is given by Odin and Fullagar (1988).

Glauconitisation Process

The glauconitisation process is related to the chemical diffusion of those elements naturally in solution in water to which are added those directly solubilised in the poral waters of the microenvironment. The presence of organic matter imposes reducing conditions that permit local-scale solubilisation of the terrigenous Fe-oxyhydroxides. This presence causes the diffusion of these elements in the direction poral water to solid substrate by retaining a redox

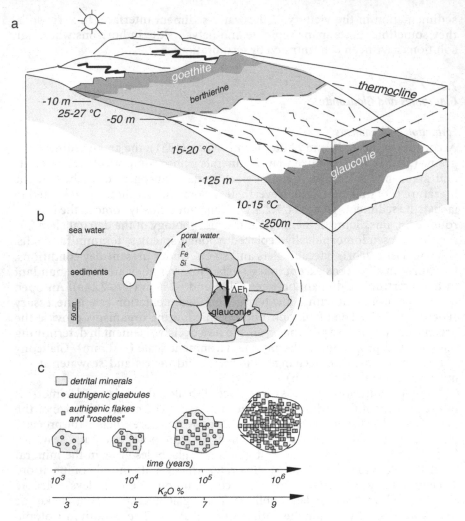

Fig. 7.14a–c. Conditions of glaucony formation. **a)** Coastal marine origin (after Porrenga 1967). **b)** Glauconitisation takes place within microsystems that are close to the sediment-seawater interface and in which redox potential gradients (ΔEh) are established. **c)** The fixation of potassium increases with time and causes changes in the morphology of glauconitic corpuscles (after Odin and Fullagar 1988)

potential gradient (ΔEh), the excess ions in the solid entering the solution. This process leads to the epigenesis of clasts that retain their outward form: calcareous shells, siliceous sponge spicules and so on.

Although containing Fe^{3+} ions principally, glauconite is sometimes associated with pyrite owing to the reducing properties of organic matter. The process is favoured but not controlled by the temperature conditions of tropical waters since glauconies are formed under high latitudes. This has been confirmed by experimental syntheses from co-precipitates by Harder (1980). The significant parameter is the activity of silica in solution: when low, berthierines are formed, when high, glauconitisation is triggered. Using the phase rule, Velde (1985) shows that only aluminium remains inert in the microsystem. Indeed, the fact that glauconite is the only formed phase imposes that all the other variables be intensive ones (controlled by the environment out of the microsystem), the role of which depends on their chemical potential. In fact, even the inert element is likely to migrate, as shown by the epigenesis of the siliceous spicules or the calcareous bioclasts.

The proportion of interstratified smectite varies with the age of sediments (Fig. 7.14c): the more recent the sediments, the greater the smectite content. This variation is controlled by the amounts of potassium fixed in glauconies. Note that the same tendency is observed both for recent sediments (Odin and Fullagar 1988) and for diagenetic series (Thompson and Hower 1975; Velde and Odin 1975). The composition of the smectite component has long been considered to be of nontronite type. A reaction of glaucony formation from a precursor nontronite has been proposed by Giresse and Odin (1973):

nontronite (16–21% Fe_2O_3) + K^+ → glauconite

Nevertheless, this reaction is not consistent with the real chemical composition domain of glauconies, which is situated between celadonite-glauconite micas and beidellites (see Sect. 2.1.4.2). By contrast, the phase diagram proposed by Velde (1985) takes into account the real chemical composition domain and imposes a reaction between an aluminous phase (kaolinite) and a ferric phase (nontronite or Fe-oxyhydroxides):

kaolinite + nontronite or Fe-oxyhydroxide + K^+

→ aluminous mixed-layer mineral + ferric mixed-layer mineral

→ glauconite

The presence of two types of mixed-layer minerals imposes a simultaneous increase in their mica content as the fixed potassium content increases. This is purely speculative and has not been shown by X-ray diffraction.

While the formation of a ferric smectite in the first stages of glauconitisation seems to be proved now, (a ferric montmorillonite in fact, Wiewiora, personal communication), nothing indicates that this precursor has the same composition as that of the smectite interstratified in the mixed-layer structure. This precursor is unstable and reacts with its environment to yield the two constitutive types of layers of glauconies: Fe-rich beidellite and glauconite mica. The reaction then becomes:

ferric montmorillonite + kaolinite + K^+

→ Fe-beidellite–glauconite mica mixed layer.

As emphasised by Velde (1985), the formation then the destabilisation of the precursor ferric smectite follows the inverse path of weathering (see

Sect. 6.1.3.4). This shows that glauconitisation is a reaction process of the Earth's surface involving true equilibria given its reversibility under the same conditions. Glauconitisation and weathering of glauconitic rocks should then be described using a single-phase diagram in the μ_{K^+}–Al–Fe^{3+} system (Fig. 7.15). Indeed, the glauconitisation diagram is very close to the weathering diagram (Fig. 6.22, Sect. 6.1.3.4); the slight differences are in fact due to the resolution of the petrographical analysis. This diagram shows that the precursor ferric montmorillonite is stable with kaolinite for very low values of μ_{K^+}. As soon as these values increase, this assemblage is no longer stable and a Fe-rich beidellite-like dioctahedral smectite is formed. This is the "smectite" component of the interstratification. For higher values of μ_{K^+}, the precursor disappears while glauconite mica is formed in the interstratification, as well as Fe-oxihydroxides if allowed by the chemical composition. Glauconitisation is not about solid-state transformations (SST) but a dissolution-recrystallisation process.

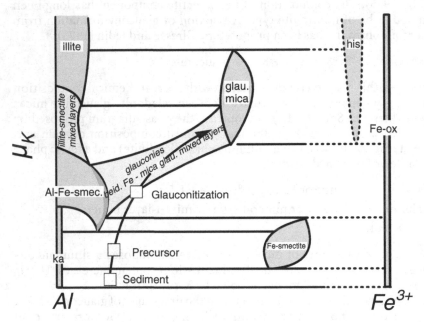

Fig. 7.15. Phase diagram in the μ_{K^+}–Al–Fe^{3+} system showing the relationships between the chemical composition of sediments and the formation of a precursor phase (ferric montmorillonite) before glauconitisation. Al–Fe–Sm: aluminous dioctahedral smectites. The composition of the mixed-layer smectite (Fe-beidellite) is stressed by the *darker area*; *ka*: kaolinite; *glau mica*: glauconite mica; *Fe-sm*: ferric smectites whose domain goes from nontronites to ferric montmorillonites; *his*: hisingerite or hydrated ferric silicates; *Fe-ox*: Fe-oxyhydroxides

7.2.4
Berthierine, Odinite and Chamosite (Verdine and Oolitic Ironstone Facies)

The "verdine" facies occurs in tropical environments where water temperature exceeds 20 °C (Porrenga 1967). This facies has not been observed in the intertropical space along the coasts bathed in cold oceanic currents (Odin and Matter 1981). Green grains are formed at shallow depth along the shores in the vicinity of the mouths of rivers carrying Fe-rich colloids (Fig. 7.16a). Coastal swamps and mangrove areas seem to play a major part as organic matter suppliers. The "greening" of sediments takes place between 10 and 60 m in depth, in the area where organic matter imposes reducing conditions that solubilise the iron of sedimentary oxides and hydroxides.

In ancient sedimentary formations, another facies is observed: oolitic ironstones. Formed of concentric deposits of iron oxides and clays, their origin is still questioned (Chamley 1989). A reliable reconstruction of the chemical processes taking place at the time of deposition is very difficult when the observed sediments have been subjected to diagenesis. The verdine and oolitic ironstone facies have in common the formation of Fe-rich phyllosilicates, which are neither glauconites nor nontronites. These are chlorites, the half-cell unit formulae of which are as follows:

1. berthierine (7 Å): [Si Al] O_5 (Al Fe^{2+}) $(OH)_4$ for the Fe-rich end member,

2. chamosite (14 Å): [Si_3 Al] O_{10} (Al Fe_5^{2+}) $(OH)_8$ for the Fe-rich end member,

3. odinite: [Si_{4-x} Al_x] O_{10} ($R_{2,2-2,8}^{3+}$ $R_{1,7-2,9}^{2+}$) $(OH)_8$ (after Odin et al. 1988).

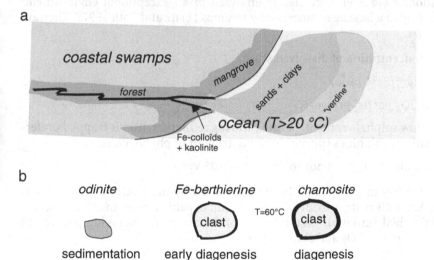

Fig. 7.16a,b. Odinite – berthierine. **a)** The verdine facies is formed at the mouth of rivers of tropical areas flowing into a hot ocean. **b)** The transformation of odinite into berthierine and into chamosite takes place by dissolution-precipitation reactions, and not only by reduction of the ferric iron

Berthierine and chamosite have a trioctahedral structure. Berthierine is characterised by high aluminium content. Odinite has a 7 Å-chlorite-like crystal structure intermediate between di- and trioctahedral structures (Bailey 1988). The typical assemblages of the verdine and ironstone facies are odinite + Fe-rich smectite and berthierine + chamosite+swelling chlorite, respectively. It looks as if the crystal structures intermediate between di- and trioctahedral are not stable under low diagenetic conditions.

Odinite can progressively transform into ferrous berthierine by reduction of iron, first yielding intermediate berthierines that retain the di-trioctahedral character: $[Si_{1.74} Al_{0.26}] O_5 (Al_{0.82} Fe^{3+}_{0.28} Fe^{2+}_{1.01} Mg_{0.46} Mn_{<0.01} \square_{0.43}) (OH)_4$ (Hornibrook and Longstaffe 1996). Nevertheless, a moderate transformation does not mean that the reaction is limited to the change in iron valency. Indeed, odinite crystallises by filling bioclasts or porous detrital grains whereas berthierine forms coatings on the surfaces of the detrital elements outlining the rock pores (Fig. 7.16b). Therefore, the reaction implies both the dissolution of odinite (and of the accompanying detrital clay mineral assemblage) and the crystallisation of berthierine. The more reducing the environment, the closer to the ferrous trioctahedral end member the composition of berthierine. Subsequently, the burial of sediments triggers diagenetic reactions that transform berthierine into chamosite. The mean conversion temperature is estimated at about 60 °C (Hornibrook and Longstaffe 1996).

The mineral sequence odinite – intermediate berthierine – trioctahedral berthierine results from the destabilisation of the detrital sediments carried by rivers of tropical areas. As regards fine particles, these sediments are mainly Fe-oxyhydroxides and kaolinite associated with organic matter. The reaction conditions have been very clearly analysed in an exceptional environment: weathering of a laterite submerged by swamps (Fritz and Toth 1997). They are as follows:

- low concentration of dissolved silica,

- low Mg^{2+}/Fe^{2+} ratio,

- high CO_2 partial pressure,

- very low sulphate content, because during reduction, iron is trapped preferentially in sulphides (pyrite) to the detriment of phyllosilicates,

- moderate reducing conditions: $Eh \approx -0.05$ V.

These conditions correspond to those imposed by the discharge of fresh waters laden with oxihydroxides, kaolinite and organic matter into a hot ocean. The microbial activity is certainly one of the main factors taking part in the regulation of the CO_2 and sulphur partial pressure.

Suggested Readings

Velde B (1985) Clay Minerals. A physico-chemical explanation of their occurrence. Elsevier, Amsterdam, 427 pp

Allen PA (1997) Earth surface processes. Blackwell, Oxford, 404 pp

Chamley H (1989) Clay sedimentology. Springer, Berlin Heidelberg New York, 623 pp

Odin GS (ed) (1988) Green marine clays. Oolitic Ironstone facies, Verdine facies, Glaucony facies and Celadonite-bearingfacies – A comparative study. Developments in Sedimentology Series, Elsevier, Amsterdam, 445 pp

Worden RH, Morad S (2003) Clay mineral cements in sandstones. Intern Assoc Sediment Special Publ 34, Blackwell, Oxford, 509 pp

References

CHAPTER 8

Diagenesis and Very Low-Grade Metamorphism

8.1
Sedimentary Series

Introduction

The physical properties and the mineralogical composition of sediments are transformed by diagenesis during burial in sedimentary basins. These physical and chemical transformations take place in the presence of complex fluids in which salt solutions, hydrocarbon compounds and gases are mixed. Burial at several kilometres in depth is possible only if the basin floor (basement) sinks gradually. This process, known as subsidence, is caused by sediment weight. The more abundant the sediment sources, the more active the subsidence.

During diagenesis, highly porous soft sediments (muds, sands) are transformed into less porous coherent rocks (shales, sandstones ...). This transformation is due to compaction (pressure effect) and cementation (mineral precipitations from over-saturated solutions). The resulting reduction in the pore volume causes the release of a great part of the water contained in the sediment. Mud and sand contain up to 80% and 30% water, respectively, whereas shales and sandstones contain a few percent only. The water released from rocks flows in high-permeability zones, notably in faults. The latter serve as drains, the activity of which being discontinuous over time (alternance of sealing and seism-induced reopening periods).

Diagenesis does not transform only the mineral matter. The organic components of sediments (river-borne soil organic matters, continental or oceanic living organisms) form hydrocarbon compounds and gases. These substances play a major part in the chemical environment, notably as regards redox conditions and solution pH. This chapter is aimed at showing the significant parameters of diagenesis, particularly those with an action on the mineral transformations of clays. The reactions will be detailed for two mineral sequences: the illite/smectite mixed layers (I/S) and the kaolin-group minerals.

Among the numerous books dealing with diagenesis, the general survey published by Larsen and Chilingar (1983), the subsidence process (Force et al. 1991), the review of organic matter evolution (Gautier et al. 1985; Gautier 1986), the methods for analysing the composition of fluids (Hanor 1988) and clay minerals (Eslinger and Pevear 1988) should be known. The reading of a few synthetic articles is strongly advised (Kübler 1984; Velde 1995).

8.1.1
Parameters of Diagenesis

8.1.1.1
Variation in Pressure and Temperature

Subsidence

The progressive sinking of the basement of sedimentary basins is due to the combined effect of slides along faults (tectonic subsidence) and volume reduction related to the cooling of this basement (thermal subsidence). The tectonic subsidence starts at the basin opening under an extension regime that induces the thinning and fracturation of the continental crust, thus forming a depression. The oldest sediments accumulate and their weight increases the downward movements (Fig. 8.1) until isostatic equilibrium has been reached. Thermal subsidence subsequently takes place (reduction in the basement volume by cooling process). The depression is sealed by increasingly recent sediments whose deposition is not disturbed by faults. Modelisation of these effects is now possible (Klein 1991). Nevertheless, the older the basins, the more complicated may be their tectonic history. Tectonic subsidence is sometimes interrupted by compression periods causing the reverse faulting of gravity faults. Finally, the basin structure itself is often greatly modified by the re-activation of gravity faults after the sedimentary filling period. Many basins are cut in "panels" in horst or graben position.

The various sinking, uplift and stability episodes experienced by the "panels" bring about variations in the pressure, temperature and permeability conditions of rock bodies. Therefore, successive processes of dissolution or precipitation of mineral phases take place in a same rock volume, yielding additional organic or inorganic chemical elements or, conversely, bringing about depletions by evacuation of some of these elements. These events are recorded in the petrographical structure of rocks, in the composition and shape of minerals, in fluid inclusions, and in the chemical and isotopic compositions of solutions collected in their environment. Accordingly, the history of diagenesis can be reconstructed gradually as shown by the North Sea example (Lacharpagne, personal communication), for which the sinking curve has been

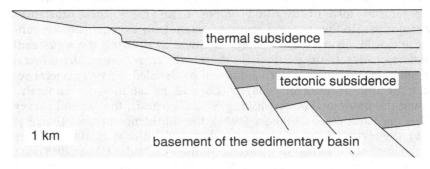

Fig. 8.1. Schematic representation of the subsidence of a sedimentary basin (Klein 1991)

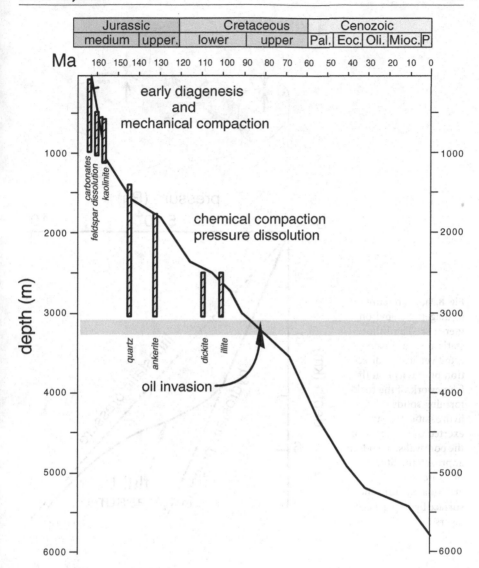

Fig. 8.2. Reconstruction of the diagenetic history of a sedimentary basin, North Sea (Lacharpagne, personal communication)

plotted as a function of time and has been marked with mineral or organic reactions (Fig. 8.2).

Lithostatic and Hydrostatic Pressures
The pressure exerted at one point of the sedimentary pile essentially depends on the weight of the rock column above that point. The higher this column (depth

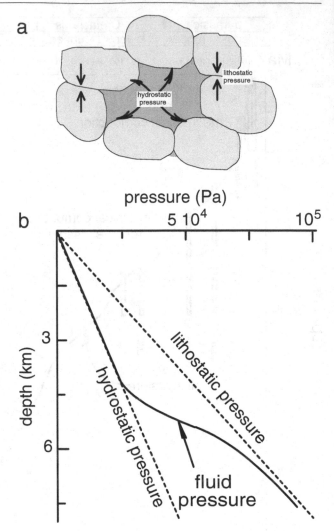

Fig. 8.3a,b. Pressure in diagenetic environments. **a)** The lithostatic pressure is exerted vertically (direction of gravity) on the framework of the rock-forming solids. **b)** The hydrostatic pressure is exerted isotropically on the pore walls. It tends to approach the lithostatic pressure when fluids are isolated from the surface by impermeable layers

of the point involved), the greater the density and the higher the pressure. Since sediment porosity is never zero in the rock column, it can be connected up to the surface. In this case, pressure can be subdivided in lithostatic pressure (which is controlled by the mass of the solids) and hydrostatic pressure (which is controlled by the mass of the pore-filling fluids), both increasing with depth (Fig. 8.3a). However, considering the density difference in between solids and solutions, the lithostatic pressure increases at least twice as fast as the hydrostatic pressure (Fig. 8.3b).

Since permeability is reduced by compaction, fluids do not form a continuous column up to the surface. The hydrostatic pressure increases with the decreasing porosity and tends to approach the lithostatic pressure (Fig. 8.3b).

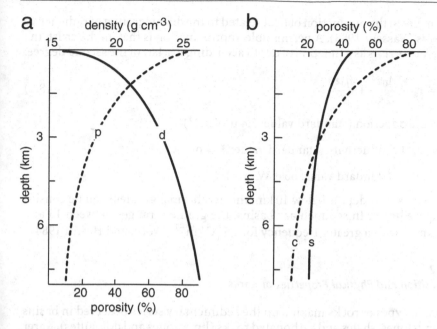

Fig. 8.4a,b. Compaction effects. a) Opposite variations in density (*d*) and porosity (*p*) of a clay sediment with increasing burial depth. b) The compaction rate (reduction of the poral volume) is greater in the clay sediment (*c*) than in a sand (*s*)

The relatively frequent occurrence of totally impermeable layers may induce a fluid overpressure, thus reducing the rock compaction rate, even at great depth. If overpressure exceeds the rock mechanical strength in its environment, the excess energy is released by hydraulic fracturation.

Compaction increases the density of sedimentary rocks by reducing their poral volume (Fig. 8.4a). Therefore, a great part of the ambient fluids is expelled, which results in the decrease of the solution/rock mass ratio. The higher the clay content in the sediment, the greater the decrease of this mass ratio. Accordingly, a mud containing 80% water yields a shale containing less than 20% water, whereas a sand containing a little more than 40% water is transformed into a sandstone containing only 30% water (Fig. 8.4b). From a geochemical standpoint, this means that during compaction, a sedimentary rock passes from an open system where fluid composition is controlled by the ambient environment to an increasingly closed system where this composition is controlled by equilibria with its mineral constituents and organic substances.

Geothermal Gradient

One of the effects of burial of a given sediment is the temperature changes it experiences with increasing depth. The average geothermal gradient for the continental crust is about 30 °C km^{-1}. Nevertheless, the temperature increase is not a linear function of depth. Indeed, temperature rises fast in the first

100 km due to the production of heat related to the disintegration of radioactive elements (Vasseur 1988). A permissible approximation is that, in the crust, the temperature rises as a function of depth according to a law of the second degree:

$$T = T_s - \frac{A_0 z^2}{2K} + \frac{q_0 z}{K} \tag{8.1}$$

A_0 heat production (standard value: $2.4\,\mu\mathrm{W\,m^{-3}}$)

K thermal conductivity (standard value: $3\,\mathrm{W\,m^{-1}\,C^{-1}}$)

q_0 heat flow (standard value: $60\,\mathrm{mW\,m^{-2}}$)

Nevertheless, for depths below 10 km, the geothermal gradient can be considered quite linear. In sedimentary basins, the gradient ranges between 15 and $35\,^\circ\mathrm{C\,km^{-1}}$ with a greater frequency for $25\,^\circ\mathrm{C\,km^{-1}}$ (Wood and Hewitt 1984).

8.1.1.2
Composition and Physical Properties of Rocks

The main types of rocks making up the sedimentary series observed in basins are sandstones, shales and carbonated rocks (limestones and dolomites). Rarer and much smaller are the salty rocks and the volcanoclastic rocks. Among the latter, bentonites are of great interest because they constitute stratigraphic markers (see Sect. 7.2.2.1) and particularly bear witness to diagenetic conditions (Sucha et al. 1993). From the standpoint of clays, the most important rocks are shales, bentonites and sandstones.

Sedimentary clays in muds or in clayey sands are generally mixtures of several species occurring in varying proportions: smectites, vermiculites, mixed-layer minerals, micas and so on. Large- and small-sized clays settle in the form of isolated particles (micas, kaolinite) and flakes (smectites, vermiculites), respectively (Fig. 8.5). The structures of these deposits are generally determined by edge-edge contacts, more rarely by edge-face contacts and never by face-face contacts (Fig. 5.22). This is due to the fact that the chemical bonds of the crystal framework of phyllosilicates are interrupted at the edges of crystals. When the pH conditions of the ambient environment are neutral or alkaline, flocculation is favoured by the unbalanced electric charges. Flocculation takes place when suspension in fresh water is dispersed in the sea.

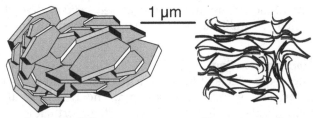

Fig. 8.5. Schematic representation of the structure of large-sized particle aggregates (mica or kaolinite) and small-sized particle flakes (smectite or vermiculite)

mica or kaolinite   smectite or vermiculite

The compaction rate increases according to the clay content of the sediment, thus introducing porosity contrasts. Accordingly, at 2 km in depth for instance, the porosity of a shale is about 10% whereas that of a clay-free sandstone is 30%. These differences not only affect the local water/rock ratio but also the solution flow. Therefore, shales behave like closed systems whereas sandstones constitute open systems. Bentonites are nearly monomineral rocks consisting of pure smectite (montmorillonite). Their extremely high compaction rate reduces their porosity and permeability more than for any other rock. Consequently, they form impermeable layers whose diagenetic evolution is less advanced than that of other clay rocks (Sucha et al. 1993).

8.1.1.3
Diagenetic Fluids

Open System – Closed System
The composition of the solutions impregnating sediments during their deposition is very close to that of seawater. The latter is an infinite reservoir that buffers the chemical activities of alkaline ions notably. Since compaction reduces porosity and permeability, then sandstones and shales form two systems with differing chemical behaviours. Accordingly, as fluids flow in sandstones, changes in their chemical composition are inferred from the chemical composition of seawater under the influence of interactions with solids (precipitation, dissolution). The marine inheritance is recorded in the isotopic compositions, as discussed below. The driving force for the displacement of solutions is the hydraulic gradient. Conversely, clay beds quickly acquire very low permeability and porosity. They contain small amounts of fluids, which flow very slowly, or not at all. In this case, the chemical activities of the dissolved species are controlled by the bulk chemical composition of the rock. The system is closed, and each clay bed undergoes changes independently of the others. Exchanges at the interfaces of these beds occur by chemical diffusion, the driving force being then the chemical potential gradients. To summarise, diagenetic reactions take place in open (Fig. 8.6a), half-closed (Fig. 8.6b) and closed (Fig. 8.6c) systems.

Isotopic Geochemistry
The isotopic composition of authigenic clay minerals is controlled by exchanges with the fluids from which they have been formed. At equilibrium, for a fluid with a given composition, the isotopic exchange rate depends on temperature (see Sect. 4.1.4). This has been shown in the Gulf Coast area where the isotopic ratios $^{18}O/^{16}O$ and D/H vary with depth for the fine fractions only ($< 0.1 \mu m$), which must be considered as representing the authigenic fraction of clays (Fig. 8.7a,b). Yeh and Savin (1977) have shown that the $\delta^{18}O$ of this fine fraction depends on the temperature measured in the drill hole (Fig. 8.7c). The sudden variation of the D/H ratio at about 3 km in depth indicates that the system is closing (Yeh 1980). The fine clay fraction reaches isotopic equilibrium with residual fluids mixed with dehydration fluids of shales.

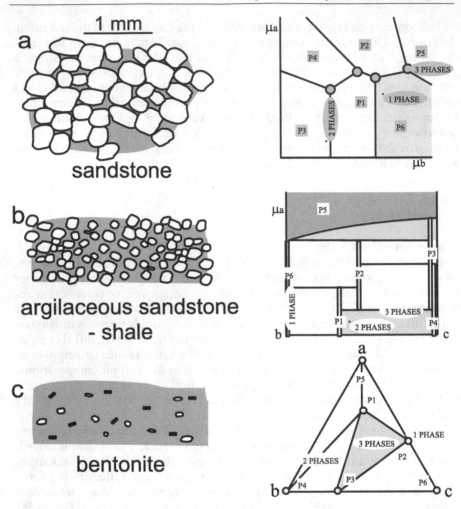

Fig. 8.6a-c. Behaviour in open **a**), half-closed **b**) and closed **c**) systems of the major clay rocks in diagenetic series. The appropriate phase diagrams are represented in front of the rocks. In an open system, the chemical potential of all elements in solution is controlled by the external environment (significant fluid flows). In a half-closed system, the chemical potential of only a few chemical elements is controlled by the external environment. In a closed system, the chemical potential of all elements in solution are controlled by the rock composition

It must be remembered that the isotopic composition of clays depends on three parameters: temperature, fluid/rock ratio and fluid isotopic composition. While the latter is readily controlled in supergene environments where it is known to depend on latitude, its determination is made much more difficult in diagenetic series. Indeed, fluids cannot be analysed most of the time, and even when possible, the fluids sampled in drill holes are not necessarily those that existed at the time of mineral formation.

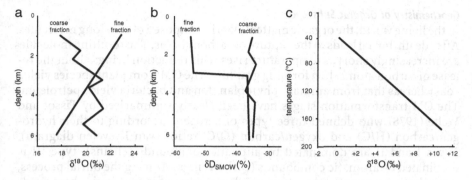

Fig. 8.7a-c. Isotopic composition of the various grain size fractions of clay minerals derived from the Gulf Coast diagenetic shales. **a)** Variation of $\delta^{18}O$ with depth according to data by Yeh and Savin (1977). **b)** Variation of the D/H ratio with depth (Yeh 1980). **c)** Relationship between the temperature measured in drill holes and the variation of the $\delta^{18}O$ ratio

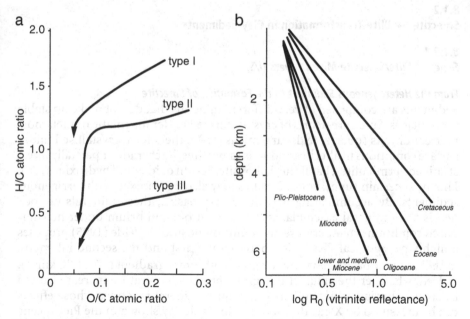

Fig. 8.8a,b. The organic matter in diagenesis. **a)** van Krevelen diagram showing the three types of OM. **b)** The reflectance of vitrinite varies according to a factor *time × temperature* (Kübler 1985)

Geochemistry of Organic Substances

In the living world, the organic matter (OM) is composed of very long molecules. After death, burial leads to the rupture of carbon chains. The resulting molecules are increasingly short as temperature rises. This reduction brings about the release of carbon atoms then forming graphite. The OM from plant species yields coal whereas that from animals, phytoplankton and bacteria yields petroleum. The OM transformation stages have been clearly summarised by Tissot and Welte (1978) who defined three types of kerogens according to their hydrogen/carbon (H/C) and oxygen/carbon (O/C) ratios (van Krevelen diagram). Types I and II are dominated by aliphatic compounds (chains), type III is dominated by aromatic compounds (carbon rings). During the burial process, these three types form separate mineral sequences of increasingly short products (reduction of the H/C and O/C ratios), thus retaining the signature of the initial OM (Fig. 8.8a). In the same time, the solid compounds (organoliths) and notably vitrinite are modified during maturation (Kübler 1984). Their reflectance increases steadily as a function of the factor *time × temperature* (Fig. 8.8b).

The presence of liquid or gaseous organic compounds influences the mineral reactions in four different ways: (1) control of CO_2 and H_2S partial pressures, (2) prevailing influence on redox conditions, (3) influence on pH conditions (OM provides protons) and (4) capacity to form organo-mineral complexes that substantially modify the solubility of elements and increases mass transfers.

8.1.2
Smectite → Illite Transformation in Clay Sediments

8.1.2.1
Series of Illite/Smectite Mixed Layers (I/S)

From the Heterogeneous Sediment to the Formation of Smectite

Sediments are composed of very different minerals that do not make up stable assemblages. Some are debris of crystals formed under magmatic or metamorphic conditions (quartz, feldspars, micas ...), others have crystallised under surface conditions in alterites and soils (smectites, intergrade or partially interstratified vermiculites, kaolinite, halloysite, Fe-Mn oxides and hydroxides ...). Finally, at certain times, volcanic ashes may also have mixed with terrigenous sediments. By accumulating in sedimentary basins, these minerals or rock debris form mineral assemblages that are out of equilibrium under new imposed conditions (temperature and fluid composition). Velde (1995) proposes a highly pedagogical distinction between the first and the second kilometre in sedimentary basins having a normal geothermal gradient (25–$30\,°C\,km^{-1}$). Indeed, whatever the value of this gradient, the temperature increase caused by shallow burials is not sufficient to trigger mineral reactions whose effects can be measured by X-ray diffraction. This is clearly shown in the Pleistocene series of the Colorado Delta (Jennings and Thompson 1987) where the sedimentary influence continues as long as a temperature of $80\,°C$ has not been reached. Once said temperature has been reached, some mineral species of the sediments react, notably clay species except micas. The reaction homogenises the clay composition by yielding pure smectite (low-charge montmorillonite). The older the sedimentary basin, the lower the reaction temperature: 80 °C for a recent age (Pleistocene), 50 °C for Cretaceous sediments. This means that smectite formation during burial is controlled by the parameter *time × temperature* ($t \times T$). This phenomenon differs from "early diagenesis" that leads to lath-shaped overgrowths of Al–Fe clays at the interface between sediments and seawater (see Sect. 7.2.2.3).

Smectite → illite Reaction

The smectite previously formed by the reaction of the clay fractions in sediments in turn transforms as burial progresses, yielding illite. In the first stages of transformation, illite forms randomly ordered mixed-layer minerals with smectite. As the reaction progresses, the stacking sequence becomes ordered and the illite content rises (Fig. 8.9a). Such sequences are known in most of the sedimentary basins that have been drilled for oil exploration (Fig. 8.9b). From a chemical balance standpoint, this reaction consumes potassium, which implies an outside source for this element insufficiently present in smectite.

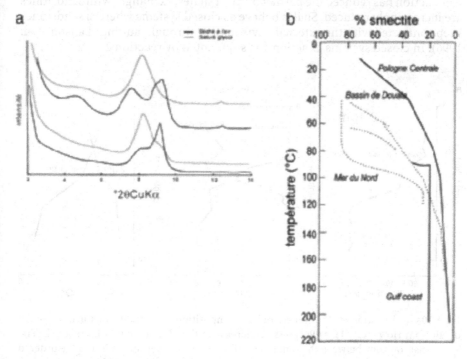

Fig. 8.9a,b. Series of illite/smectite mixed layers (I/S). **a)** Diffraction patterns (air-dried and ethylene glycol-saturated oriented samples) of two examples of I/S minerals derived from the Paris Basin (Lanson 1990). **b)** Variation in the smectite content of I/S series derived from several sedimentary basins (Srodon and Eberl 1987)

The source is derived either from the dissolution of some detrital minerals that had remained inert during the previous stage (smectite formation): micas and K-feldspars or from external solutions brines). The smectite → illite reaction can then be written in two different ways according to whether aluminium is considered (1) a mobile element (Hower et al. 1976) or (2) an immobile element (Boles and Franks 1979):

1. $1\,\text{smectite} + K^+ + Al^{3+} \rightarrow 1\,\text{illite} + Na^+ + Ca^{2+} + Si^{4+} + Fe^{2+} + Mg^{2+} + H_2O$

2. $1.6\,\text{smectite} + K^+ \rightarrow 1\,\text{illite} + Na^+ + Ca^{2+} + Si^{4+} + Fe^{2+} + Mg^{2+} + O^{2-} + OH^-$
   $+ H_2O$

The first reaction does not entail any loss of mass (hence of volume) whereas the second reaction entails the loss of more than 35% of mass.

Both reactions can take place in all types of clay sediments as long as compaction has not established any permeability contrast between sandstones and shales. The composition of the sedimentary solutions is altered by the release of water and dissolved ionic species. Dissolved silica causes the formation of the first overgrowths on quartz. According to the CO_2 partial pressure value, the Ca^{2+}, Mg^{2+} and Fe^{2+} cations form carbonates or Fe-rich chlorites. Once compaction has reduced the permeability of shales, exchanges with sandstones are increasingly reduced. Shales behave as closed systems whereas sandstones keep connected with the external environment through faulting (Lanson et al. 1996). In closed systems, reaction 1 is supplanted by reaction 2.

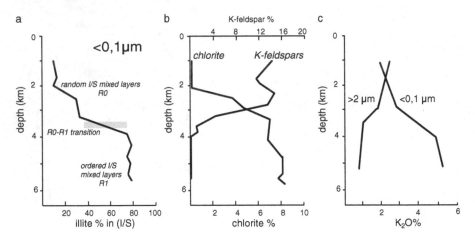

Fig. 8.10a–c. Variation in the mineralogical composition of sediments as a function of depth from Hower (1981). **a)** Progressive increase in the illite content of I/S mixed layers. The transition zone between the randomly ordered mixed-layer minerals and the ordered mixed-layer minerals is indicated. **b)** Progressive decrease in the K-feldspar content and increase in the chlorite content. **c)** Progressive increase in the K_2O content of the clay fine fraction dominated by authigenic minerals (I/S minerals) as the amounts of the coarse fraction dominated by detrital minerals (K-feldspars, micas) decrease

Under diagenetic conditions, the illite resulting from one or the other reaction has never been shown to exist in the first place as individual crystals or particles. It comes in the form of layers interstratified with smectite (montmorillonite) in randomly ordered stacking sequences (R0 I/S minerals). As reactions progress, the illite content of R0 I/S minerals progressively increases up to values exceeding 50% (Fig. 8.10a). They are then replaced by ordered mixed-layer minerals (R1 I/S minerals) and illite crystals (see Sect. 1.2.4.2). In the same time, the proportion of chlorite increases and that of K-feldspars decreases (Fig. 8.10b). The K_2O content of authigenic fractions increases (Fig. 8.10c). The simultaneity of these variations has been shown by Hower (1981).

Series of Mixed-Layer Minerals: 2 Components (I/S) or 3 Components (I/S/V)?

A direct reading of the diffraction patterns reveals a seemingly sudden transition between randomly ordered mixed-layer minerals (R0 I/S minerals) and ordered mixed-layer minerals (R1 I/S minerals) (Fig. 8.11a). Indeed, this transition corresponds to two criteria: (1) disappearance of the 17 Å band whose background in small angles increases with the illite content (increasing saddle/peak ratio), and (2) occurrence of a band near 13.4 Å typical of R1 structures (Fig. 8.9b). This is in fact a slower change. Decomposition of XRD patterns (Fig. 8.11a) and mostly the fit by calculated curves (Fig. 8.11b) show that R0 and R1 I/S minerals co-exist over greater depth than what was first supposed.

Comparative Reactivity of Sedimentary Components

Mineral components of terrigenous sediments do not all react at the same rate when burial depth increases (Fig. 8.12). The most reactive mineral components (unstable under diagenetic conditions) are clays formed under surface conditions (soils and alterites). All species (smectites, di- and trioctahedral vermiculites, mixed-layer minerals, illite, halloysite, allophanes and imogolites) mostly react to the change in the chemical composition of fluids (Velde's first kilometre). Kaolinite itself, which is known to occur down to significant depths (at least 5 km), recrystallises and loses its supergene characteristics such as the presence of Fe^{3+} ions and a high density of crystal defects.

Debris of magmatic or metamorphic minerals persist over longer periods of time. They take part in the mineral reactions as providers of some chemical elements. Their dissolution produces a secondary porosity that adds to or replaces the primary porosity of the sediment. Quartz only dissolves on microsites under compression whereas overgrowths are formed in pore free space. Abercrombie et al. (1994) have shown that the dissolved silica content is first controlled by low-temperature amorphous silica, then by the dissolution of feldspars and finally by high-temperature quartz (Fig. 8.13).

Some minerals remain totally inert under the physicochemical conditions imposed by diagenesis. They then form a detrital inheritance that survives up to the very low-grade metamorphic domain. This is the case of garnets, zircons and magnetite-like oxides. Glauconite remains stable in practically the whole diagenetic domain. The only change is the progressive decrease in the proportion of interstratified smectite they may contain since their formation

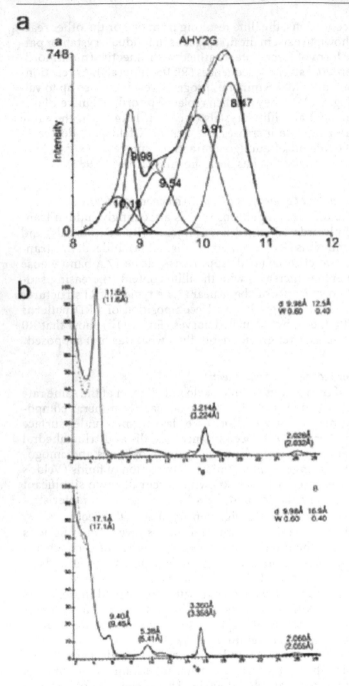

Fig. 8.11a,b. Methods for interpreting X-ray diffraction patterns of illite/smectite mixed layers. **a)** Example of decomposition applied to a diagenetic bentonite sample (Aquitaine Basin, France): smectite + R0 and R1 IS MLMs + mica (°2θ Cu Kα). **b)** Calculation of the closest fit between the theoretical and the experimental pattern (from Drits et al. 1999)

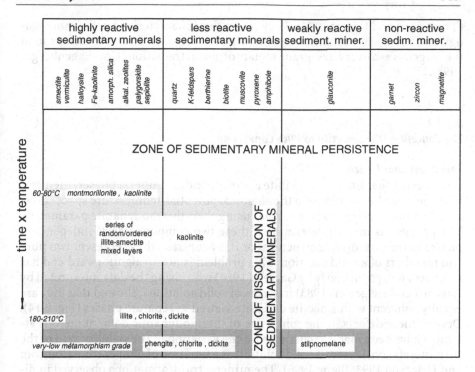

| | | | | | |
| highly reactive sedimentary minerals | | | less reactive sedimentary minerals | | weakly reactive sediment. miner. | non-reactive sedim. miner. |

Fig. 8.12. Reactivity differences between the mineral components of the terrigenous sediments under diagenetic conditions

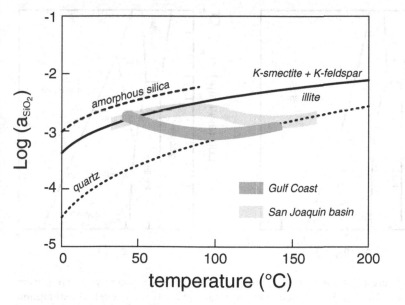

Fig. 8.13. Variation in the silica activity of solutions derived from the reservoirs of two sedimentary basins: San Joaquin and Gulf Coast offshore, U.S.A. (Abercrombie et al. 1994)

(Thompson and Hower 1973). They retain their characteristic pellet morphology and finally recrystallise yielding Mg-rich phyllites and stilpnomelane in the diagenesis-to-very low-grade metamorphism transition (Goy-Eggenberger 1998).

8.1.2.2
The Concept of the Smectite to Illite Conversion

Thermodynamic Models

Considering that smectite and illite are independent mineral phases, their stability field can be calculated in the chemical potential-temperature space (pressure seemingly playing a minor role) using their thermodynamic parameters. The problem is then to determine if these two components are independent phases in the mixed-layer structures or if, by contrast, they represent two pure end members of a solid solution. This problem is not straightforward and has been debated, particularly by Garrels (1984) who, using the data interpreted by Aagaard and Helgeson (1983) in terms of solid solutions, showed that they are equally coherent with smectite and illite viewed as separate phases (Fig. 8.14). Despite these defects in the coherence of thermodynamic data, numerous attempts have been made to characterise the variation in the stability field of the I/S minerals considered as solid solution as a function of temperature (Ransom and Helgeson 1993; Blanc 1996). The mineral transformations observed in di-

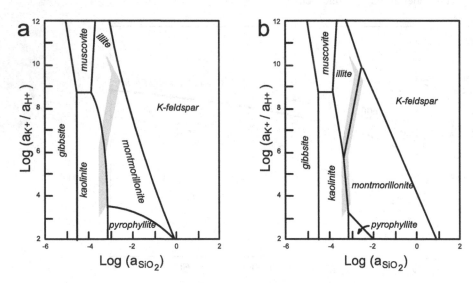

Fig. 8.14a,b. Representations of the solution composition in the sedimentary basin reservoirs in diagram Log (a_{K^+}/a_{H^+}) as a function of Log (a_{SiO_2}) at 25 °C (*grey zone*). **a)** Assumption of a solid solution between smectite and illite (Aagaard and Helgeson 1983). **b)** Assumption of independent phases (Garrels 1984)

agenesis are not properly described by these attempts because crystal growth along a and b directions (see Sect. 1.2.4.2) is ignored.

Kinetic Models

Based upon kinetic laws (see Sect. 3.2.3), several models have been proposed either from observation of natural I/S sequences (Bethke and Altaner 1986; Pytte and Reynolds 1989; Velde and Vasseur 1992) or from experimentation (Eberl and Hower 1976; Huang et al. 1993). Values of reaction order, activation energy and pre-exponential factor vary according to whether illitisation is considered as a simple reaction (smectite $\rightarrow$ illite) or as a stage reaction (smectite $\rightarrow$ R0 I/S minerals $\rightarrow$ R1 I/S minerals ...).

In the smectite $\rightarrow$ illite simple reaction, the law of mass action is considered to affect the reaction rate. Huang et al. (1993) propose a rate formulation in which the relative K^+ content acts as a linear factor:

$$\frac{-dS}{dt} = A \exp\left(\frac{-E_a}{RT}\right) \times [K^+] \times S^2 \qquad (8.2)$$

S % smectite in the I/S minerals

S^2 indicates a second-order reaction

t time (s)

A frequency factor ($8.08 \times 10^{-4}\,s^{-1}$)

E_a activation energy ($28\,kcal\,mol^{-1}$)

R perfect gas constant ($1.987\,cal\,°K^{-1}\,mol^{-1}$)

T absolute temperature (°K)

K^+ concentration (molarity) of K^+ in solution

By comparing the results of experiments carried out under the same temperature conditions (180 °C) for different K^+ concentrations (Eberl and Hower 1976; Robertson and Lahann 1981; Howard and Roy 1985; Whitney and Northrop 1988; Meunier et al. 1998) have shown that a linear relationship appears between the potassium content and the reaction progress (dS/dt).

The major difficulty in this type of approach (simple reaction) is the selection of the reaction order, which is determined by its mechanism. What does a second-order (Huang et al. 1993) or a sixth-order reaction (Pytte and Reynolds 1989) mean? To avoid this problem, other kinetic approaches consider that illitisation is a stage reaction. Bethke and Altaner (1986) have tried to describe the I/S sequences observed in shales by successive reactions forming SSI then ISI structures from smectite S. Each of the forms S, SSI and ISI is characterised by a specific reactivity coefficient: 0.5, 1 and 1.1. The three reaction constants k_1, k_2 and k_3 are calculated as follows:

$$k_1 = A \exp\left(\frac{-E_1}{RT}\right) \tag{8.3}$$

$$\frac{k_0}{k_1} = \exp\left(\frac{-[E_0 - E_1]}{RT}\right) \tag{8.4}$$

$A = 10^{-3} \, \text{s}^{-1}$

$E_0 - E_1 = 0.5 \, \text{kcal mol}^{-1}$

$E_2 - E_1 = 1.7 \, \text{kcal mol}^{-1}$

This velocity rule has been successfully applied to shales which have undergone a deep burial but fails to describe properly the less advanced diagenetic transformations.

Velde and Vasseur (1992) have proposed an empirical kinetic model based on I/S sequences derived from seven basins of differing ages (1 to 200 Ma). This model rests upon the assumption of a two-stage smectite-to-illite transformation corresponding to the formation of two mineral structures with changing compositions (% illite). The first reaction describes the transformation of smectite into a randomly ordered I/S mineral (R0) whose percentage in smectite component ranges from 100% to 50%. The second reaction yields I/S minerals whose percentage in illite component ranges from 50% to 100%. The second reaction is dependent on the first one because, for each lost smectite layer in a R0 I/S mineral (relative increase in the illite content), a R1 50% illite crystal is formed. Subsequently, this crystal progresses towards the 100% illite composition, end of the reaction (Fig. 8.15). New R1 I/S crystals are produced as long as there are R0 I/S minerals with a composition tending towards the end of reaction 1 (50% illite). Therefore, the proportion of R0 crystals decreases as the reaction progresses.

The kinetic equations are as follows:

Reaction 1 (randomly ordered I/S (R0) whose percentage in smectite component ranges from 100 to 50%):

$$\frac{dS}{dt} = -k_1 S \text{ with } \log(k_1) = \log(A_1) - \frac{E_1}{RT} \tag{8.5}$$

$$\log(A_1) = 24.4 \, \text{Ma}^{-1} \text{ and } E_1 = 76.8 \, \text{kJ mol}^{-1} \tag{8.6}$$

Reaction 2 (regularly ordered I/S mineral (R1) with 50 to 0% smectite)

$$\frac{dM}{dt} = k_1 S - k_2 M \text{ with } \log(k_2) = \log(A_2) - \frac{E_2}{RT} \tag{8.7}$$

$$\log(A_2) = 7.2 \, \text{Ma}^{-1} \text{ and } E_2 = 37.4 \, \text{kJ mol}^{-1} \tag{8.8}$$

This model gives a rather good description of the diagenetic transformations of the I/S minerals in shales, and even in bentonites. Its major drawback lies

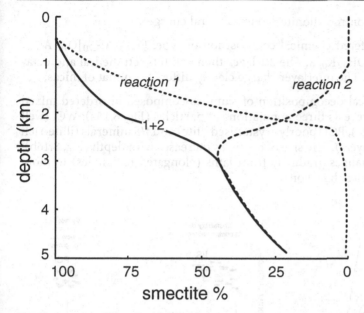

Fig. 8.15. Empirical kinetic model of the smectite-to-illite transformation through two reactions (Velde and Vasseur 1992). Reaction 1: smectite→randomly ordered I/S mineral (R0) with 100 to 50% smectite (*dotted curve*); reaction 2: R0 → regularly ordered I/S mineral (R1) with 50 to 0% smectite (*dashed curve*); the sum of both reactions (*full line*)

in a too early prediction (too shallow depth) of the end of the reaction end at 100% illite (Varajao and Meunier 1995).

The empirical model by Velde and Vasseur (1992) based on a mineral reaction between two phases is supported by recent works (Claret 2001). Indeed, the classical approach to the series of illite-smectite mixed layers in diagenesis has been modified by the fit of experimental XRD patterns by calculated ones using 3-component mixed layer minerals. The study of the diagenetic series in the eastern part of the Paris Basin and in the Gulf Coast area (USA) shows that the sample diffraction patterns result from the sum of the diffraction patterns of the discrete phases, of which the composition remains constant but the proportions vary with depth. This more rigorous but unfortunately very cumbersome new approach will undoubtedly become the most efficient procedure for interpreting diffraction patterns in the future, once the trial and error work performed by hand to date is somehow automated.

8.1.2.3
The Concept of Illite Growth on a Montmorillonite Nucleus

Basis of the Concept
The analytical data of a great number of series amount to three essential facts (Meunier et al. 2000):

1. smectite is a montmorillonite (no tetrahedral charge),

2. illite has a buffered chemical composition of type: $[Si_{3.30}Al_{0.70}]O_{10}(Al_{1.80}Fe^{3+}_{0.05}Mg_{0.15})(OH)_2K_{0.90}$. The 2:1 layer then exhibits tetrahedral and octahedral charges. The interlayer charge clearly differs from that of micas,

3. the mathematical decomposition of samples composed of ordered mixed layers to illite reveals three populations of particles (Fig. 8.16a): WCI (well crystallised illite), PCI (poorly crystallised illite) and I/S minerals (illite-rich R1 I/S mixed layers). The size of particles increases with depth (Fig. 8.16b). Their shape changes gradually from laths (elongated rectangles) to more isotropic particles (hexagons).

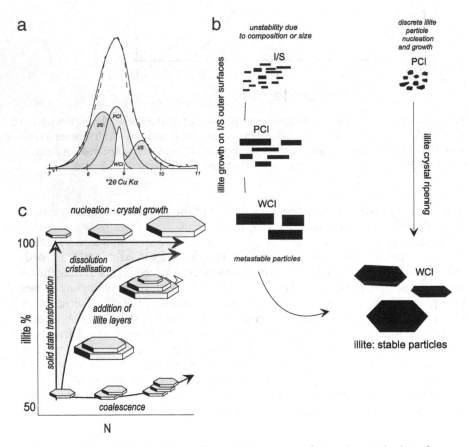

Fig. 8.16a–c. The R1 I/S minerals in diagenesis (Lanson and Meunier 1995). **a)** Mathematical decomposition of the broad band near 10 Å. *WCI*: well crystallised illite; *PCI*: poorly crystallised illite; *I/S*: the two bands correspond to an illite-rich I/S mineral. **b)** Schematic representation of the relationships between morphology, size and composition of I/S particles under increasing diagenetic conditions. **c)** The different possibilities of illitisation for regularly ordered I/S minerals (*N*: number of layers in the coherent domain)

The illite content of I/S crystals progressively increases as they grow. This growth takes place simultaneously with the formation and growth of discrete illite particles. This means that the surrounding fluids are oversaturated with respect to illite (buffered composition) and not to mixed-layer minerals (illite+smectite). The montmorillonite component does not disappear from the growing I/S crystals, but is "fossilised" within the structure (see Sect. 1.2.5.1). Accordingly, the size and proportion of illite in regularly ordered I/S minerals can increase in four ways (Fig. 8.16c): (1) smectite to illite solid state transformation (N constant); (2) coalescence of I/S crystals of similar composition (limited increase in the illite content through interfaces); (3) nucleation and direct growth of illite crystals; (4) addition of illite layers on I/S crystals.

Model of the I/S/V Crystal Structure

At a first glance, the identification by X-ray diffraction of three types of layers in the I/S minerals (Drits et al. 1997) – montmorillonite (2 glycol layers), vermiculite (1 glycol layer) and illite (0 glycol layer) – does not seem consistent with the chemical compositions corresponding to a two component (montmorillonite + illite) mixture. This is only an outward discrepancy if these 3-component mixed-layer minerals are considered chemically identical to mixtures of illite and montmorillonite. In this case, layers with vermiculite behaviour are partly illites from the standpoint of tetrahedral and octahedral charges, but the interlayer zone is devoid of potassium (Meunier et al. 2000). Indeed, a I/S/V mixed-layer mineral corresponds to a succession of interlayers bordered with tetrahedral sheets without substitution (Si_4O_{10}) compensating for octahedral charges (montmorillonite) and interlayers compensating for tetrahedral and octahedral charges (illite). If tetrahedral substitutions are added, the charge is increased and a 2:1 layer polarity is induced (Altaner and Ylagan 1997).

The growth of I/S/V minerals by addition of illite layers requires that the composition of the crystal outer surfaces be such that cumulative interlayer charges correspond to +0.90 per Si_4O_{10}. The outer interfaces of each I/S/V crystal have the chemical composition of a half illite layer (Fig. 8.17). When crystals are superposed in an oriented sample for instance, interfaces are saturated by the surrounding cations. If the latter are cations with a strong hydration power (Ca^{2+}, Mg^{2+}, Na^+), the joining interfaces form a pseudo interlayer sheet having a swelling power similar to that of vermiculite. Consequently, when examined by X-ray diffraction, mixed-layer minerals exhibit three types of layers: illite, montmorillonite and vermiculite. This model in which growth phenomena are accounted for is still hypothetical to date.

Formation and Growth of I/S/V Minerals

The addition of illite layers over a montmorillonite structure used as an embryo poses the problem of the nature of the solid-solution interfaces. To obtain an I/S mixed-layer structure similar to the one described in Fig. 8.17, it is necessary that low-charge montmorillonite interfaces be transformed into vermiculite interfaces whose charge is partly derived from tetrahedral substitutions

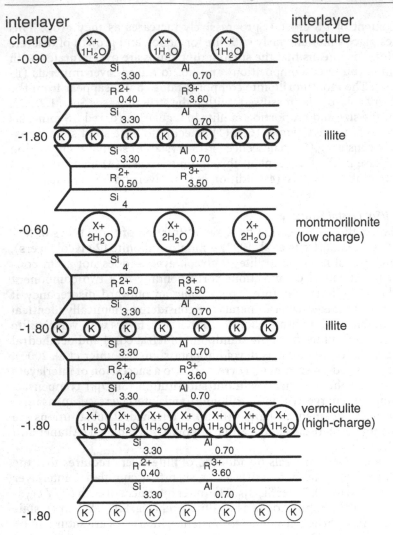

Fig. 8.17. The symmetrical distribution of electric charges in the 2:1 layers about the interlayer zones corresponds to illite and montmorillonite in an I/S/V mixed-layer structure (Meunier et al. 2000)

(Fig. 8.18a). This type of transformation has been shown in experimental works (Howard 1981; Whitney et Northrop 1988). The reaction can be written as follows:

low-charge montmorillonite + Na–Ca solutions

→ vermiculite + Mg-smectite + quartz

The formation of quartz and magnesian smectite is due to the occurrence of substitutions of Al for Si in tetrahedral sheets and to the decrease in the sub-

a Vermiculitization of the crystal outer interfaces

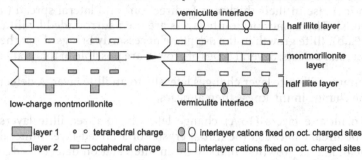

b the illite growth stage

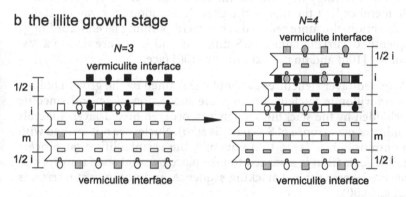

Fig. 8.18a,b. I/S/V mixed layers. a) Transformation of the specific interfaces of a low-charge montmorillonite into vermiculite by occurrence of charges in the tetrahedral sheet. b) Addition of illite layers. In the growth process considered here, the charge of the interface is equal to half the charge of an illite layer (0.45 per Si_4O_{10})

stitution rate of Mg for R^{3+} in octahedral sheets, which releases Si^{4+} and Mg^{2+} ions into solutions. Under higher temperature conditions, magnesian smectite is replaced by chlorite. In the case of high CO_2 pressure, magnesian phyllosilicates are replaced by dolomite. The increase in the size of I/S particles then takes place by addition of illite layers over vermiculite interfaces (Fig. 8.18b).

Illite layers grow over vermiculite interfaces such that the charge of the interlayer zone so created is equal to 0.9 per Si_4O_{10} (Fig. 8.18b). The crystal then gains an additional layer (growth along the c direction) and the proportion of illite increases in the mixed-layer structure. In its new state, the crystal still exhibits vermiculite interfaces likely to fix a new layer. The process can take place continuously by spiral growth from a screw dislocation emerging on the (001) faces, as shown in Fig. 1.30. This type of growth, well known in micas, has been revealed in I/S crystals by Inoue and Kitagawa (1994). This process is much less energy consuming than direct nucleation of illite layers on the I/S (001) faces. This means that growth can be triggered for a much lower fluid oversaturation rate.

The growth of crystals or particles takes place in the three space directions, resulting in an increase in their thickness (*c* direction) and lateral spread (*a* and *b* dimensions). The montmorillonite layers are then surrounded by illite layers (Fig. 8.19a,b). Illite growth in the *a-b* plane necessarily brings about the occurrence of crystal defects. Several types of growth are possible in this plane:

1. the initial montmorillonite layer changes laterally to an illite layer. A defect related to the change in thickness then appears;

2. two montmorillonite layers (15 Å) change laterally to three illite layers (10 Å). The boundary between the initial crystal of a mixed-layer mineral and the zone of lateral growth is marked by an intercalation defect of edge dislocation type (Fig. 8.19c).

3. the crystal morphology depends on the crystal growth directions (Güven 2001). Remember that the higher the growth rate along a given crystallographic direction, the more reduced is the corresponding face. Accordingly, the frequently occurring laths in I/S minerals and illites are due to a fast growth along [100] and do not exhibit any (100) face.

All I/S/V mixed-layer structures cannot be explained by the growth of illite over a montmorillonite nucleus, because the succession of several smectite layers is forbidden by the stacking sequence in ordered mixed-layer minerals (the probability of occurrence of S–S pairs is zero). Nuclei are not formed with a single montmorillonite layer. Consequently, the vermiculitisation process preceding illite formation is assumed to take place by solid state transformation of smectite layers within a stacking sequence, and not only at interfaces (Meunier et al. 2000).

8.1.2.4
Direct Precipitation of Illite and I/S Minerals

In sandstones, illite is not always formed by transformation of smectite. It may nucleate and grow directly on pre-existing solid supports, such as kaolinite or quartz notably. This has been shown by the study of the Rotliegende sandstone reservoir in the Broad Fourteens Basin, North Sea (Fig. 8.20). From the time of sedimentary deposition (275 Ma) to the Cimmerian orogeny (155 Ma), the burial diagenesis caused kaolinite to recrystallise into dickite. The faulting process during orogeny allowed hot and salted fluids derived from underlying formations (Zechstein) to flow. The sudden chemical disequilibrium imposed in the reservoir triggered the precipitation of illitic minerals (pure illite and I/S minerals), the morphology and crystallochemical characteristics of which did not depend on the stage of advancement of the classical reaction smectite $\rightarrow$ illite (Lanson et al. 1996). These characteristics (illite content of I/S minerals, illite / I/S minerals ratio, tridimensional crystal structure) reflect the temperature conditions established at the time of the crystallisation of these minerals. This is typically the result of a sudden hydrothermal phenomenon due to the invasion of the reservoir by K^+-rich brines.

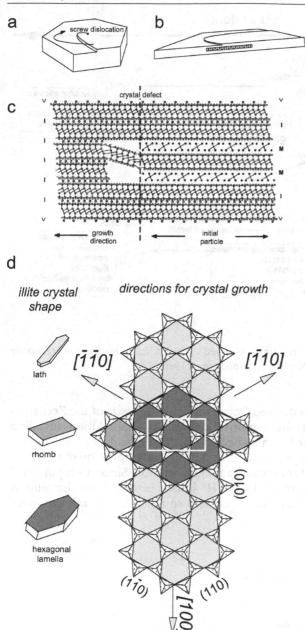

Fig. 8.19a–d. Possible growth process of I/S crystals. **a)** Spiral growth from a screw dislocation emerging on the (001) face of a I/S crystal. The growth direction is given by the arrow. **b)** Growth along the three space directions. Illite surrounds the montmorillonite layers that were used as a nucleus. **c)** Growth takes place from a crystal defect of edge dislocation type. *I*: illite; *M*: montmorillonite; *V*: vermiculite. **d)** Control of illite (or mica morphology by the crystal growth directions (Güven 2001)

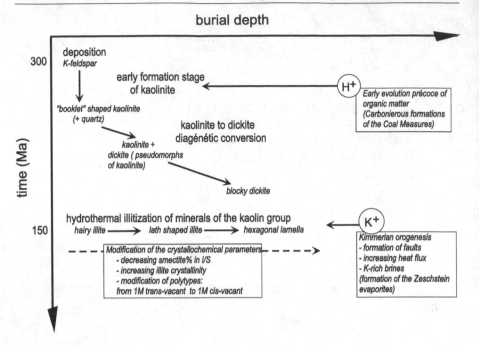

Fig. 8.20. History of the mineral reactions observed in the Rotliegende sandstone reservoir in the Broad Fourteens Basin, North Sea (Lanson et al. 1996)

The potassium source is derived from evaporitic deposits of the Zechstein Formation and not from the dissolution of K-feldspars. The solutions invading sandstones are oversaturated with respect to illite and K-feldspars. The higher oversaturation rate with respect to illite explains the precipitation of the latter whereas K-feldspars do not show any trace of dissolution. Since the input of K^+ ions is not compensated for by the loss of H^+ ions as is the rule for dissolution reactions, the a_{K^+}/a_{H^+} activity ratio decreases so that I/S minerals are more easily formed than illite.

8.1.3
Transformations of Other Clay Minerals

8.1.3.1
The Kaolinite-to-Dickite Transition

Kaolin-group minerals are $Si_2O_5Al_2(OH)_4$ polymorphs. Kaolinite forms book-shaped triclinic flat crystals whereas dickite, although of monoclinic symmetry, exhibits rhomboidal prisms (Fig. 8.21a,b). Kaolinite (1-layer polymorph) comprises a layer stacking sequence whose vacancy is in the B position. Dickite (2-layer polymorph) is formed by the regular alternation of type B- and type C-layer stacking sequence (Fig. 8.21c). Nacrite is a monoclinic 2-layer poly-

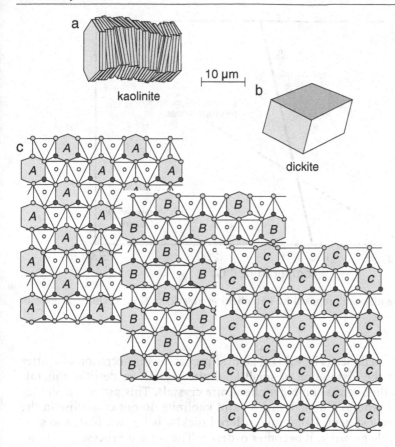

Fig. 8.21a–c. Crystallographic characteristics of kaolinite and dickite. **a,b)** Crystal shapes in diagenetic environments. **c)** The three types of vacancy position in the 1:1 layer (*light grey*: octahedral vacancy in A, B or C position; *heavy lines*: the three octahedra as seen through the hexagonal cavity of the tetrahedral sheet)

morph. Relatively rare in diagenesis, nacrite is known to be formed in rocks containing bitumen or coal.

The dimensions of the dickite unit cell are smaller than those of two kaolinite unit cells superposed along the c^* direction. Therefore, its molar volume is smaller and dickite is viewed as the stable polymorph at high pressure. Ehrenberg et al. (1993) have proposed a kaolinite-dickite isograde in the pressure-temperature space, kaolinite transforming into dickite between 100 and 160 °C for pressures in the range 0.5–0.8 kbar (Fig. 8.22). Therefore, according to these authors, if pressure is known (calculated by the weight of the sedimentary column), the kaolinite-to-dickite transition can be used as a geothermometer.

In reality, isograde is a simplistic view of dickite crystallisation. Beaufort et al. (1998) have shown that dickite crystallisation takes place according to two simultaneous processes in the Rotliegende sandstone reservoir, Broad Four-

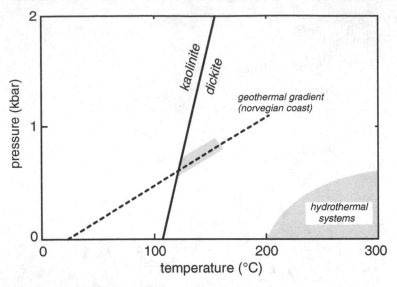

Fig. 8.22. Phase diagram proposed by Ehrenberg et al. (1993). *Grey*: occurrence fields of dickite

teens Basin, North Sea (Fig. 8.23). The first process is the accretion of matter derived from the dissolution of small kaolinite crystals or detrital minerals (K-feldspars then micas) on bigger kaolinite crystals. This process is similar to a ripening process although dickite and kaolinite do not crystallise in the same symmetry system. Randomly ordered dickite is formed first, and subsequently, while growing, it becomes ordered. The second process is a direct neoformation of ordered dickite from the solutions.

8.1.3.2
Corrensite and Magnesian Chlorite

Corrensite and minerals described as chlorite/smectite mixed layers are formed either in volcanic environments (volcanoclastic rocks, basalts) or in detrital or carbonated sedimentary series. Volcanic rocks will be discussed in a later paragraph. Since corrensite and magnesian chlorite are not directly formed by precipitation in present day evaporitic environments (Bodine and Madsen 1987), they were classically assumed to derive from the transformation of a magnesian smectite. The latter was believed to assume the role of precursor in the diagenetic crystallisation sequence: saponite → corrensite → chlorite (Fig. 8.24a). This sequence is not that commonly found because corrensite and magnesian chlorite occur in carbonated rocks (Hillier 1993) or in the fractures of the sedimentary basin basement. In fact, diagenesis of sedimentary series leads to the formation of these minerals only when brines impose very special local chemical conditions of pH and dissolved magnesium activity.

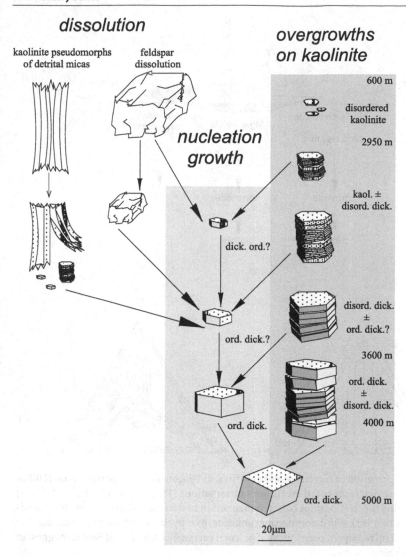

Fig. 8.23. Schematic representation of the two processes of dickite formation in the Rotliegende sandstone reservoir (Broad Fourteens Basin, North Sea): growth of dickite on kaolinite crystals, nucleation and growth of dickite crystals

The source of chemical elements necessary for the crystallisation of corrensite or magnesian chlorites in carbonated rocks is double: dolomite for Fe and Mg; dioctahedral detrital clays and quartz for Si and Al. Two mineral reactions are considered:

5 dolomite + 1 kaolinite + 1 quartz + 1 H_2O
$\rightarrow$ 1 chlorite + 5 calcite + 5 CO_2 (8.9)

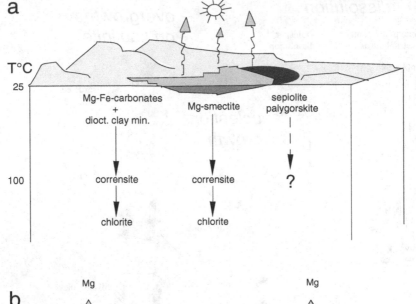

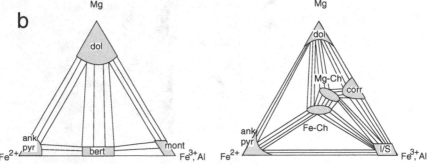

Fig. 8.24a,b. Formation of corrensite and chlorites. **a)** Diagenesis of evaporitic series (Hillier 1993). **b)** Phase diagrams based on Hillier's observations (1993) in the Mg–Fe^{2+}–(Fe^{3+},Al) system showing the relationships between magnesian and ferrous mineral sequences (modified from Velde 1985). *dol*: dolomite, *ank*: ankerite, *pyr*: pyrite, *mont*: montmorillonite, *I/S*: illite-smectite mixed layers, *bert*: berthierine, *corr*: corrensite, *Mg-chl* and *Fe-chl*: magnesian and ferrous chlorites, respectively

(Zen 1959)

$$15\,\text{dolomite} + 2\,\text{illite} + 3\,\text{quartz} + 11\,H_2O$$
$$\rightarrow 3\,\text{chlorite} + 15\,\text{calcite} + 2\,K^+ + 2(OH)^- + 15\,CO_2 \qquad (8.10)$$

(Hutcheon et al. 1980)

In Devonian lacustrine clays of the Orcadian Basin (Scotland), Hillier (1993) has shown two essential diagenetic effects: (1) the amount of chloritic minerals increases as the amount of dolomite decreases, (2) a magnesian mineral sequence (corrensite, chlorite and corrensite-chlorite intergrowths) is formed along with a chlorite having a higher iron content (Fig. 8.24b). This observation is important because crystallisation of chlorite is shown to take place according to two different processes depending on its iron content. The co-existence of two chemically different chlorite populations shows that equilibrium under the P, T, μx conditions is not reached. The predominating kinetic effects eliminate any possibility of searching for geothermometers based on the evolution of composition within the compositional range of a solid solution. The rock chemical composition is obviously of paramount importance for the formation of mineral assemblages in diagenesis.

Flows of hot brines at the interface between sediments and basement in sedimentary basins bring about mineral reactions in their flowing zones: porous sedimentary formations, fractures in basements. The example of the scientific drill hole at Sancerre-Couy (Géologie Profonde de la France programme) has shown that hot and highly salted fluids have infiltrated into the basement fractures and have reacted with the wall rock (gneiss and amphibolites). Their residence time has been long enough for them to reach the isotopic equilibrium with rocks, and for the formed secondary minerals to be exceptionally well crystallised (Beaufort and Meunier 1993). The formation of these minerals is controlled by the rock local chemical composition and not by temperature: saponite in pyroxene bearing rocks, corrensites in amphibolites, chlorites in gneiss.

The formation of corrensite, like the formation of magnesian chlorite, does not always take place at the same temperature. It depends on the local chemical composition, the geothermal gradient and the duration of diagenesis. Drill holes in which present-day temperature has been measured give a few indicators for the formation of corrensite (Table 8.1).

Table 8.1. Estimated temperatures of corrensite formation in various geological contexts

Temp. (°C)	Geological formation	Age	References
60–70	Detrital sandstones	Cretaceous	Chang et al. (1986)
	and shales	Tertiary–Upper Cretaceous	Pollastro and Barker (1986)
84–91	Marine sediments–tuffs	Tertiary	Iijima and Utada (1971)
100	Carbonates	Jurassic	Kübler (1973)
	Volcanoclastic rocks	Palaeocene–Oligocene	Chudaev (1978)
	Volcanoclastic rocks	Miocene	Inoue and Utada (1991)
120–130	Volcanoclastic rocks	Upper Eocene–Lower Oligocene	Stalder (1979)
	Volcanoclastic rocks	Tertiary	Kisch (1981)
	Molasse	Tertiary	Monnier (1982)
160	Arkoses	Palaeocene	Helmold and van de Kamp (1984)

Corrensite and magnesian chlorite are often interlayered in composite crystals. The proportion of chlorite increases as a function of the *time × temperature* parameter. Pure magnesian chlorite does not exist below 100 °C; it is of polytype IIb (β = 97°) and forms mostly platelets assuming a honeycomb pattern and covering the grain surface. It retains the porosity of sandstones by inhibiting the precipitation of quartz overgrowths on detrital grains (Hillier 1994). These coverings are frequently found in sandstones formed from Eolian sands (dunes, sebkas) reflecting desert climates and evaporitic conditions.

8.1.3.3
Berthierines, Glauconites and Ferrous Chlorite

Berthierine-like iron-rich chlorites the half-cell formula of which is $Si_{2-x} Al_x$ $R_a^{2+} R_b^{3+} \square_c O_5(OH)_4$ are formed at the beginning of diagenesis (Walker and Thompson 1990; Hillier 1994; Ehrenberg 1993) or even under surface conditions in a reducing environment (laterite alteration under peat or lignite, shallow marine sedimentation). They can be found at varying depths (temperatures) according to the conditions imposed by the environment (presence of hydrocarbon compounds, carbonates, sulphides etc). They are subsequently transformed into a chamosite-like 7 Å chlorite. Temperature at the beginning of transformation is variable (Table 8.2).

The berthierine-to-chamosite transformation takes place through the interstratification of 7 Å layers with 14 Å layers. The number of serpentine-type layers decreases with temperature through a transformation process that is still poorly known (Ryan and Reynolds 1996). This transformation might take place by inversion of tetrahedra without variation in chemical composition (Brindley 1961; Jiang et al. 1994). The presence of 7 Å layers can be detected in ferrous chlorites up to 200 °C. These chlorites are of polytype Ib (β = 90°) at first, subsequently transforming into polytype IIb (β = 97°) above 200 °C. Ferrous chlorites are indicators of tropical sedimentary environments (great river deltas) in which ferric ooliths are formed (Hillier 1994).

Those glauconites scattered in sandstones or limestones were beidellite-glauconite mica mixed layers at the time of their formation (see Sect. 2.1.4.2). The proportion of smectite depends on the local chemical conditions and particularly on the potassium concentration of the environment. Accordingly,

Table 8.2. Temperature of berthierine formation in various geological contexts

Precursor	Fe^{2+} source	Temperature (°C)	References
Kaolinite	Siderite	160	Iijima and Matsumoto (1982)
Kaolinite	Volcanic clasts	70	Hornibrook and Longstaffe (1996)
		100	Jahren and Aagaard (1989)
		195	Walker and Thompson (1990)
		100	Hillier (1994)
Odinite			Odin (1988)

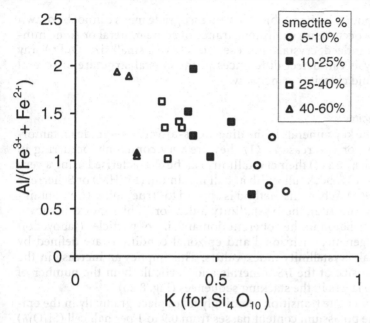

Fig. 8.25. Evolution of the chemical composition of glauconites as a function of their interstratified smectite content (data from Thompson and Hower 1975)

their heterogeneous composition derived from the sedimentary episode makes the diagenetic effects (*time* × *temperature*) difficult to grasp. Nevertheless, the series of samples studied by Thompson and Hower (1975) shows that the smectite content decreases with the increasing age of the formations: 5 to 10% for the Cambrian – Ordovician Period, 10 to 25% for the Jurassic – Cretaceous Period, 25 to 40% for the Eocene Period and 40–60% for the Eocene Period-to-present. Figure 8.25 shows that the K content is inversely correlated with the smectite content and with the $Al/(Fe^{3+} + Fe^{2+})$ ratio. This confirms that, when interstratified with glauconite mica, smectite is an aluminous dioctahedral smectite and not a nontronite.

8.1.4
From Diagenesis to Very Low-Grade Metamorphism

8.1.4.1
Geology of the Diagenesis-to-Very Low-Grade Metamorphism Transition

The transition from diagenesis to the very low-grade metamorphism can be observed in four types of geological contexts: (1) extension tectonic basins (West of England), (2) subduction zones (California), (3) collision zones (The Alps), and (4) thermal metamorphism around magmatic bodies. The relationships between geodynamics and transformation facies of clays have been summarised by Merriman and Frey (1999). The very low-grade metamorphism has long been a poorly known subject because isograde maps cannot be drawn by observing the occurrence or disappearance of some mineral or some mineral assemblage. Indeed, crystals in these rocks have a small size. Only X-ray diffraction analysis can detect differences in the crystal structure of several species – illites and chlorites principally.

From Illite to Phengites

Illites proved to be key minerals in the diagenesis-to-very low-grade metamorphism transition for two reasons: (1) they are very commonly occurring in sedimentary rocks, and (2) their crystallinity can be characterised using a very simple index: the 10 Å peak full width at half maximum (FWHM) or Scherrer's width (Kübler 1964). When this method is applied to "true" illites (i. e. containing less than 5% smectite), the crystallinity index (or Kübler index) is shown to describe the variation in the coherent domain size of particles (Jaboyedoff et al. 1999). Diagenetic, anchizonal and epizonal conditions are defined by increasingly small crystallinity index values. This implies an increase in the coherent domain size of the I/S minerals and particularly in the number of consecutive illite layers in the stacking sequences (Fig. 8.26).

The illite-to-phengite transition seemingly takes place gradually in the epizone. Indeed, the potassium content passes from 0.9 to 1 per half cell (Si_4O_{10}) whereas aluminium and bivalent elements ($R^{2+} = Fe^{2+} + Mg^{2+}$) contents do no longer vary independently but are controlled by the Tschermak substitution: $(Mg, Fe^{2+})^{VI}, Si^{IV} \Leftrightarrow Al^{VI}, Al^{IV}$. However, electron microprobe analyses cannot be blindly relied upon because observations by transmission electron microscopy show that mica and chlorite often constitute intergrowths (Lee et al. 1984; Ahn et al. 1988), which alters the evaluation of the phengite substitution. The progressive transition from illite to phengites occurs through the increase in the coherent domain size, which passes from 15–25 layers to more than 60. Although the polytype change from $1M$ to $2M_1$ is not as system-

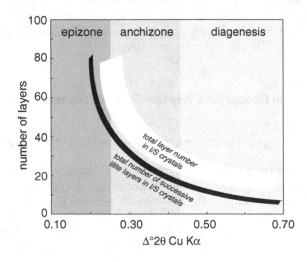

Fig. 8.26. Variation in the crystallinity index of illites (Kübler index) in the diagenesis – anchizone – epizone transition. The increasingly small index values show an increase in the coherent domain size of the I/S minerals and in the number of successive illite layers in the stacking sequence

atic as admitted for many years (Dong and Peacor 1996), $2 M_1$ still remain the prevailing polytype in the epizone.

Chlorites and "Metamorphic Vermiculites"

The chemical composition of chlorites changes during diagenesis. They become increasingly aluminous and the number of vacancies in the octahedral sheet decreases (Hillier and Velde 1991). Nevertheless, the variations are small and difficult to determine owing to contaminations. By compiling thousands of data, Zane et al. (1996) have shown that the chemical composition of chlorites and their host rocks are correlated. Therefore, any attempt to establish geothermometers from the composition of chlorites is fruitless. The only unquestionable fact is that the compositional range of solid solutions decreases with increasing temperature. "Metamorphic vermiculites" (Velde 1988) have optical properties and compositions intermediate between those of biotites and chlorites. Although low, their swelling property is unquestionable and makes them closer to randomly ordered chlorite/smectite mixed layers (Beaufort 1987).

A crystallinity index of chlorites has been modelled on that of illites, using the 14 and 7 Å diffraction peak full width at half maximum (Arkai 1991). The values of this index are rather well correlated with those of illite, which makes its use possible to establish a zoneography of the diagenesis-to-metamorphism transition. However, the coherent domain size of chlorites increases more slowly than that of illites. This is due to the greater deformability of chlorites when subjected to an increase in pressure. The coherent domain size is reduced by the resulting greater number of crystal defects.

8.1.4.2
Mineral Zoneography of the Diagenesis-to-Metamorphism Transition

The diagenesis-to-very low-grade metamorphism transition has been described in numerous geological sites (Merriman and Frey 1999). It has been thoroughly explored for many years in the Alps (Frey et al. 1980), and more particularly in the Morcles thrust sheet south-west of Switzerland (Kübler 1969; Goy-Eggenberger 1998). The intensity of the deformations in the structure of this thrust sheet regularly increases from the front zone (diagenesis) towards the root zone (epizone). The diagenesis–anchizone and anchizone–epizone boundaries are plotted from the crystallinity index of illite, 0.33 and $0.22 °2\theta$ Cu Kα respectively (Fig. 8.27a). The mineralogical study of the various types of rocks (limestones, sandstones, shales) shows that the transition to metamorphism is emphasised by several reactions (Fig. 8.27b):

1. progressive transformation of glauconites into stilpnomelane in the anchizone;

2. disappearance of kaolinite (or dickite) in the epizone while pyrophyllite starts forming in the anchizone;

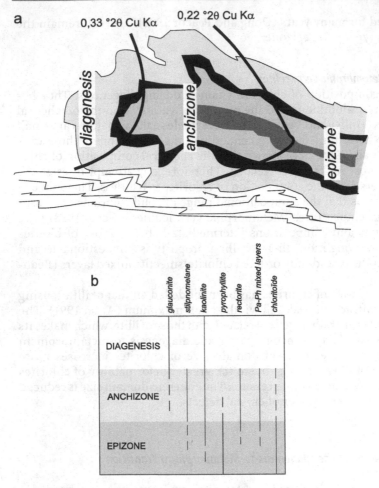

Fig. 8.27a,b. Diagenesis-to-metamorphism transition. **a)** Sketch profile of the Morcles thrust sheet with the diagenesis-anchizone and anchizone-epizone boundaries determined by the crystallinity index of illite (Goy-Eggenberger 1998). **b)** Mineral sequence observed in the thrust sheet. *Pa-Ph*: paragonite – phengite mixed layer

3. crystallisation of rectorite, of a Na-mica – K-mica mixed layer and of chloritoid in the anchizone. These minerals subsequently disappear in the epizone, except chloritoid.

Rectorite, which has long been considered a regularly ordered smectite-paragonite mixed layer, is in reality a mineral phase sensu stricto like corrensite, its trioctahedral equivalent. Rectorite is definitely different from its potassic equivalent that should not bear the same name. In ancient literature, this mixed-layer mineral was referred to as *allevardite*. This mineral commonly occurs in I/S sequences typical of diagenesis. An improved knowledge of the growth processes of these two minerals shall permit us to update terminology. The paragonite–phengite (Pa–Ph) mixed layer, contrary to rectorite, is a transient crystal structure that disappears by formation of separate phengite and paragonite crystals (Frey 1969).

The classical interpretations of the diagenesis-to-metamorphism boundary described above are substantially modified by the following observations:

1. chloritoid and pyrophyllite may occur in the anchizone in case of favourable chemical composition of rocks (aluminous shales);

2. kaolinite and pyrophyllite sometimes co-exist in the anchizone and epizone.

Sudoite is a relatively rare mineral whose occurrence is related to the reaction kaolinite + chlorite $\rightarrow$ sudoite + quartz + H_2O (Fransolet and Schreyer 1984). This reaction is triggered at about 200 °C. Nevertheless, sudoite may occur under diagenetic conditions of lesser intensity by transformation of smectites or I/S minerals in bentonite beds in the presence of organic matter (Anceau 1993).

8.2
Volcanic Rocks

Introduction

Volcanic rocks exhibit a significant distinguishing feature when compared to clastic sediments: they contain glass. Owing to the high reactivity of glass in the presence of fluids, the activities of the major chemical components of silicates, silica notably, are controlled by its dissolution. Rocks of volcanic origin are commonly found in series of detrital sediments in the form of relatively thin beds. Nevertheless, in the vicinity of island arcs, sedimentation is dominated by these glass-bearing rocks. They are transformed during the burial process; glass and some magmatic minerals disappear whereas zeolites and clays crystallise forming mineral sequences that vary with depth.

Such sequences can result from diagenesis (temperature rise due to burial by heat diffusion) or from hydrothermal alterations (temperature rise due to the flow of hot fluids: thermal convection). The respective roles assumed by diffusion and convection is not easy to define because volcanic areas are essentially regions with high geothermal gradient where hydrothermal flows are very active. In fact, their distinction cannot rest upon the local and instantaneous temperature because, in this case, the period of time during which this temperature has been stabilised is not known. Indeed, the reaction processes are controlled by kinetic laws. The significant parameter in diagenesis as well as in hydrothermal alterations is *time* × *temperature*. The difference between diagenesis and hydrothermal alterations is rather based on the importance of hot fluid flows, which can be measured by the water/rock ratio. Consequently, even when the thermal gradient is higher than the mean value of 30 °C km^{-1}, the process will be referred to as diagenesis provided that the water/rock ratio remains low as compared to the values reached in geothermal fields.

8.2.1
Diagenesis of Ash and Vitreous Rock Deposits

8.2.1.1
Diagenesis of Bentonites

Bentonites are formed by alteration of volcanic ashes under conditions that are still poorly known (see Sect. 7.2.2.1). These are practically monomineral rocks composed nearly exclusively of montmorillonite-like smectite. They form reference stratigraphic layers and bear witness to the volcanic activity in the vicinity of the sedimentary basin. The thickness of beds ranges between a few millimetres to one metre (rarely more) owing to several factors, such as the distance to the ash-producing volcano or the flow turbulences in the atmosphere. The number of beds inserted in the sedimentary pile may be significant in some basins (e. g. the Slovak Basin described by Sucha et al. 1993).

Smectites persist in very old bentonites (Ordovician Period) even if the surrounding shales are composed of I/S minerals with a very high illite content

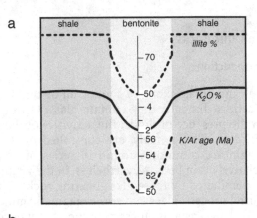

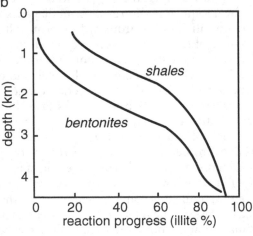

Fig. 8.28a,b. Diagenesis of bentonites. **a)** Illitisation of the bed rims controlled by the diffusion of potassium (Altaner et al. 1984). **b)** Comparison of the advancement stage of the smectite → illite reaction as a function of depth between shales and bentonites in the eastern part of the Slovak Basin (Sucha et al. 1993). Bentonites react more slowly than shales

(Velde 1985). Nevertheless, mineralogical differences can be observed within bentonite beds according to their thickness and to the nature of the rocks with which they are in contact. When beds are 1 m thick or thicker, a zonation is clearly perceptible between the illitised rims and the centre that retains a high smectite content (Huff and Türkmenoglu 1981; Altaner et al. 1984; Brusewitz 1986). The potassium enrichment modifies the measured K/Ar radiogenic ages (Fig. 8.28a). This type of zonation takes place by chemical diffusion (Pusch and Madsen 1995).

When bentonite beds are thin (a few centimetres), their mineralogical composition is homogeneous and no zonation is perceptible. Sucha et al. (1993) have compared the advancement stage of the illitisation reaction in a sequence of bentonites and shales in the eastern part of the Slovak Basin (Fig. 8.28b). They have shown that the I/S series are seemingly identical in both types of rocks but that, for a given depth, the illite content is always higher in shales than in bentonites. The reaction is then slower in bentonites.

8.2.1.2
Diagenesis of Volcanoclastic Deposits

The abundance of zeolites is the signature of volcanoclastic deposits under diagenetic conditions (Iijima and Utada 1966; Coombs 1970 among others). Some basics of zeolite crystallochemistry may be recalled here. There are two main mineral sequences: Ca-zeolites (gismondine, laumontite, heulandite, lawsonite, scolecite, chabazite, epistilbite) and Na-K zeolites (natrolite, analcite, clinoptilolite). Several species are common to both mineral sequences: phillipsite, stilbite, mordenite and erionite. These two mineral sequences differ by the value of their $(Na + K + 2Ca)/Al$ ratio (Fig. 8.29a). The stability of zeolites in the pressure-temperature space has been studied by experimental synthesis of a few species. These works set the general lines of the diagenesis-to-metamorphism transition (Ghent et al. 1979), which spans the stability fields of analcite and laumontite (Fig. 8.29b). For low pressures (shallow depths), the hydrothermal sequences (high geothermal gradient) result in wairakite, which can be viewed as the indicator of the boundary with thermal metamorphism.

The sequences observed in the deep drill holes of New Zealand volcanoclastic areas (Fig. 8.29c) show that heulandite and analcite are the two main species characterising the upper part of the series where plagioclases are strongly albitised (low temperatures). Heulandite and analcite disappear to the benefit of laumontite and non-zeolitic minerals: prehnite and pumpellyite (Surdam and Boles 1979). Zeolites are commonly observed along with alkaline feldspars (albite, adularia) and illite/smectite mixed layers. Nevertheless, the complexity of zeolite solid solutions does not enable an accurate zoneography to be established, owing to the significant – if not dominating – effects of the local chemical composition.

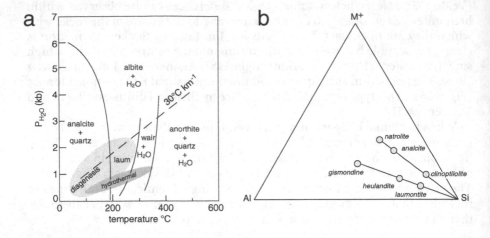

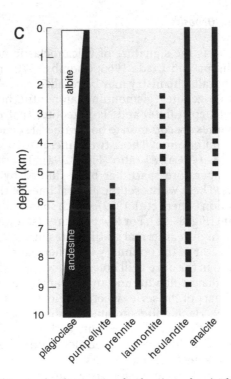

Fig. 8.29a–c. Zeolites in the diagenesis of volcanic rocks **a)** The main zeolite species are represented in the M$^+$–Al–Si chemical system (M$^+$ = Na$^+$ + K$^+$ + 2 Ca^{2+}). **b)** Analcite and laumontite stability fields in the pressure-temperature space (from Ghent 1979). **c)** Mineral sequence observed in volcanoclastic sediments in New Zealand (from Surdham and Boles 1979)

8.2.2
Diagenesis – Very Low-Grade Metamorphism of Basalts

8.2.2.1
Distinguishing Features of the Basaltic Series

Basaltic rocks are the most widely occurring rocks at the Earth's surface. They make up the entire ocean floor and may form huge stackings on continents (trapps or flood basalts). Despite their relatively extended chemical compositional range (Juteau and Maury 2000), their differences are negligible from the standpoint of diagenetic alterations because their prevailing feature lies in high Mg and Fe^{2+} contents. Therefore, they differ from silicoclastic rocks, which exhibit higher Si, Al, Fe^{3+} and K contents. Diagenetic reactions are characterised by the prevalence of trioctahedral clay minerals and zeolites.

The observation of mineral transformations in continental basaltic series is for the most part made possible in natural outcrops and geothermal boreholes. Geothermal fields, apart from the intensely fractured zones where great amounts of hot fluids flow (heat transport by convection), exhibit sequences of zeolites and clay minerals that are close to those formed in zones with lower thermal gradients (heat diffusion by thermal conduction). These sequences are simply shortened. Accordingly, they will be described in chapter 9 along with the alterations of deep-sea floors (see Sect. 9.1.4.2). To avoid boring repetitions, only those aspects which are unquestionably related to burial under conditions of geothermal gradient and fluid flows typical of diagenesis will be considered here: $30-40\,°C\,km^{-1}$, water of meteoric or marine origin.

Besides their chemical composition, basalts exhibit another characteristic that distinguishes them from sediments: the contrasted porosity within a single lava flow. The cooling magma flowing out to the open air degasses. Gas bubbles nucleate, go up in the lava flow and are vented to the atmosphere as long as permitted by magma viscosity ($T > 1,150\,°C$). Below this threshold, they remain trapped and form vesicular levels at the top and bottom of the lava flow that surround a massive zone in between (Fig. 8.30). These massive zones are more or less roughly cut into prisms by fractures that spread once temperature has dropped below $700\,°C$. During burial, the vesicular or massive levels of basaltic flows form zones of contrasted permeability that behave like open or nearly closed systems, respectively (Levi et al. 1982).

8.2.2.2
Mineral Zonation at Various Scales

Regional Scale
The burial of basaltic series causes the crystallisation of zeolites and clay minerals whose assemblages form a large-scale mineral zonation. Whatever the age and the emplacement mode of basaltic series, the main facies are systematically observed: Achaean (Schmidt and Robinson 1997), late Precambrian (Bevins et al. 1991), Cretaceous (Lévi et al. 1982), and Tertiary (Neuhoff et al. 1999).

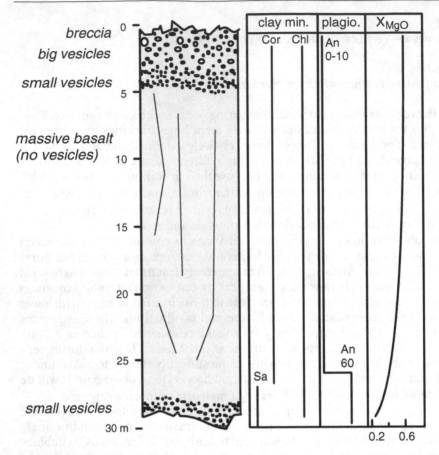

Fig. 8.30. Structure of a basaltic flow and mineral and chemical zonation (from Schmidt and Robinson 1997)

Four typical mineral assemblages can be distinguished in order of increasing intensity of metamorphism (Alt 1999):

1. smectite + alkaline zeolites and heulandite,

2. corrensite + laumontite,

3. chlorite + epidote ± prehnite,

4. chlorite + actinolite.

Local variations may appear according to P_{CO_2} and P_{S_2} whose values enable or impede the formation of carbonates and sulphides. The alteration of deep-sea floor basalts exhibits an additional phyllosilicate–celadonite, which persists in the corrensite facies.

Lava Flow Scale

Owing to the permeability contrast between the vesicular and massive levels, each lava flow exhibits an internal mineral zonation. In the basalts of the Chilean Andes cordillera (Jurassic, Cretaceous, Tertiary), Levi et al. (1982) have shown an increase in the albitisation of plagioclases and in the ratio of the epidote amount to the prehnite + pumpellyite amounts from the massive levels to the vesicular and brecciated levels. Similar phenomena have been observed in other basaltic series, particularly in Iceland (13 Ma, Neuhoff et al. 1999) and in the metabasalts of the Keweenaw Peninsula, Minnesota, U.S.A. (Precambrian, Schmidt and Robinson 1997). Generally, smectitic minerals better withstand the condition changes imposed by burial in the massive parts of lava flows, which causes a shift of facies with the vesicular levels. Accordingly, for the same lava flow, saponite persists in the massive part whereas corrensite and chlorite are formed in vesicles. This shows that, within the greatest volume of basaltic series, thermodynamic equilibria are reached in microsystems only. The internal zonation subsists practically up to the actinolite zone.

Sample Scale

Clay minerals can be observed mainly in two types of sites: either they fill vesicles in which they form zoned deposits (Fig. 8.31a) or they replace a mesostasis classically considered to be formed of glass (Fig. 8.31b). Two essential pieces of information have been brought by the recent detailed petrographical studies (Mas 2000) on mesostasis: (1) glass is formed only on the lava flow margin, which is subjected to a sudden quenching, and (2) clay minerals are not always formed by alteration of phenocrystals along microfissures: they can precipitate directly on the unaltered crystal surfaces of these phenocrystals (Fig. 10.6). Their post-magmatic origin is now contemplated (see Sect. 10.2.1).

The physicochemical conditions imposed during diagenesis and very low-grade metamorphism differ from those imposed during the magmatic episode. Particularly, the circulation of fluids triggers interactions between solution and minerals, the products of which depend on the activity of ions in solution and on the water/rock ratio. Some of the initial mineral components become unstable under such conditions and react at different velocities. The nature of the secondary parageneses is controlled by the solid porosity: parageneses in the vesicular level differ from those in the massive level. Generally, whatever the reaction progress, the chemical compositions of altered and unaltered rocks are similar, except for water contents. Accordingly, cations are mobile over very short distances (a few millimetres), and only water is brought by the external environment. Neuhoff et al. (1999) have shown in Icelandic trapps that the filling rate of vesicles increases with the intensity of metamorphism (Fig. 8.31c). They distinguish two burial stages: replacement of 8% of the porosity by precipitations of celadonite and quartz (stage 1); hydrolysis of olivines, plagioclases and glass, feeding the crystallisation of chlorite/smectite mixed layers whose chlorite content increases with increasing depth, and of zeolites that seal 40% of the porosity (stage 2). Stage 3 is a hydrothermal alteration stage related to the emplacement of basaltic dykes. Zeolites precipitate and occupy 10% of the initial porosity.

Alteration patterns of primary components and conservation of rock composition are the petrographical and geochemical signatures of the metamorphic process. The mineralogical signature will be discussed later. The formation of clay minerals is fed by dissolution of primary components. However, since basaltic glass occurs only in lava flow chilled margins, its alteration remains problematical. The glass assumed to be present in the mesostasis is the final product of the fractional crystallisation and must be of rhyolitic composition, namely very different from the composition of its alteration products (saponite, nontronite and chlorite). There is no unquestionable petrographical evidence of the occurrence of this glass, except in the form of solid inclusions in plagioclase phenocrystals. Mas (2000) has shown that clays in the mesostasis are probably the result of a direct precipitation from residual fluids enriched with incompatible and volatile elements. Accordingly, these "initial clays" located in the porous zones of the rock–and not glass–would be transformed by metamorphism.

8.2.3
Di- and Trioctahedral Mixed-Layer Minerals

8.2.3.1
Celadonite – Glauconite – Nontronite

Celadonite is systematically described in the vesicles of deep-sea floor basalts (Humphris et al. 1980; Laverne 1987; Kaleda and Cherkes 1991; Alt 1999) and more rarely in the vesicles of continental basalts (Neuhoff et al. 1999). This mineral always forms within the rock at the external interface of the vesicle (Fig. 8.31a). The first mineral sheet inside the vesicle comprises microcrystalline quartz. The following sheets show a zonation: they comprise trioctahedral minerals (saponite and C/S minerals) whose chlorite content is increasingly high inwards. The vesicle may be totally sealed by precipitation of zeolites.

The composition of celadonites is variable, ranging from the domain of sedimentary glauconites to the compositions of the theoretical end members of celadonite and glauconite micas (Fig. 8.32). This dispersion of values is due to three factors: (1) variation of the measured K_2O contents in the range 6–9%, (2) presence of Al ions in tetrahedral position, and (3) presence of Fe^{2+} ions although the oxidation state of iron remains undetermined most of the time. Note that the compositional range of solid solutions rather extends towards the beidellite domain, and not towards that of nontronites. The latter have a limited chemical composition range close to the theoretical end member.

Celadonites and nontronites form in a wide range of temperatures. Indeed, low-pressure experimental syntheses have shown that the celadonite – biotite joint is reached at about 400 °C (Wise and Eugster 1964; Velde 1972; Klopproge et al. 1999). The higher the temperature, the closer the solid solution to the theoretical end member, and the lower the temperature, the greater the proportion of expandable layers. Stable isotopes show that celadonite can form at temperatures below 40 °C in marine environment (Stakes and O'Neil 1982

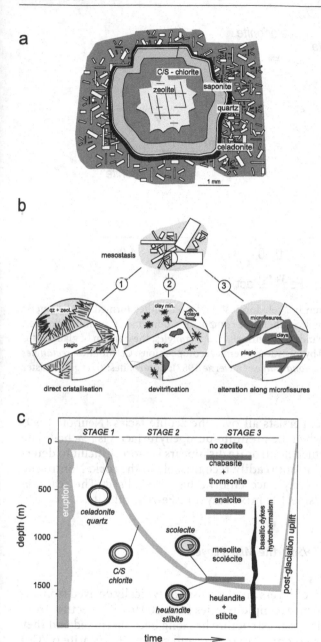

Fig. 8.31a–c. Clay minerals of basalts. **a)** Schematic representation of a vesicle totally sealed by deposits of clay minerals and zeolites. **b)** The three possible origins of mineral clays in the mesostasis: direct precipitation from post-magmatic fluids, glass ageing, alteration of glass and phenocrystals by external fluids. **c)** Temporal sequence of parageneses observed in Icelandic trapps (Neuhoff et al. 1999)

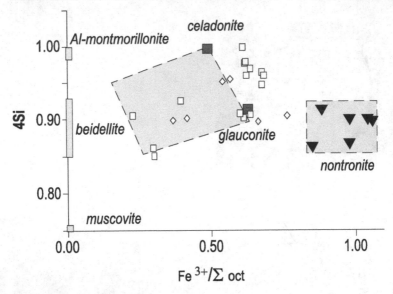

Fig. 8.32. Chemical composition of "celadonites" and nontronites formed by diagenetic alteration of basalts. The theoretical chemical composition domains of montmorillonites, beidellites and muscovites are indicated with *shaded rectangles*; the domain of glauconites and nontronites is determined by measured chemical compositions. Celadonites: *rectangles* (Humphris et al. 1980) and *diamond shapes* (Laverne 1987); nontronites: *triangles* (Koster et al. 1999)

among others). Celadonite persists all over the zeolite facies (Neuhoff 1999) and probably disappears when the prehnite-pumpellyite facies is reached. The metamorphism stage at which nontronite disappears is more difficult to determine because this mineral is not readily recognisable in the rock. Nontronite probably transforms very early once chlorite has crystallised. The reaction produces either a chlorite/smectite mixed layer or Fe-rich chlorite.

8.2.3.2
Chlorite – Corrensite –Chlorite/Smectite Mixed Layers

Chemical Composition Domains
Chlorite/smectite (C/S) mixed layers are described as typically derived from diagenesis of basalts. They form a complete series from 100 to 0% smectite (from saponite to chlorite). Their crystal structure has long been discussed and they are considered today to be saponite-corrensite and corrensite-chlorite (C/Co) mixtures rather than true smectite/chlorite mixed layers. However, the chemical compositions measured with the electron microprobe in basalt vesicles are not consistent with either model (Fig. 8.33). Indeed, a few compositions only are situated between the compositional ranges of saponites and those of chlorites, as should theoretically be the case. Many of them are situated in the saponite-nontronite-chlorite space. Unfortunately, in the absence of accurate

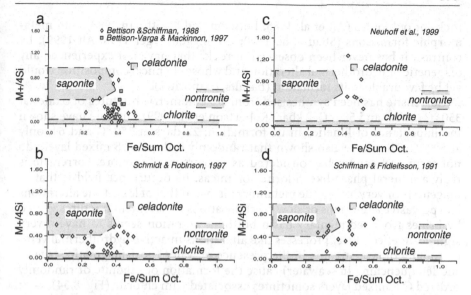

Fig. 8.33a–d. Representation of the chemical compositions of several series of smectite/chlorite mixed layers derived from the very low-grade metamorphism of basaltic rocks. **a)** Ophiolite from Point Sal, California (Bettison and Schiffman 1988; Bettison and Mackinnon 1997). **b)** Metabasalts from Keweenaw, USA (Schmidt and Robinson 1997). **c)** Basaltic trapps from Iceland (Neuhoff et al. 1999). **d)** Basalts from the Nesjavellir geothermal field, Iceland (Schiffman and Fridleifson 1991)

crystallographic data, the actual co-existence of nontronite with C/S or C/Co cannot be established.

Regardless of how the metamorphic sequence is described (saponite – corrensite – chlorite or a complete series of C/S minerals between end members saponite and chlorite), the chemical balance of the reaction is based on saponite disappearance and chlorite formation. Considering the theoretical compositions of these minerals (saponite: $Si_{3.7}Al_{.3} O_{10} R_3^{2+} (OH)_2 M^+._3$; chlorite: $Si_3Al O_{10} R_3^{2+} (OH)_2 - AlR_2^{2+} (OH)_6$), and knowing that quartz and zeolites are to precipitate too, the global reaction can be written as follows:

$$1\,Saponite + 1.7\,Al^{3+} + 2\,R^{2+} + 6(OH)^- \rightarrow 1\,Chlorite + 0.7\,Si^{4+} + 0.3\,M^+ \tag{8.11}$$

The Si^{4+} and M^+ ions released by the reaction are consumed by the precipitation of zeolites. This is consistent with the structure of the concentric deposits in vesicles.

Corrensite: a Mineral Indicator of Metamorphism

Corrensite has been identified in various natural environments spanning a large range of P, T, t conditions: diagenetic evaporitic series (April 1981), geothermal fields (Inoue and Utada 1991), hydrothermal alteration of basaltic

rocks or ophiolites (Alt et al. 1985; Bettison and Schiffman 1988), and metamorphic formations (Shau et al. 1990; Goy-Eggenberger 1998; Alt 1999). By
contrast, it has never been observed in rocks that have not experienced any
diagenetic or hydrothermal alteration and whose chemical composition would
yet be favourable to its formation (basalts, evaporites).

Corrensite has been synthesised from stoichiometric mixtures of oxides at
350 °C–2 kbar and 500 °C–2 kbar (Robertson et al. 1999). Authors have shown
the importance of kinetics in its formation: 22 days at 350 °C and 6 h only
at 500 °C. They have also shown that randomly ordered C/S mixed layers do
not form and cannot be considered as precursors. Therefore, corrensite is
truly a mineral phase like chlorites or micas. Its occurrence field is that of
diagenesis or very low-grade metamorphism and that of long-time alterations
like pervasive alterations or slow crystallisations in basement fractures visited
by brines from sedimentary basin bottoms. Duration seems to have a very
significant effect. Brief processes like alterations in active hydrothermal veins
or cooling of lava streams in the presence of water (post-magmatic stage or
sudden quenching in seawater) cause the formation of saponite or randomly
ordered C/S mixed layers sometimes associated with chlorite (Fig. 8.34).

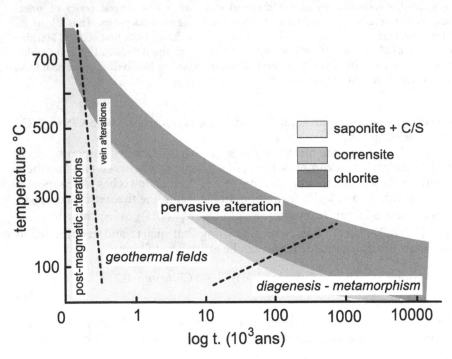

Fig. 8.34. Influence of the factor *time* × *temperature* (kinetics) on the stability fields of
saponite (saponite-rich C/S minerals), corrensite and chlorite in the main geological processes

Suggested Readings

Frey M, Robinson G (1999) Low-grade metamorphism. Blackwell, Oxford, 313 pp

Velde B (1995) Origin and mineralogy of clays. Clays and the environment. Springer, Berlin Heidelberg New York, 334 pp

Larsen G, Chilingar GV (1979) Diagenesis in sediments and sedimentary rocks. Developments in Sedimentology 25A, Elsevier, Amsterdam, 579 pp

McIlreath IA, Morrow DW (1990) Diagenesis. Geoscience Canada, reprint series 4, Geological Association of Canada, 338 pp

Somerton WH (1992) Thermal properties and temperature-related behaviour of rock/fluid systems. Developments in Petroleum Science 37, Elsevier, Amsterdam, 257 pp

Montanez IP, Gregg JM, Shelton KL (1997) Basin-wide diagenetic patterns: integrated petrologic, geochemical, and hydrologic considerations. SEPM special publication 57, Society for Sedimentary Geology, Tulsa, 302 pp

CHAPTER 9

Hydrothermal Process – Thermal Metamorphism

9.1
Fossil and Present-Day Geothermal Fields

Introduction

Continental hydrothermal systems have been known since Antiquity for their potential ore deposits (copper, tin) and the "miraculous" properties of hot springs. However, their scientific exploration only began in the second half of the 20th century. Many works were undertaken in both fossil systems and active geothermal fields with the aim of searching for mineral zonations comprising sulphides, oxides and silicates. These zonations have been described in numerous books or reviews in which are explained the fundamental mechanisms of interactions between hot fluids and rocks: fossil systems (Lovering 1950; Bonorino 1959; Creasey 1959; Lowell and Guilbert 1970; Titley 1982; Beane 1992; Pirajno 1992; Barnes 1997) and active geothermal fields (Steiner 1968; Ellis and Mahon 1977; Browne 1978; Elder 1981; Henley and Ellis 1983; Reyes 1990; Browne 1998).

The hydrothermal systems affecting the deep-sea floor basalts and ophiolites have been discovered recently even though the Cyprus copper was the primitive Greeks' ore resource. Better known today thanks to oceanic research programs, these systems have been abundantly described. Their structure is rather well known from their roots in the dyke system to the floor surface from which they arise in the form of the famous "black smokers" (see Alt's review 1999).

From the standpoint of clay minerals, fossil and active geothermal fields offer a great variability in their formation conditions: temperatures, composition and pH of solutions, fluid/rock ratio. Two processes operate simultaneously or successively depending on the location and activity period of the system: alteration of primary minerals and direct precipitation from oversaturated solutions. Considering the rapid changes in these conditions over time, several clay formation episodes may be recorded in the same rock. The result is then a complicated combination of different clay species that cannot be considered to form a single paragenesis, the analysis of which would permit the reconstruction of the accurate physicochemical conditions. Particularly, their use as geothermometers should be definitely avoided.

9.1.1
Geological and Dynamic Structure of Geothermal Fields

9.1.1.1
Conduction and Convection

Mineral Zonations in Hydrothermal Systems

The first classifications of alterations were based on the prevailing parageneses of secondary minerals. They have been used to define zones around the porphyry copper deposits: potassic, phyllitic, propylitic, and argillic zones (Creasey 1959). The classification proposed by Utada (1980) uses combinations of two variables: temperature and ratio of cation activities in hydrothermal solutions. This classification enables the major types of hydrothermal systems to be described (Table 9.1). Three groups are so distinguished according to the value of the cations/H^+ ratio: acid-type system (low ratio), intermediate-type system (medium ratio) and alkaline-type system (high ratio).

The deep roots of geothermal fields as they have been observed in the porphyry copper deposits are the seat of interactions of high-temperature (and mostly high-salinity) fluids and intrusive rocks with their immediate wall rocks. These interactions have been described as "phyllic alteration" and "potassic alteration" (Creasey 1959; Lowell and Guilbert 1970). The resulting secondary silicated minerals are not clays but white micas (muscovite and phengite) and K-feldspars. Nevertheless, it is important to know how to identify these early alterations, which can either be superimposed on an earlier propylitic alteration or be themselves followed by late low-temperature alterations. Since phyllic and potassic alterations do not produce clay minerals, they are not the subject of this book. On the other hand, the propylitic alteration will be described.

Table 9.1. Zonal arrangement of parageneses in the three types of hydrothermal systems defined by the value of the cations/H^+ ratio in solutions (Utada 1980)

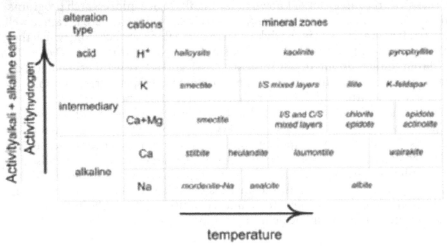

Dynamics of Continental Systems

It is widely admitted now that active geothermal fields are the surface expression of a hydrothermal activity related to the emplacement of magmatic intrusions. These geothermal fields form the upper part of a greater system whose deep roots are the porphyry copper deposits (Henley and Ellis 1983) and whose intermediate parts are the epithermal ore deposits, as shown by Hedenquist et al. (1998) in the Far Southeast and Lepanto deposits in the Philippines. All these structures represent huge volumes of rocks submitted to interactions with hot fluids (Fig. 9.1).

Whether fossil or active, geothermal fields are geological structures that enable the heat stored in the Earth's crust to be dissipated. Two processes take place simultaneously or alternately: conduction and convection (Fig. 9.2). The emplacement of the magmatic intrusion is a slow process that enables the formation of thermal aureoles by conduction. The seismic activity opens faults, thus forming high-permeability networks that drain a major part of hot fluids. The most efficient process for dissipating the heat is then convection. The chemical and mineralogical interactions accompanying these circulations cause the sealing of the networks through deposits of secondary minerals (clays, carbonates, sulphates, sulphides, oxides). The permeability of the networks is reduced thus bringing about the predominance of conduction over convection. Such alternations may repeatedly take place during the activity period of the geothermal field, which itself depends on the amount of heat to

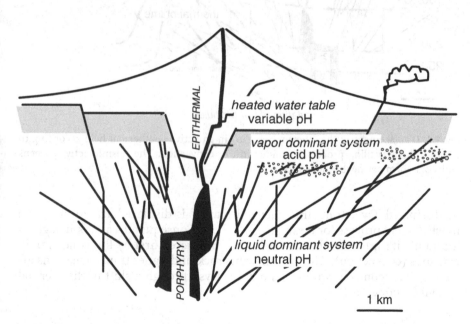

Fig. 9.1. Schematic representation of the organisation of a geothermal field around a stratovolcano. *"Bubbles"* represent the vapour-dominant levels. The fumarole symbolises the active geothermal field at the Earth's surface

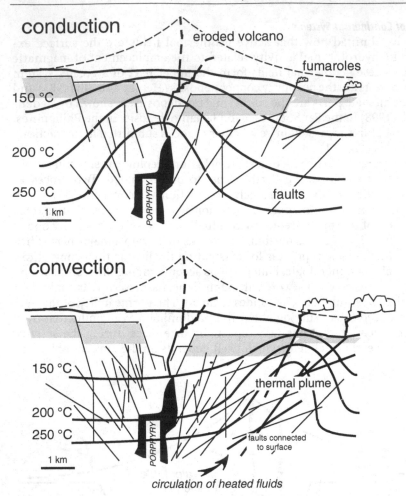

Fig. 9.2. Displacement of the "thermal plume" in an active geothermal field according to whether the prevailing process is conduction or convection. The seismic activity opens faults that form permeable networks

be dissipated. Over time, the "thermal plume" is displaced according to the most efficient process for dissipating the heat. Consequently, a rock in a given location within the geothermal field may be altered first by fluids flowing in fractures (convection). After sealing, this rock undergoes transformations affecting the secondary products, notably clays, through a shift in the thermal system (conduction).

Propylitic Alteration
The greater the amount of heat (i. e. the size of the magmatic body), the greater the rock volumes in which temperature is increased by conduction. Thermal

models show that heat dissipation is a slow phenomenon owing to the very low thermal conductivity of rocks (Jaeger 1968; Cathles 1977; Norton and Knight 1977). The fluids entering these rocks through fractures and microfissures (intergranular joints, intramineral fissures) trigger mineral reactions: dissolution of primary minerals, precipitation of secondary ones. An aureole in which wall rocks are transformed forms around the heat source. These mineral transformations are described under the term *propylitic alteration*.

The mineral parageneses typical of the propylitic alteration have been abundantly described (Creasey 1959; Lowell and Guilbert 1970; Titley et al. 1986 etc). They are formed owing to local equilibria (microsystems) and they retain the rock initial microstructure (Meunier et al. 1984). Mass transfers take place over short distances of the order of magnitude of the size of primary minerals. The main mineral reactions are interdependent in that one reaction provides the elements in solution needed by another:

$$\text{plagioclase} + K^+ \rightarrow \text{illite} + Al^{3+} \tag{9.1}$$

$$\text{biotite} + \text{amphibole} + Al^{3+}$$
$$\rightarrow \text{chlorite} \pm \text{epidote} \pm \text{calcite} \pm \text{titanite} + K^+ \tag{9.2}$$

Epidote or calcite is formed according to the value of the CO_2 partial pressure. Quartz crystallises in microfractures by consuming excess silica. The amounts of fluids involved are so low that the composition of secondary phases is essentially controlled by that of primary phases (Berger and Velde 1992).

The study of the Saint-Martin fossil system, Lesser Antilles (Beaufort et al. 1992) has shown that the assemblages and chemical compositions of secondary minerals in the propylitic aureole vary with distance to magmatic body (Fig. 9.3a). The maximum temperature recorded by rocks (fluid inclusions in quartz + epidote veins) decreases as a complex function of distance. The ferro-magnesian minerals form an actinote – chlorite – chlorite-rich chlorite-smectite mixed layers series. Chlorites experience gradual changes in their chemical composition; their Fe content increases as the distance to the magmatic body increases. There is great temptation to consider that these compositions are controlled by temperature and that, as a consequence, they can be used as geothermometers. The reality is more complex because temperature is not the only parameter subjected to variation with the distance to the magmatic body. Particularly, the oxygen partial pressure (f_{O_2}) changes as attested by the presence of magnetite at short distance and of hematite farther. This parameter that modifies the oxidation state of iron is very important for determining the Fe^{2+} amounts taken up by the crystal lattice of chlorites. The chemical composition of the latter changes according to the minerals with which they are combined (Fig. 9.3b).

The temperature curve is not the best way for evaluating the amount of energy released by heat conduction. Indeed, not only is temperature higher near the magmatic body, but the duration of this "heating" before temperature has reached equilibrium with the wall rock temperature is longer. The *time* × *temperature* parameter is the one that best accounts for transfers of energy. It

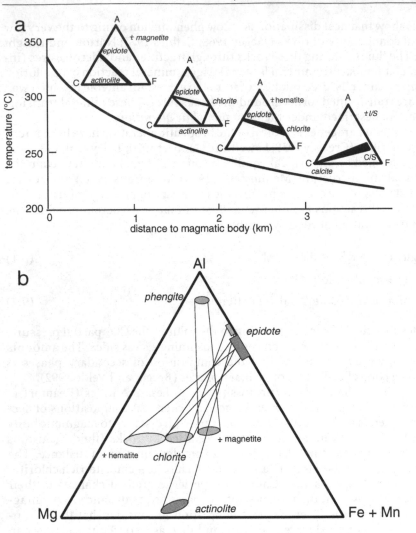

Fig. 9.3a,b. Alteration parageneses of the Saint Martin fossil geothermal field (Antilles).
a) Variation as a function of the distance to the magmatic intrusion (hence of temperatures).
b) Variation in the chemical composition of chlorites as a function of the minerals with
which they are combined

is a determining parameter particularly for the crystallisation state: number of
defects (order-disorder degree) and crystal size. Patrier et al. (1990, 1991) have
shown that the order-disorder degree of epidotes increases with increasing
distance to the magmatic body whereas their size decreases. Similar results
have been obtained for chlorites from the width of their diffraction peaks on
oriented samples (unpublished data).

9.1.1.2
Petrography: Example of the Chipilapa Field (El Salvador)

Geological Structure
The Chipilapa Field (El Salvador) belongs to the intermediate-type hydrothermal system as defined by Utada (1980). It developed in recent calcoalkaline
volcanic formations (Pliocene to present day) where andesitic lava flows and
dacitic pyroclastic deposits alternate. It seems that the hydrothermal activity
has been established for about 16,000 years. It has experienced recent paroxys-
mal crises as attested by phreatic explosion craters. This field is fed by meteoric
fluids that are heated up to 250 °C in reservoirs. The hot fluids are drained by
faults that delineate a "thermal plume" (Fig. 9.4). Two reservoirs are exploited
600 to 800 m deep and 1,150 to 1,400 m deep. Drill hole CH 8 is far from the production area; it is used as a reference for the study of hydrothermal alterations
in low-permeability areas where thermal conduction is the prevailing heat dissipation process. Drill hole CH 7bis is located at the centre of the "plume". It

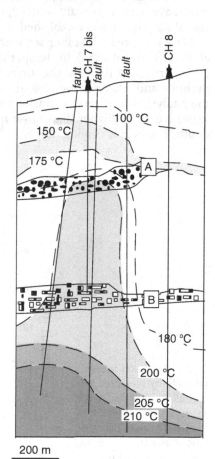

Fig. 9.4. Diagram of the geological structure
of the Chipilapa geothermal field, El Salvador
(Beaufort et al. 1995). The "thermal plume" is
framed by three faults. Two reservoirs (*A* and
B) are exploited by a series of drill holes

200 m

is used as a reference for hydrothermal alterations in high-permeability areas where convection is the prevailing heat dissipation process.

The surface geological formation (0–500 m) comprises rocks whose clay content increases with increasing depth. They form an impermeable layer (clay cap) that strongly slows down surface heat losses. Therefore, the thermal gradient is very high between 300 and 550 m: 60 to 80 °C over 250 m only. Surface reservoir A immediately beneath the clay cap contains high-temperature fluids (> 185 °C) with a significant vapour phase (steam cap). From 650 to 1,100 m, the permeability of rocks is low, as is the thermal gradient. The vapour amount of the deeper aquifer (reservoir B) between 1,100 and 1,400 m is small (prevailing liquid phase). Today, the boiling point of fluids ranges between 700 and 800 m in depth.

Early Episode: Propylitic Alteration (16,000 to 4,000 Years)

Referring to the distribution of calcium silicates, the vertical thermal gradient contemporaneous with the propylitic alteration episode in the Chipilapa Field must have been significant. Indeed, the clinoptilolite – stilbite – laumontite – wairakite sequence is established over 500 m in thickness. Epidote occurs at 750 m. Compared with other sequences in the world (Utada 1970; Cavaretta et al. 1982; Bird et al. 1984), the temperature must have exceeded 300 °C at 2,500 m and approached 100 °C at the surface. Anhydrite occurs from 1,000 m on, and prehnite and adularia from 2,400 m on. The mean temperatures estimated by the analysis of the fluid inclusions trapped in quartz and calcite (Bril et al. 1996) are much higher than those measured in present-day fluids at similar depth. The field is cooling down.

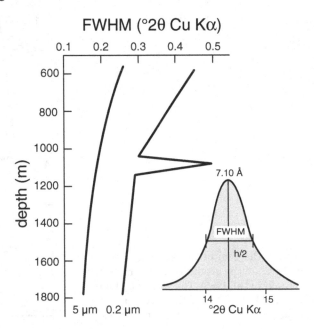

Fig. 9.5. Variation in the crystallinity measured by the full width at half maximum intensity (FWHM) of the (002) peak of the chloritic phases in the > 5 μm and < 0.2 μm fractions

The phyllosilicate phases of the propylitic alteration are mostly illite and chlorite. These minerals can be found all over the altered zone; only their crystallinity changes. Measured in the $> 5\,\mu m$ fraction by the full width at half maximum intensity (FWHM) of the (002) peak, the crystallinity of chlorite steadily increases with depth (Fig. 9.5). This means that the coherent scattering domain size (dimension along the c direction) increases whereas the number of crystal defects decreases; this is also so with the very fine fraction ($< 0.2\,\mu m$), except at about $- 1,100\,m$ (reservoir B) where crystallinity clearly decreases. The mathematical decomposition of the (002) peak shows that the FWHM increases by addition of two effects: the presence of chlorite-rich chlorite/smectite mixed layers and the chlorite peak broadening. The minerals crystallising under conditions imposed by present-day hot fluid circulations are concentrated in the fine fraction whereas the coarse fraction corresponds to older crystallisations in the propylitic domain.

Late Episode: "Clays – Carbonates" Alteration (4,000 Years – Present Day)

Crystals from the illite – chlorite – epidote propylitic paragenesis show dissolution patterns when located in highly fractured rocks. Microfractures form very dense networks and are sealed by deposits of calcite, hematite and clay minerals (Fig. 9.6). Intense fracturation is due to the explosive expansion of fluids. The presence of hydrothermal breccia and the noteworthy compositional homogeneity of fluid inclusions in carbonates whatever the depth show that a sudden boiling phenomenon caused the hydraulic fracturation of the propylitised rocks. The new paragenesis sealing fractures is typical of the clays – carbonates alteration, as described by Steiner (1968), Browne and Ellis (1970), Lowell and Guilbert (1970), McDowell and Elders (1980), Beaufort and Meunier (1983), Beaufort et al. (1990).

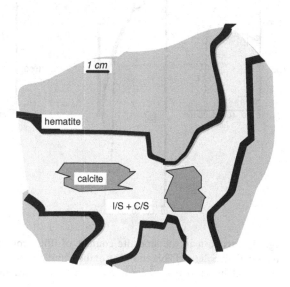

Fig. 9.6. Diagram (from photograph) of microfractures sealed by deposits of carbonates, hematite and clays

Taking place at temperatures lower than those of the propylitic alteration (230–240 °C in the present case), the clay-carbonates alteration transforms pre-existing minerals (primary or propylitic) by modifying the physicochemical conditions. The system becomes open and oxidising, fluid/rock ratios are much higher. Consequently, the mineral reactions are no longer controlled by the rock, but by fluids (Berger and Velde 1992). Fluids have a low salinity and a meteoric origin, as shown by the biphased fluid inclusions in carbonates (salinity contained between 0.8 and 1.8 NaCl mass equivalent) and by the $^{18}O/^{16}O$ and D/H ratios of the neoformed minerals.

The clay minerals neoformed under these conditions are illite/smectite (I/S) and chlorite/smectite (C/S) mixed layers whose composition varies with depth. Their smectite content decreases with increasing depth (Fig. 9.7). In both cases, the variation is not progressive, but occurs in stages. For the I/S minerals, two different series appear depending on whether the drill holes go through the thermal anomaly (CH 7bis) or are far away from it (CH 8). Far away from the anomaly, the I/S minerals are randomly ordered (R0) and rich in smectite (> 90%) in the upper portion of the field. The major change occurs between 400 and 600 m, where temperature suddenly passes from 60 to 80 °C. The smectite-rich R0 I/S minerals transform into regularly ordered (R1) I/S minerals (50% smectite – 50% illite) then quickly into R1 I/S minerals with 60% illite. The illite content then increases steadily up to over 95% at −1,750 m. In the thermal anomaly (CH7 bis), transformations of the I/S minerals take place differently, with pure smectites and R1 IS minerals with 60 and 90% illite coexisting. There is no steady enrichment in illite.

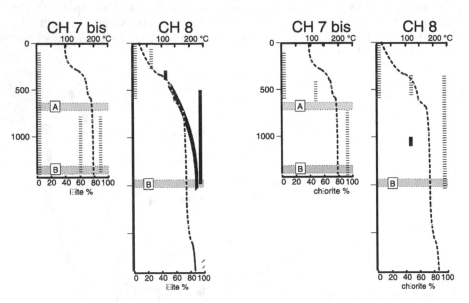

Fig. 9.7. Variation in the smectite content of illite/smectite and chlorite/smectite mixed layers in drill holes CH 7bis and CH 8 of the Chipilapa geothermal field. Measurements were performed by mathematical decomposition of the diffraction patterns into Gaussian curves

The smectite (saponite) – C/S series is similar in both situations: the upper portion of the field contains only saponite, then the gradient suddenly increases (400 to 600 m) and saponite is found to coexist with corrensite and C/S minerals with more than 95% chlorite. Saponite and corrensite disappear below 600 m. Corrensite reappears sporadically in low-permeability basaltic rocks between 1,000 and 1,100 m.

In the surface zone, down to −350 m, the clay cap of the field constitutes a low-permeability environment in which temperature is lower than 130 °C. Several smectite species co-exist: saponite, nontronite and a dioctahedral smectite whose composition is intermediate between beidellite and montmorillonite. R0 I/S minerals appear.

Present-Day Episode: High-Temperature Smectites

The alterations related to present-day fluid circulations can be observed in reservoir fractures only. They cause the formation of smectites, sometimes in combination with kaolinite and hematite, that are deposited as more or less continuous coatings on propylitic minerals or on minerals relating to the clays– carbonates episode (Fig. 9.8). Fluids locally impose physicochemical conditions that are similar to those of the clays – carbonates episode: temperature between 185 and 210 °C, low salinity, boiling state, oxidising environment. The differences stem from permeability because fractures are not sealed by carbonated deposits. Ebullition brings about the oversaturation of the fluid phase, an increase in the oxygen fugacity and a decrease in pH.

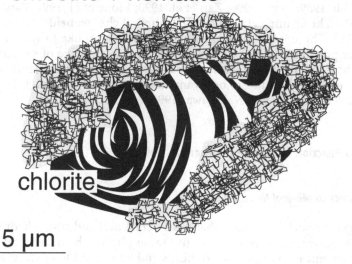

smectite + hematite

chlorite

5 µm

Fig. 9.8. Schematic representation of the growth of smectites in combination with hematite over large crystal rosettes of chlorites of propylitic origin (from a micro-photograph, Beaufort et al. 1995)

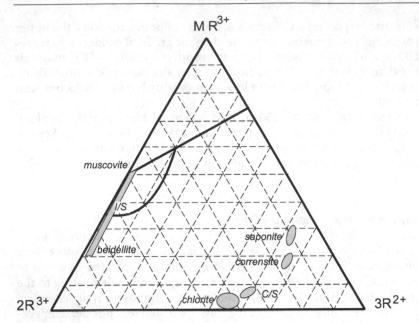

Fig. 9.9. Chemiographical representation of the chemical compositions of the di- and trioctahedral smectites, I/S minerals, illite, corrensite and chlorite from the Chipilapa geothermal field in the MR^3–$2R^3$–$3R^2$ system (Velde 1985)

Oversaturation is the driving force of the nucleation and growth of high-temperature smectites that are hallmark features of liquid-vapour reservoirs (Reyes and Cardile 1989; Reyes 1990; Zan et al. 1990; Inoue et al. 1992; Papa-panagiotou 1993). The forming clays have a tetrahedral charge: beidellite and saponite (Fig. 9.9). There is no montmorillonite component, like in the clay cap in the upper portion of the geothermal field. The assemblage of di- and trioctahedral phases is identical to what is produced by the experiments of high-temperature mineral syntheses by Iiyama and Roy (1963), Eberl (1978), Yamada et al. (1991), Yamada and Nakasawa (1993) (see Sect. 10.1.1.1).

9.1.2
Precipitation and Reaction of Clays in Geothermal Fields

9.1.2.1
From Field Dynamics to Mineral Reaction Kinetics

The Chipilapa geothermal field shows that the dioctahedral and trioctahedral phyllosilicates have syncrystallised under the same physicochemical conditions (identical chemical compositions of fluids and equivalent temperature) at two periods of the hydrothermal activity about 4,000 years apart. Despite these identical conditions, the two generations exhibit differing mineralogical and crystallochemical properties. Indeed, the minerals that are currently crystallising (t_0) between 180 and 240 °C in the field's active fractures are tetrahedrally charged smectites: beidellite and saponite. The clay minerals whose formation took place at about 240 °C about 4,000 years ago ($t_{4,000}$) are today's saponite-chlorite and beidellite-illite mixed layer series. Their respective transformation rates towards the end members chlorite and illite increase with depth (Fig. 9.10). This shows that the crystallochemical properties of "ancient" clays do not reflect the temperature of their occurrence in rocks but rather the period of time during which the transformations of smectites (metastable)

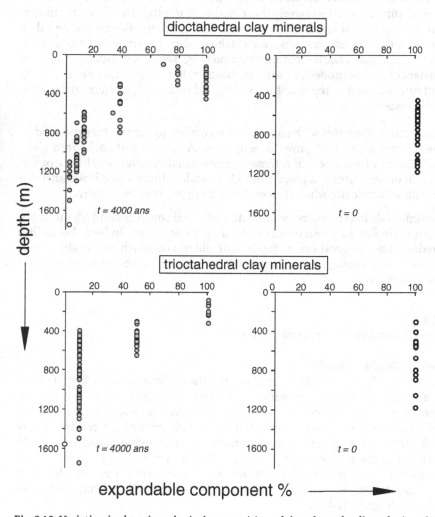

Fig. 9.10. Variation in the mineralogical composition of clays from the di- and trioctahedral series as a function of depth. Time t_0 corresponds to the precipitation of clays in fractures from boiling solutions. *Time 4,000 years* represents the duration of the local thermal anomaly (4,000 years at Chipilapa) that caused high-gradient diagenetic transformations in the rock

into illite and chlorite have been thermally activated in various locations of the geothermal field over the last 4,000 years.

The progressive disappearance of the expandable components to the benefit of chlorite or illite at increasing depth indicates that during the considered time interval (4,000 years), the conductive thermal effect (mean geothermal gradient) prevailed over the convective effect (local thermal anomaly due to hydrothermal fluid circulation). Therefore, in active geothermal fields, clay minerals must be considered as indicators of the *time × temperature* parameter rather than as geothermometers.

The first data on geothermal fields (Steiner 1968) showed that the illite content of I/S minerals progressively increased with depth. Therefore, hydrothermal series were considered equivalent to a high-gradient diagenesis. Based on this assumption, kinetic models were established on the smectite (montmorillonite) to illite conversion yielding a complete I/S sequence (Huang et al. 1993 for instance). These models, as sophisticated as they may be (see Sect. 8.1.2.2), do not fully succeed to represent the geological reality of geothermal fields for several reasons:

1. conduction alternates with convection according to the fractures opened by the seismic activity or phreatic eruptions. A plurality of rocks in a given location may be the seat of intense hydrothermal circulations, hence of formation of smectites at a given time. Once sealed, these smectites transform into illite or chlorite when the conductive regime is re-established.

2. clay minerals in reservoirs, notably dioctahedral smectites and I/S minerals, are not similar to those observed in diagenetic series. Indeed, beidellites predominate in geothermal fields and illites are much more aluminous (close to muscovite). Accordingly, the montmorillonite→illite reaction is inappropriate.

9.1.2.2
I/S and C/S Minerals: Discontinuous Series

I/S Series in Geothermal Fields
In geothermal fields, beidellite – and not montmorillonite – is the initial smectitic mineral whose transformation into illite as a function of the *time × temperature* parameter yields a series of I/S mixed layers, contrary to what takes place in the diagenetic context. The total absence of the random interstratification (R0) stage in the series is noteworthy. Montmorillonite occurs in the clay cap of the geothermal field where R0 I/S minerals are equally observed. It seems that the random interstratification only occurs if smectite does not exhibit any tetrahedral charge. This fact raises questions that have not been satisfactorily answered yet.

Out of this surface portion, the I/S sequence begins with a potassic rectorite phase (regularly ordered mixed-layer mineral). A sudden transition makes them disappear to the benefit of R1 I/S minerals whose illite content and coherent domain size increase steadily with depth (Fig. 9.11). The crystal

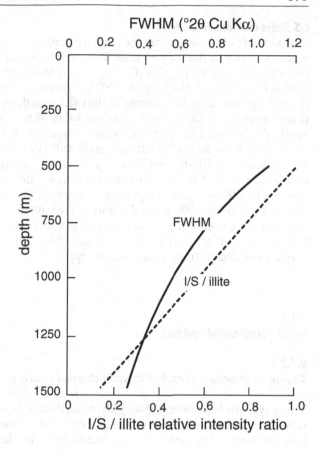

Fig. 9.11. Variation in the full width at half maximum intensity (FWHM) of the illite peak and in the intensity ratio of I/S and illite diffraction bands as a function of depth in the Chipilapa geothermal field. These parameters have been determined after decomposition of the diffraction patterns into Gaussian curves

morphology passes from small-sized lath-shaped particles (60–80% illite) to more isomorphic and thicker large-sized particles (> 85% illite). In geothermal fields, the illitic component is close to muscovite. It greatly differs from that of diagenesis, which exhibits a high octahedral charge (see Sect. 8.1.2.3).

Unfortunately, the discrimination between coexisting beidellite particles and regular I/S minerals is difficult. There is seemingly no significant difference of size and morphology between them. Therefore, the standard reaction process is probably of STT type (solid state transformation). The adsorption of K^+ ions brings about the irreversible collapse of high-charge layers. Subsequently, the increase in the illite content of I/S minerals with depth accompanied by size and morphology changes can only be due to a crystal growth phenomenon by addition of illite layers over I/S particles that is similar to the one described in diagenesis. Inoue and Kitagawa (1994) have shown that, in geothermal systems, the growth of I/S minerals and illites produces spiral steps originating from the emergence of a dislocation on the (001) faces (see Sect. 1.2.3.1).

C/S Series in Geothermal Fields

It was classically thought that the transformation of saponite into chlorite took place similarly to that of montmorillonite into illite through a continuous series of mixed-layer minerals (Chang et al. 1986; Bettison and Shiffman 1988; Bettison-Varga and Mackinnon 1997). However, some observations relating to active geothermal fields showed that the transformation rather took place discontinuously with randomly ordered C/S minerals (100 to 80% saponite) suddenly shifting to corrensite (50 to 40% saponite) then to "swelling chlorites", which are in reality C/S minerals with 15 to 0% saponite (Inoue et al. 1984; Inoue and Utada 1991). As a general rule, saponite or R0 C/S minerals and corrensite on the one hand and corrensite and chlorite on the other hand co-exist in wide-ranging temperature domains. The successive disappearance of saponite (or R0 C/S minerals) and corrensite at certain depths enable the following reaction sequence to be written: saponite → corrensite → chlorite. Corrensite is in fact a true mineral phase (Beaufort et al. 1997). Reactions are probably of dissolution-crystallisation type.

9.1.3
Acid Hydrothermal Systems

9.1.3.1
Alunite – Kaolinite – Pyrophyllite Hydrothermal Systems

These systems have been thoroughly described in Japan because they relate to recent volcanism and are sometimes exploited for kaolin. Figure 9.12a shows how the mineral zonation is established between deep propylitic zones and surface silica deposits. In the case of the Ugusu Mine, the post-Pliocene volcanic activity triggered circulations of acidic hydrothermal fluids (pH < 4 at 20 °C) that deeply altered the volcanic formations over a few hundred meters. Under these pH conditions, aluminium becomes more soluble than silicon. Therefore, silica forms porous deposits in the centre of the altered zones where hydrothermal springs emerge. The first altered zone in the vicinity of these deposits is formed of kaolinite, dickite and nacrite. Pyrophyllite appears near alunite zones $(KAl_3(SO_4)_2(OH)_6)$. On the outside, the altered rocks contain dioctahedral smectites. This zonation is explained by hydrolysis reactions of type $2\,KAlSi_3O_8 + 2\,H^+ + 9\,H_2O \rightarrow Si_2Al_2O_5(OH)_4 + 2\,K^+ + H_4SiO_4$ occurring during the decrease in the temperature of hydrothermal fluids. The total leaching of cations, including aluminium from rocks near the vents, is due to the pH stabilisation in the very acid domain by H_2S oxidation. Therefore, the physico-chemical conditions (temperature – cation activity/proton activity log) probably change following the path indicated in Fig. 9.12b. The alunite precipitation is due to the H_2S oxidation. Muscovite has not been observed. However, the presence of dioctahedral smectites near unaltered propylitic rocks shows that the cation concentration of solutions and pH values are higher away from the vents.

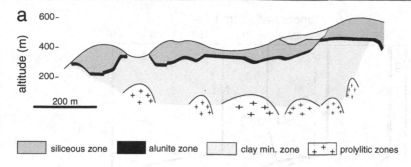

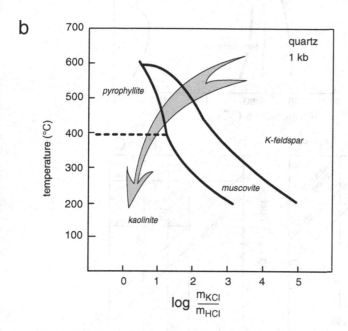

Fig. 9.12a,b. Acid hydrothermal alteration. **a)** Sketch profile of the acid hydrothermal system of the Ugusu Mine, Shizuoka Prefecture, Japan (Nagasawa 1978). **b)** Schematic representation of temperature and ion activity conditions (Montoya and Hemley 1985). The *arrow* symbolises the probable path of the temperature-composition conditions of fluids in the hydrothermal system

9.1.3.2
Halloysite-Kaolinite Transition in Acid Lakes

Clay minerals react in acid and hot environments by dissolution of certain species and precipitation of others. The acid lake of the Ohyunuma crater (Japan) provides a unique opportunity to study these reactions because it is supplied by rivers that renew solutions (Inoue and Aoki 1999–2000). The mean residence time of water is 25.5 days. Two environments are superimposed (Fig. 9.13a):

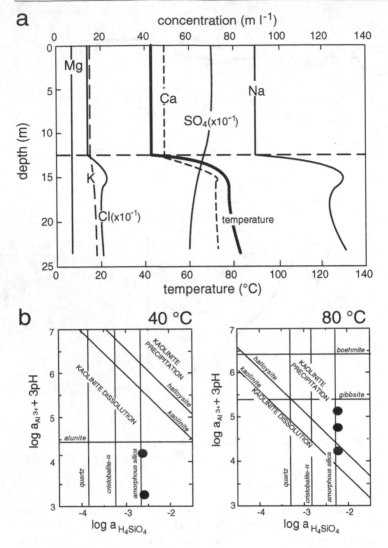

Fig. 9.13a,b. The acid lake of the Ohyunuma crater (Japan) from Inove and Aoki (1999–2000). **a)** Variation in chemical parameters and temperature in the lake. **b)** The chemical composition of waters from the upper part (40 °C) and the lower part (80 °C) of the lake are represented by *black dots*. The slopes of solubility lines are determined from thermodynamic parameters calculated by Helgeson et al (1978)

1. a low-salinity, oxidising, well homogenised upper part, with a high sulphate content, a low chlorine content and with pH = 2.4, where temperature does not exceed 40 °C;

2. a high-salinity lower part, with high sulphide and chlorine contents, with pH = 2.7, where temperature increases from 40 to 121 °C at the bottom.

The distribution of the clay minerals composing the sediments at the bottom of the lake is controlled by several factors, such as the location of hydrothermal discharge zones, the composition of river-borne suspensions, the bottom topography, the flow direction. Nevertheless, these sediments are strictly composed of halloysite and kaolinite. Quartz and α-cristobalite are of allochthonous origin. Sulphur and pyrite precipitate in the entire water column; they are related to the very hot vents and are kept in suspension by turbulent flows. Alunite forms in the lower part.

The upper part is kept in an undersaturation state with respect to halloysite and kaolinite owing to the renewal of solutions by river inputs. Accordingly, the reactions relating to silicates amount to the dissolution of all suspended clays in rivers (smectite, halloysite and supergene kaolinite). The compositions of the solutions at various depths are represented in the diagram $\log a_{Al^{3+}} + 3$ pH vs. $\log a_{H_4SiO_4}$ (Fig. 9.13b). The slopes of the solubility lines are determined from the thermodynamic parameters calculated by Helgeson et al. (1978). At $40\,°C$, they are greatly within the undersaturation domain.

In the lower part of the lake (mean temperature $80\,°C$), the compositions remain undersaturated with respect to halloysite but become oversaturated with respect to kaolinite. Accordingly, only the largest halloysite particles are not dissolved and settle at the bottom of the lake (Inoue et al. 2000). Kaolinite is formed first through a two-dimensional nucleation process due to the high oversaturation degree. The growth takes place by coalescence of small crystals agitated by lake currents. However, since the major part of the aluminium dissolved in solutions is consumed by sulphate precipitation (alunite), the growth of kaolinite is limited.

9.1.3.3
Donbassite and Tosudite in Granitic Cupola

The emplacement of granitic batholiths is a complex phenomenon that comprises different stages taking place over long periods. The physicochemical processes of the magmatic differentiation sometimes result in the formation of small granitic bodies having a high alkaline and volatile element content. The granitic cupola of Echassières (Massif Central, France), studied in the scope of the Programme Géologie Profonde de la France, is an example of this type of intrusion into a two-mica granite and micaschists. The presence of breccia adjacent to micaschists shows that the emplacement took place at very shallow depths (brittle domain). The magmatic body is divided into two entities: a phaneritic upper part and a lower part with a fluidal microstructure (Fig. 9.14a). Its mineralogical composition is dominated by albite (40–70%); other components are quartz (15–25%), K-feldspar (5–15%), lepidolite (10–25%) and topaz (1–5%). The rocks have been affected by two successive episodes of hydrothermal alteration:

- early alteration stages in veins, chronologically: quartz, muscovite, pyrophyllite, donbassite, tosudite, and kaolinite;

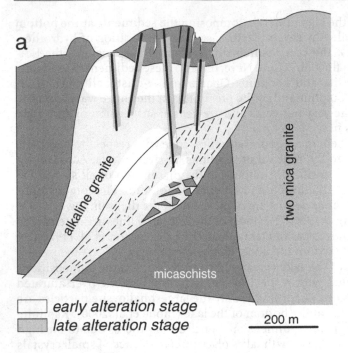

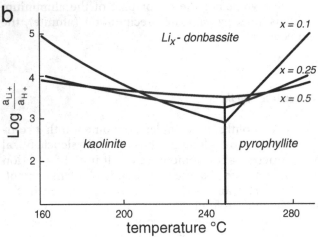

Fig. 9.14a,b. Hydrothermal alterations in the granitic cupola of Echassières, Massif Central, France (Merceron et al. 1992). **a)** Geological profile of the granitic cupola showing the intrusion of two magmatic bodies (phaneritic upper part, fluidal lower part) and the early alteration stages related to their cooling. The late alteration stages transform the fault wallrocks (post-magmatic tectonic event). **b)** Phase diagram in the Si–Al–Li–H$_2$O system, showing the extension of the stability field of donbassite as a function of its Li content ($0 \leq x \leq 1$)

– late alteration stages in veins, chronologically: fluorite + quartz, illite + quartz, illite/smectite mixed layer + quartz.

Donbassite seals small-sized veins (0.2 to 0.7 mm) and replaces some pre-existing minerals: primary feldspars and lepidolites, secondary muscovites. It contains up to 1.6% Li_2O (in weight). Its mean unit formula is as follows:

$$[Si_{3.81}Al_{0.19}]\,O_{10}\,(Al_{3.81}Li_{0.52}Fe^{2+}_{0.01}Mg_{0.01}Mn_{0.01})\,(OH)_8Ca_{0.02}Na_{0.07}K_{0.04} \tag{9.3}$$

The lithium content of tosudite (regularly ordered dioctahedral donbassite/smectite mixed layer) is a bit lower: 0.72%. Its unit formula is as follows:

$$[Si_{3.50}Al_{0.50}]\,O_{10}\,(Al_{2.95}Li_{0.22}Fe^{3+}_{0.01}Ti_{0.01})\,(OH)_5Ca_{0.01}Na_{0.15}K_{0.18} \tag{9.4}$$

The salinity and temperature of solutions measured by fluid inclusions and by isotopic analyses of quartz and associated minerals are controlled by the cooling of the magmatic body after its emplacement. They range between 10 and 1% NaCl equivalent and between 400 and 150 °C, respectively. These fluids are acid and exhibit high chlorine and fluor contents. Upon reaction with alkaline granite, they yield a sequence of secondary minerals from 400 to 200 °C: muscovite – pyrophyllite – donbassite-$Li_{0.5}$ – tosudite – kaolinite. Donbassite and tosudite form at the end of this early episode between 300 and 200 °C preferentially to kaolinite owing to the presence of lithium in the system (Merceron et al. 1992). Indeed, the higher the Li content, the greater the temperature range of the stability field of donbassite (Fig. 9.14b). Late alteration stages relate to the opening of fractures and vertical faults that cut across the enclosing micaschists. Fluids are of meteoric origin. Their Fe and Mg contents are increased by their passage in the micaschist cover. The precipitated illite and I/S minerals have compositions that greatly differ from those of the minerals generated by early alterations.

9.1.4
Geothermal Systems with Seawater (Alkaline Type)

9.1.4.1
Active Geothermal Fields

The geothermal fields supplied with seawater are located in volcanic islands. The best known have developed in basaltic structures like Iceland, others are related to andesitic volcanic arcs (Bouillante, Guadeloupe), and still others form in fractured basements (Milos, Greece). Whatever their geological setting, they are characterised by the abundance of magnesium-rich trioctahedral phyllosilicates in zones of thermal diffusion (conduction) as well as in fractured systems where hot fluids flow (convection). Anhydrite is frequently observed (Tomasson and Christmannsdottir 1972).

Thermal diffusion in active geothermal fields establishes a zonation of zeolites and clay minerals similar to that developing in vast regions by diagenesis or metamorphism. The only singularity is the shortening of zones owing to the high geothermal gradient. The sequence as a function of depth (increasing temperature) summarised by Alt (1999) is based on the mineral transitions observed in Icelandic active fields. In this sequence the transitions in ferromagnesian silicates (phyllosilicates and amphiboles) and in Ca–Na silicates are paralleled (Kristmannsdottir 1975, 1977, 1978; Schiffman and Fridleifsson 1991): (1) smectite – chlorite-smectite mixed layers – corrensite – chlorite – actinote + epidote; (2) low-temperature zeolites (mordenite, stilbite, clinoptilolite, heulandite) – laumontite – wairakite – prehnite (Fig. 9.15a).

More or less similar parallel zonations are observed in active fields supplied with seawater. The example of the reservoir of the Milos geothermal field (Greece) enables this organisation to be verified in a non-basaltic context. The low-permeability upper portion of the field exhibits a mineral zonation related to heat diffusion by conduction. Low-temperature clays are montmorillonite, which disappears at 150 m in depth ($T < 150\,°C$), and saponite, which persists beyond 250 m. Corrensite forms between 200 and 350 m in depth, from where it is replaced by chlorite. A non-expandable mixed-layer mineral of talc-chlorite type occurs in places. This regular zonation contrasts with clay minerals that are currently forming in permeable zones at temperatures over $300\,°C$.

The reservoir is located in the basement composed of micaschists (blue schist and green schist metamorphic facies). The primary phyllosilicates of schist rocks are muscovite (phengite) and biotite. This dominating vapour reservoir is located in the seawater-boiling zone. All phyllosilicates forming at $300\,°C$ are of trioctahedral type. Saponite reappears, combined with a talc/saponite mixed layer (T/S), talc and actinote (Fig. 9.15b). This combination is totally out of thermodynamic equilibrium. It results from the kinetics of the nucleation of these various phases and only depends on the composition of fluids. This effect of kinetics has been confirmed by experiments of basalt alteration between 375 and $425\,°C$ that have shown the coexistence of smectite with chlorite-epidote or chlorite-actinolite-tremolite parageneses (Berndt et al. 1988)

9.1.4.2
Hydrothermal Alteration of Oceanic Basalts

Pillow Lavas (Seawater at 2–4 °C)
The rims of pillow lavas are formed by very thin exfoliation structures produced by the thermal shock with seawater (Fig. 9.16a). These structures transform gradually into green breccia (Scott and Hajash 1976; Juteau et al. 1979). The interaction between seawater and glass yields an alteration micro-profile a few millimetres thick (Fig. 9.16b), which is not to be confused with the black halo caused by sea-water weathering (see Sect. 6.1.3.5). The external part is formed of green palagonite (altered glass) that gradually transforms (variolitic zone 2 to 3 mm thick) into an apparently unaltered brown basaltic glass. The green palagonite has a perlitic structure, each "pearl" being zoned concentrically. The breccia is formed of green palagonite pieces cemented together. Voids are sealed

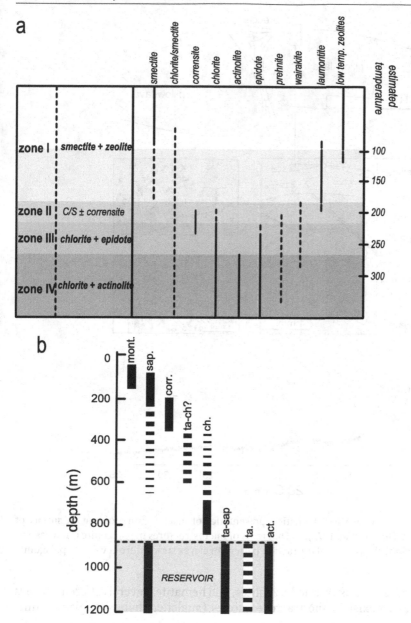

Fig. 9.15a,b. Active geothermal fields supplied with seawater. **a)** Mineral zonation related to heat diffusion by conduction (Alt 1999). **b)** Contrast between the diffusion-related zonation in the upper portion of the Milos geothermal field (Greece) and the forming paragenesis in the reservoir (Beaufort et al. 1995)

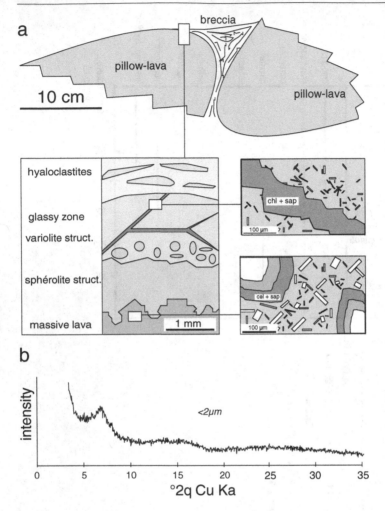

Fig. 9.16a–c. Pillow lava. **a)** Schematic representation of an alteration profile at the surface of a pillow lava (Juteau et al. 1979). **b)** Diagrams of microstructures in palagonite fractures and massive basalt. **c)** Example of diffraction pattern of clays extracted from veins in palagonite

by polyphased deposits: calcite, zeolites, and hematite. Several different species of zeolites co-exist in the fractured zones (analcite, chabasite, gismondine, faujasite).

Phyllosilicates in the glassy zones of pillow lavas are often difficult to identify because their diffraction patterns exhibit only broad and low-intensity bands (small size X-ray scattering coherent domains) while their chemical composition, even measured with the electron microprobe, always corresponds to phase mixtures (Fig. 9.16c). Most compositions so measured are located in the domains of celadonite, saponite and chlorite although only smectite can be identified by X-ray diffraction.

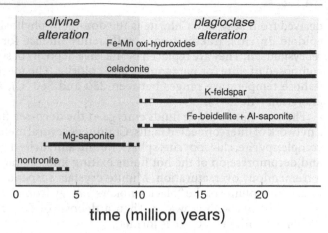

Fig. 9.17. Simplified sequence of secondary minerals formed by the low-temperature hydrothermal alteration (< 50 °C) of deep-sea floor basalts (Alt 1999)

Alteration of the Deep-Sea Floor (Sea Water 10–50 °C)

Hydrothermal alterations of the deep-sea floor have been extensively studied. Alt (1999) has published an excellent data synthesis. During the first two million years, the seawater-basalt interactions produce a black halo composed of a mixture of nontronite, celadonite and Fe- and Mn-oxihydroxides. Clay phases have very small-sized coherent scattering domains as is the case for the palagonitised zone of pillow lavas. This halo ranges between a few millimetres and 1 or 2 centimetres in thickness. The basalt vesicles remain empty. This is seawater weathering; see Sect. 6.1.3.5.

Older samples (2 to 10 My) show the presence of a saponite zone between the unaltered rock and the halo. The presence of this clay is interpreted as the superimposition of a hydrothermal alteration subsequent to the formation of the black halo by fluids derived from seawater and heated (< 50 °C) in fractures. The hotter the fluids, the more intense the alteration. At this stage, olivines are totally altered. Saponite remains the dominant phyllosilicate but talc may occur in vesicles in combination with pyrite and calcite. The vesicles are then totally sealed.

The hydrothermal alteration under similar conditions (seawater, 50 °C) of rocks that have been exposed for 10 to 23 Ma still exhibits the black halo. The zone intermediate with the fresh rock is characterised by an aluminous saponite + ferriferous beidellite + K-feldspar mineral assemblage. The sequence of occurrence of alteration minerals as a function of age (Fig. 9.17) shall not be interpreted as the replacement of the magnesian saponite by an aluminous equivalent. The change in composition is not a question of relative stability but of chemical composition of the altered environment. The presence of secondary minerals relates to the destabilisation of plagioclases.

Alteration of the Dyke Zone (Seawater, < 150 °C)

Underneath basalts, the deep-sea floor structure comprises a layer formed by an agglomeration of dykes corresponding to the feeding conduits of submarine volcanoes. The intensely fractured rocks are crossed by heated fluids (> 50 °C) derived from seawater. Chlorite is the dominant phyllosilicate combined with epidote. In rifts, fracturation is sufficiently intense for basalts to be totally recrystallised. They are replaced by massive deposits of sulphides and by a rock composed of illite (or paragonite) and chlorite. The more abundant the fluids (whose temperature ranges between 250 and 360 °C), the more intense the alteration (Alt 1999).

The very hot sulphide fluids emerge at the deep-sea floor surface thanks to a network of interconnected faults. Chimneys several meters high are formed by complex pyrite, chalcopyrite, sphalerite and anhydrite deposits. Thermal shock and decompression of the hot fluids exiting into seawater at 2 °C bring about a tremendous oversaturation. Minute crystals suspended in these fluids emphasise the plume of the "black smokers". Away from the vents, the hydrothermal fluids mix with seawater. Then a deposit of Fe- and Mn-oxihydroxides combined with nontronite is formed.

Hydrothermal Alterations in Guyots (Sea Mounts)

The general structure of guyots is typically organised as follows: a summit formed by air lava flows, a base formed by submarine lava flows and an inner framework of dykes. The Mururoa and Fangataufa atolls have been widely studied thanks to the numerous drill holes exploring their mechanical stability (Dudoignon et al. 1989; Dudoignon et al. 1997). The hydrothermal systems that have developed there have experienced several activity periods leading to successive mineral deposits. During the most intense activity at Fangataufa, the thermal gradient was $300\,°C\,km^{-1}$. Alterations have led to the formation of a sequence of clay minerals: from top to bottom, beidellite, saponite, saponite + randomly ordered C/S minerals, randomly ordered C/S minerals + corrensite, chlorite. The last activity period ($150\,°C\,km^{-1}$) has caused the sealing of fractures by calcite deposits.

The sequence of trioctahedral minerals exhibits a general organisation that is similar to those described in the deep-sea floor or in Icelandic-like systems. However, their distinguishing feature is the formation of beidellite replacing plagioclases (Fig. 9.18). Beidellite co-exists with randomly ordered C/S minerals (190–240 °C) and corrensite (260–320 °C). The occurrence of illite/smectite mixed layers might be expected in these temperature ranges. Their absence is due to a chemical microsystem effect: plagioclase does not contain potassium.

9.2
Small-Sized Hydrothermal Systems

Introduction

Small-sized hydrothermal systems are often used as natural analogues in human works for the study of chemical component transfers or mineral reactions owing to similar size and duration scales. When the order of magnitude is a hundred years or more, any experimental reproduction of the phenomena is unthinkable. These systems are considered as natural laboratories and the

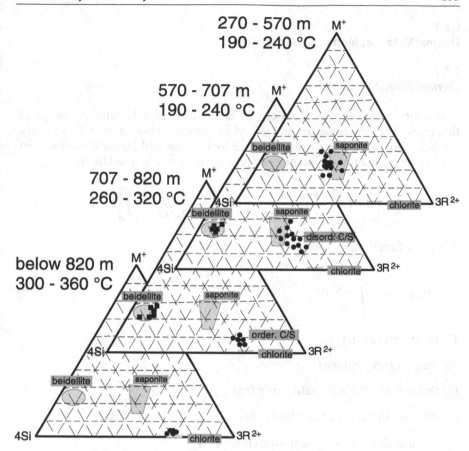

Fig. 9.18. Sequence of clay minerals in the hydrothermal system of the Fangataufa guyot (Sea Mounts), French Polynesia (Dudoignon et al. 1997). Palaeotemperatures have been determined by the stable isotope composition of phyllosilicates and carbonates

observations are treated like experimental data. Notably, the safety of the radioactive waste storages requires that the thermal and chemical behaviours of the various barriers be modelled over relatively long durations. Two points have been particularly developed:

1. the long-term stability of the bentonite engineered barrier has been modelled using natural situations in which this material underwent transformations triggered by the local temperature increase around volcanic rock intrusions;

2. the mechanisms involved in the transfers of chemical elements related to the circulations of hot fluids in fractures have been deciphered by the study of hydrothermal veins crosscutting different types of rocks.

9.2.1
Thermal Metamorphism of Clay Formations

9.2.1.1
Thermal Diffusion in Shales

The solution of complex equations of heat transfers is beyond the scope of this book. The principles are published in numerous books now. To simplify, the one-dimensional solution proposed by Carslaw and Jaeger (1959) and its application to magmatic intrusions (Jaeger 1964) will be used here:

$$T = T_0 \sum_{n=0}^{\infty} \left[\mathrm{erf}\left(\frac{(2n+1)L+x}{2\sqrt{\kappa t}}\right) - \mathrm{erf}\left(\frac{(2n+1)L-x}{2\sqrt{\kappa t}}\right) \right] \tag{9.5}$$

erf: error function:

$$\mathrm{erf}(x) = \frac{2}{\sqrt{\pi}} \int_0^x e^{-\xi^2} \, d\xi \tag{9.6}$$

T temperature (°K)

T_0 magma temperature (°K)

L thickness of the magmatic body (m)

x distance to the magmatic body (m)

κ thermal diffusivity of bentonite (mm^2 s^{-1})

t time(s)

This type of modelling has been applied by Pytte (1982) to a contact metamorphic zone produced by basaltic intrusions in smectite-rich Cretaceous shales (Fig. 9.19a). The κ value for shales is 0.64 mm^2 s^{-1}. The result is expressed graphically in Fig. 9.19b, which shows that the thermal peak is reached faster when the considered point is closer to the magmatic body. The intensity of this peak decreases with distance. In other words, these graphs show that the amount of absorbed heat is proportional to the *time × temperature* parameter: the closer the point to the heating body, the greater this parameter.

9.2.1.2
Kinetics of Mineral Reactions

Smectite → Illite Transition
The progressive increase of the illite content and the decrease of the smectite content in illite – smectite mixed layers (I/S) in diagenetic or hydrothermal formations are classically thought to be due to a single mineral reaction of

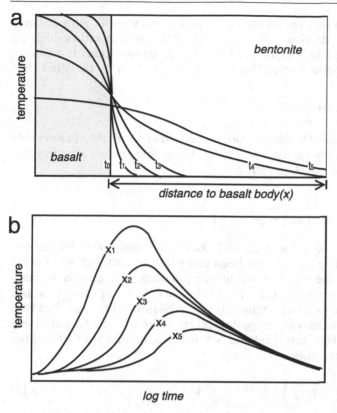

Fig. 9.19a,b. One-dimensional schematic representation of the effects of thermal diffusion around a basaltic vein crosscutting smectite-rich shales (Pytte 1982). **a)** Variation in temperature as a function of distance for different increasing time values from t_0 (date of the magmatic intrusion) to t_5. **b)** Variation in temperature as a function of time for different points of the system located at increasing distance values from x_1 to x_5

smectite $\rightarrow$ illite type. Therefore, the reaction progress can be measured by the variation in the smectite ratio S. The kinetic expression is as follows: $-dS/dt = kS^a$ where a is a constant determining the reaction order. The rate constant k is given by the Arrhenius equation (see Sect. 3.2.2.1): $k = A \exp(-E_a/RT)$ where A is the frequency factor, E_a is the activation energy of the reaction and T is the temperature expressed in K. The general kinetic equation is written as follows:

$$-\frac{dS}{dt} = S^a \left(\frac{a_K}{a_{Na}} \right)^b A \exp^{-\left(\frac{E_a}{RT} \right)} \tag{9.7}$$

The transformation of smectite into illite brings about potassium consumption. As a consequence, the reaction does not only depend on temperature but also on the potassium activity a_K in solution. In the case of the thermo-metamorphism of shales described by Pytte, an indirect measurement of a_K was performed from the observation that the amount of K-feldspars increases with distance whereas that of albite decreases. Therefore, the K/Na activity ratio calculated by the orthose-albite equilibrium at all considered temperatures is a correct approach of the potassium role. The ratio a_K/a_{Na} is given by the Van't Hoff equation:

$$a_K/a_{Na} = 74.2 \exp^{(-2490/T)} \tag{9.8}$$

Integers a and b are added to determine the reaction order. Pytte and Reynolds (1989) give the following values: $a = 5$, $b = 1$. Using these values after integration of the differential equation yields:

$$S^4 = \frac{S_0^4}{1 + 4 \times 74.2t\, S_0^4 A \exp\left(-\frac{2490}{T} - \frac{E_a}{RT}\right)} \tag{9.9}$$

Parameters $A = 5.2\,10^7\,s^{-1}$ and $E_a = 33\,kcal\,mol^{-1}$ have been determined by fitting methods. The results have been compared to data relative to other environments where the smectite→illite reaction occurs (diagenesis, metasomatism). Pytte and Reynolds (1984) have calculated A and E_a using time and temperature values for which I/S minerals with 80% illite form more than 90% of the clay fraction under various geological conditions (Table 9.2). If $S_0 = 1$ (clay fraction with 100% smectite) and $S = 0.2$ (I/S mineral with 80% illite), the above equation becomes:

$$\ln\left(t/ = \left(\frac{1}{T}\right)\left(2490 + \frac{E_a}{R}\right) - \ln(A) + 0.743 \tag{9.10}$$

The resulting numerical values are consistent with those extracted from the line presented in Fig. 9.20a. A and Ea values are $5.6 \times 10^7\,s^{-1}$ (mineral reactions assumed to be first-order ones) and $33.2\,kcal\,mol^{-1}$, respectively. They

Table 9.2. Approximate time and temperature values required for the presence of more than 90% of illite/smectite mixed layers with more than 80% illite in the clay fraction of various types of rocks (from Pytte and Reynolds 1984)

Estimated time	Estimated maximum Temperature value (°C)	Geological conditions	References
10 years	250	Contact metamorphism	Reynolds (1981)
10,000 years	150	Geothermal fields	Jennings and Thompson (1986)
1 My	127	Burial diagenesis	Perry and Hower (1972)
10 My	100	Burial diagenesis	Perry and Hower (1972)
300 My	70	Burial diagenesis	Srodon and Eberl (1984)
450 My	70	K-bentonite	Huff and Turkmenoglu (1984)

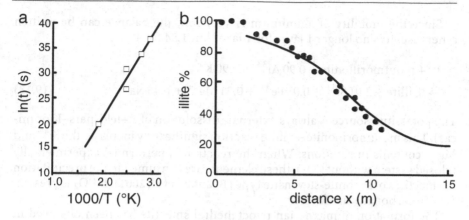

Fig. 9.20a,b. Thermometamorphism of shales (Pytte and Reynolds 1984). a) Time – temperature relationship of the formation of more than 90% of illite/smectite mixed layers with more than 80% illite in the clay fraction (values extracted from Table 2). b) Variation in the illite content in shales with distance to the basaltic vein (*full dots*). The *solid line* represents the curve calculated from the kinetic equation using the values indicated in the text

are very close to those determined by means of the fitting method. The simulation (continuous curve) is relatively consistent with the measurements of the proportion of expandable layers in the I/S minerals (Fig. 9.20b).

The Montmorillonite → Saponite + Illite Reaction

The smectite → illite transformation such as classically admitted is based upon the assumption that a smectite layer yields an illite layer. Although the true transformation processes of one into the other are still poorly known (dissolution – crystallisation, solid state transformation, growth of illite over smectite), the chemical balance of the reaction cannot be drawn up. Assuming that smectite (montmorillonite) and illite have the following compositions:

$$Si_4O_{10}\left(Al_{1.65}Fe^{3+}_{0.05}Mg_{0.30}\right)(OH)_2Na_{0.30} \quad \text{and}$$

$$[Si_{3.25}Al_{0.75}]\,O_{10}\left(Al_{1.80}Fe^{3+}_{0.05}Mg_{0.15}\right)(OH)_2K_{0.90} \tag{9.11}$$

the reaction is written:

$$1 \text{ montmorillonite} + 0.90\,Al^{3+} + 0.90\,K^+$$

$$\rightarrow 1 \text{ illite} + 0.75\,Si^{4+} + 0.15\,Mg^{2+} + 0.30\,Na^+ \tag{9.12}$$

The source of K^+ and Al^{3+} ions is generally related to the dissolution of K-feldspars or primary micas.

Since the mobility of aluminium is very low, the balance can be written otherwise. It is no longer 1 layer for 1 layer but 1.54 for 1:

$$1.54 \text{ montmorillonite} + 0.90 \text{ Al}^{3+} + 0.90 \text{ K}^{+}$$

$$\rightarrow 1 \text{ illite} + 2.91 \text{ Si}^{4+} + 0.03 \text{ Fe}^{3+} + 0.31 \text{ Mg}^{2+} + 0.46 \text{ Na}^{+} \qquad (9.13)$$

The potassium source is always external (dissolution of K–feldspars–K or micas). This montmorillonite → illite reaction significantly increases the Si^{4+} and Mg^{2+} contents in solutions. When the reaction is performed experimentally (closed system without CO_2), these elements are consumed in the precipitation of smectites of saponite-stevensite type (Beaufort et al. 2001). If CO_2 is present, Fe–Mg carbonates are formed.

The formation of magnesian trioctahedral smectite has been observed in experiments simulating nuclear waste storages in the Stripa site (Sweden) where a clay barrier formed of a kaolinite/smectite mixed layer (K/S) has been heated for many years. Despite the low magnesium content of these K/S minerals, the chemical diffusion near the heating body has caused the precipitation of saponite (Bouchet et al. 1992). Without potassium, illite does not form. It is replaced by a dioctahedral smectite whose layer charge is mostly located in the tetrahedral sheet. The saponite + beidellite assemblage has been observed in alteration experiments (Yamada and Nakasawa 1993) and natural hydrothermal systems (Beaufort et al. 1995).

9.2.2
Hydrothermal Veins

9.2.2.1
Mineral Zonations

The circulation of hydrothermal fluids in fractures causes two types of mineral reactions:

- the precipitation of minerals that finally seal the fracture, referred to as a "vein";
- the alteration of wall rocks whose colour changes over a variable distance to the fracture, referred to as an alteration "halo".

Even reduced to a few microns in thickness, the halo is most often zoned. Mineral assemblages change with the distance to the vein. These zonations were described very early in mineralised systems (Lovering 1949; Bonorino 1959). Each zone is characterised by an assemblage or by a dominating mineral whose proportions vary with distance (Fig. 9.21). These two authors had already established that zonations resulted from a chemical diffusion process under isothermal conditions (the heat diffusion rate being higher by several orders of magnitude than the chemical diffusion rate). They considered that the width of the halo was dependent on the system temperature, among other factors.

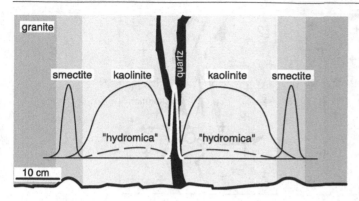

Fig. 9.21. Schematic representation of the mineral zonation of a vein from the Caribou Mine, Idaho (Bonorino 1959). The now obsolete term "hydromica" refers to illite/smectite mixed layers

The origin of zonation was subsequently reconsidered as resulting from crystallisation kinetics (Page and Wenk 1979). These authors based their conclusion on the observation of a sequence of phyllosilicates from the vein to the altered rock: $2M_1$ phengite, $1M$ then $1Md$ illite, illite/smectite mixed layers (I/S). Assuming isothermal conditions, they considered that I/S minerals are metastable precursors of illite and phengite.

The generally admitted isothermal conditions have been questioned in recent studies. Indeed, thermal gradients at various scales have been shown using different techniques:

- 200 °C over 100 m in the Amethyst vein (Horton 1985),

- 60 °C over 6 m in uranium veins (Al Shaara 1986),

- 75 °C over 11 mm in a phengite vein (Turpault et al. 1992b).

The detailed study of the mechanisms of alteration propagation in the vicinity of the phengite veins crosscutting the La Peyratte granite (Deux Sèvres, France) has shown that the thermal gradient is stabilised by a pulsed flow of hot fluids in the fracture. The hydrodynamic regime was reconstructed from temperatures measured in secondary fluid inclusions in the wall rock quartz crystals. The residence time of fluids in the fracture and the expulsion frequency are of the same order of magnitude as those observed in surface geysers (Turpault et al. 1992a).

The statistical analysis of veins 20 to 300 µm in width in the La Peyratte granite shows that the wall rock alteration obeys the two following processes:

1. formation of mineral reaction "fronts" bounding chlorite, albite or plagioclase dissolution (oligoclase) zones;

2. propagation of fronts such that the various zones do not vary independently (Fig. 9.22). This means that the alteration propagation is controlled by

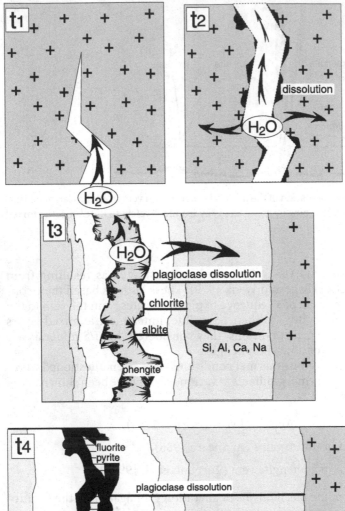

Fig. 9.22. Schematic representation of the propagation of alteration fronts in the wall rocks of fractures crosscutting the La Peyratte granite, Deux Sèvres, France (from Turpault et al. 1992a,b). t_1: mechanical strains open a fracture that brings granite minerals and hydrothermal fluids into contact; t_2: dissolution of the fracture wall rocks; t_3: wall rock alteration maintained by the establishment of chemical potential gradients. The crystallisation of phengites in the fracture starts at this stage as the last link in the chain of mineral reactions; t_4: alteration propagation by displacement of the mineral reaction fronts

a single chemical mechanism maintained during the entire "life time" of the fracture.

This alteration propagation phenomenon in wall rocks cannot be explained by the dissolution of primary minerals at the contact of fluids alone. Once the chemical equilibrium between hydrothermal solutions and the mineral has been reached, the alteration rate becomes infinitely low. Therefore, in the absence of a thermal gradient maintaining a chemical potential difference for every element, mass transfers are believed to be reduced to values that cannot be measured in hydrothermal veins.

The chemical balance of alteration is achieved as follows: (1) transfer of H_2O from the fracture to the wall rock, (2) transfer of Si^{4+}, Al^{3+}, K^+, Ca^{2+}, Na^+ cations from the rock to the fracture and (3) total consumption of cations in the mineral reactions with the exception of Si^{4+} and Ca^{2+}. The mobility of low-solubility cations like Al^{3+} is maintained by the chemical potential difference of this element between the "source" site (dissolution front of plagioclases) and the "consumption" site (biotite $\rightarrow$ chlorite and oligoclase $\rightarrow$ albite reaction front). The variation in the size of the plagioclase dissolution holes accounts for the progression of alteration fronts (Meunier 1995).

Alteration involves several mineral reactions thus forming a "trophic chain" of which the ultimate consumer is the crystallisation of phengites in the open space of the fracture. This chain starts operating at the opening of the fracture and stops when one of the constituent reactions stops. In the present case, the crystallisation of phengites ceases when the whole open space is occupied (Berger et al. 1992). The vein is definitively sealed by the precipitation of fluorite ± pyrite. The bulk rock chemical balance (fracture + altered wall rocks) shows that only Si^{4+} and Ca^{2+} are exported outward. Oddly, the alteration process took place in a nearly closed system from the cation standpoint. Considering the quartz solubility at the alteration temperature (about 300 °C), one can calculate the number of litres of fluids that have flowed per volume unit of fracture to drain off the silica lost by the rock. The retrograde path (cooling of the system) is recorded in the rock by the precipitation of illite/smectite mixed layers that are increasingly rich in smectite near the vein in the plagioclase dissolution holes.

9.2.2.2
Hydrothermal Alteration of Clay Rocks

Clay rocks may be fractured and hydrothermalised under certain geological conditions. They form small-sized systems, which exhibit also a mineral zonation composed of mixed-layer mineral assemblages whose composition varies with distance to the vein. The recent argillised dacitic rocks from the Trois Ilets quarry, Martinique (Bouchet et al. 1988), are mainly composed of I/S minerals the smectitic component of which comprises more than 80% of high-charge layers (irreversibly fixing K^+ ions). The fumarolic alteration (< 100 °C) causes the gradual decrease in the proportion of these layers down to zero in the vein

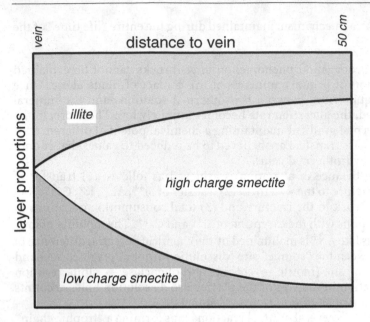

Fig. 9.23. Schematic representation of the alteration of a clay rock by hydrothermal fluids in the Trois Ilets quarry, Martinique (Bouchet et al. 1988)

(Fig. 9.23). The mineral reaction is of type:

$$\text{high-charge smectite} + Si^{4+} + K^+ \rightarrow \text{low-charge smectite} + \text{illite} \qquad (9.14)$$

Si^{4+} and K^+ cations are brought by hydrothermal fluids.

At the Pointe de la Caravelle site (Martinique), the rock is a dacitic conglomerate belonging to the ancient island arc (36–22 Ma). The clay fraction is dominated by randomly ordered I/S minerals with 55–60% smectite. A network of hydrothermal veins caused an intense alteration over about ten centimetres. The clay fraction of the wall rocks comprises kaolinite and two types of I/S minerals: randomly ordered with 80% smectite and ordered with 5% smectite. The proportion of ordered I/S minerals increases near the vein whereas that of high-charge smectite layers in the randomly ordered I/S minerals decreases.

Zonations in the vicinity of veins can be disturbed or erased during the hydrothermal activity period. Indeed, several successive reopenings of fractures initially sealed by mineral deposits are frequently observed. This is a commonly occurring phenomenon in seismically active geothermal fields, well known in metamorphism ("crack-seal"). Each episode imposes new interactions between fluids–whose chemical composition has most often changed, and the previously altered rock. The superimposition of alterations brings about the juxtaposition practically in the same sites of minerals that have formed under very different conditions. These assemblages do not constitute a paragenesis.

Suggested Reading

Velde B (1995) Origin and Mineralogy of Clays. Clays and the Environment. Springer, Berlin Heidelberg New York, 334 pp

Henley RW, Truesdell AH, Barton PB, Whitney JA (1984) Fluid-mineral equilibria in hydrothermal systems. Reviews in Economic Geology 1, Society of Economic Geologists, Littleton, CO, 267 pp

CHAPTER 10

Clays Under Extreme Conditions

10.1
Experimental Conditions

Introduction

Clay minerals are systematically considered as formed under surface conditions (low temperature and pressure), which is mostly the case as shown in the preceding chapters. Nevertheless, clays can form under much more extreme conditions: interstellar medium, subduction zones, lava flows etc. These witnesses to non-supergenic conditions raise fundamental questions about what is called the stability field. In this respect, experimental syntheses provide useful landmarks in the pressure-temperature space, despite the effects of kinetics.

Among the different species of mineral clays, smectites are considered the most representative of the low-temperature and low-pressure conditions. This chapter is aimed at presenting *clays under extreme conditions* through the investigation of their occurrence fields. Abundant data are provided by experimental syntheses. The first systematic works date back to the years 1940–1950. Fortunately, most data are available in a recent compilation in which both synthesis procedures and stability conditions of the resulting products have been summarised (Kloprogge et al. 1999).

10.1.1
High-Temperature and High-Pressure Clays

10.1.1.1
Dioctahedral Smectites

General Phase Diagram
It seems that the lower the tetrahedral charge of dioctahedral smectites, the higher the temperatures at which they are formed (Fig. 10.1a). At a pressure of 1 kbar, the maximum temperature is reached at 450 °C for a tetrahedral charge of 0.1 per Si_4O_{10} (Ames and Sand 1958; Koizumi and Roy 1959). In reality, tetrahedral and octahedral substitutions in smectites only take place in a limited chemical composition domain that is much less extended than the one represented by experimental gels and glasses. The charge variation is probably due to phase mixtures (Harward and Brindley 1966).

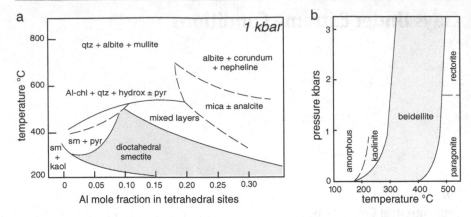

Fig. 10.1a,b. Compilation of data from the experimental syntheses of dioctahedral smectites (from Kloprogge et al. 1999). **a)** Phase diagram at a pressure of 1 kbar showing that the stability field of dioctahedral smectites depends on the aluminium content of the system. **b)** Phase diagram in the pressure-temperature space showing that beidellites can form at temperatures exceeding 450 °C for pressures above 1 kbar

The greater the proportion of Al ions, the lower the formation temperature of mica, at first as a mixed-layer mineral. The syntheses presented in Fig. 10.1a have been carried out with gels or sodic glasses; accordingly, mica is of paragonite type. When the mole fraction of Al^{3+} ions is lower than 0.1, smectite is theoretically of montmorillonitic type and is associated to kaolinite or pyrophyllite. In high temperature ranges ($T > 450\,°C$), other phyllosilicates are formed (aluminous chlorite). Above 600 °C, only anhydrous silicates crystallise: albite, mullite, nepheline.

Montmorillonite and beidellite are thus formed below 400 °C at 1 kbar. Yamada et al. (1991) have synthesised these phases in a pure state from gels having a variable Si/Al ratio. However, the formation of regular montmorillonite/beidellite mixed layers only occurs for a particular value of this ratio, thus indicating that montmorillonite has a fixed charge. For higher pressures (2 kbars), these mixed-layer minerals are formed with cristobalite at higher temperatures (Kloprogge et al. 1993).

Montmorillonites – Beidellites
The formation of montmorillonite necessitates the presence of bivalent cations such as Mg^{2+}, Fe^{2+}, Zn^{2+} … According to their proportion in the initial gels or glasses, dioctahedral smectite is pure or combined with a trioctahedral smectite (Haward and Brindley 1966; Nakasawa et al. 1991). The distribution of these bivalent elements must comply with the rules of the thermodynamic equilibrium between solid solutions. This means that the proportion of R^{2+} (hence the octahedral charge) in the montmorillonite crystal lattice varies with temperature for a given pressure. This has not been verified to date.

The formation of beidellites is favoured by high pH conditions that increase the probability of 4-fold coordination for Al^{3+} ions: AlO_4^- (Frank-Kamenestkii 1973; Kloprogge et al. 1990). The compilation of experimental data by Kloprogge et al. (1999) shows that beidellite has been synthesised at temperatures exceeding 450 °C for pressures above 1 kbar (Fig. 10.1b). Rectorite, which contains 50% of beidellite layers regularly interstratified with paragonite layers, is stable above 1.5 kbars. This shows that the higher the fluid pressure, the higher the formation temperature of hydrated minerals. This is consistent with the data on the thermal stability of smectites (Fig. 5.4b).

Dioctahedral smectites are very reactive and transform into illite (or mica) through a series of I/S mixed layers. Therefore, temperatures close to 400 °C correspond to very short synthesis times (a few days). For longer times and in the presence of K^+ ions, smectites transform into I/S minerals then into mica (Eberl 1978). Obviously, the chemical kinetics has a prevailing effect in the high-temperature synthesis.

10.1.1.2
Trioctahedral Clays

Trioctahedral clays, particularly saponite, have been synthesised at temperatures markedly higher than their dioctahedral equivalents (Koizumi and Roy 1959; Iiyama and Roy 1963). At equivalent pressure (1 kbar), they are formed above 800 °C in combination with anthophyllite and/or talc. These are the only products – up to about 600 °C for a very limited chemical composition domain – whose Al mole fraction is contained between 0.10 and 0.15 per Si_4O_{10} (Fig. 10.2). These syntheses at very high temperatures are carried out over very short periods of time (20 days maximum) in the absence of K^+ ions that favour the formation of micas. Whitney (1983) has shown that the stability tempera-

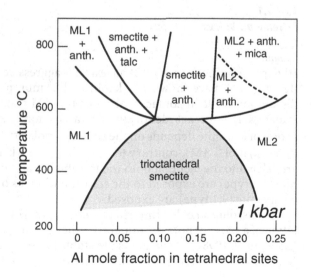

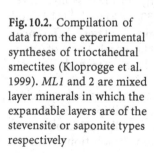

Fig. 10.2. Compilation of data from the experimental syntheses of trioctahedral smectites (Kloprogge et al. 1999). *ML1* and 2 are mixed layer minerals in which the expandable layers are of the stevensite or saponite types respectively

tures of saponite are much lower when the reaction time is multiplied by 10 (200 days): 400 °C. The rapid formation at high temperatures is obviously related to the chemical kinetics and does not represent the thermodynamic equilibrium of phases at these temperatures. However, many lessons can be drawn from these experiments since they clearly show that trioctahedral smectites are the first phases to form in the $SiO_2-Al_2O_3-MgO-(Na, 1/2 Ca, K)O-H_2O$ systems, even if they are unstable over time.

10.1.1.3
Experiments at Very High Pressures and Temperatures

In order to meet certain industrial needs for lamellar minerals, montmorillonites with an exceptional crystallinity have been synthesised under extreme conditions of pressure and temperature: 5.0 GPa–1,600 °C and 5.5 GPa–1,500 °C (Nakasawa et al. 1992). These syntheses of short duration (less than one hour) yielded an "exotic" mineral assemblage: montmorillonite + coesite + jadeite + kyanite. Even though the identified montmorillonite did not form simultaneously with the other minerals but rather after decompression (Bai et al. 1993), its exceptional crystallinity and the synthesis velocity suggest that it formed under higher P, T conditions than those usually used in synthesis experiments.

Na-beidellite has been synthesised under lower temperature conditions: 360 to 420 °C (Vidal 1997). Smectites form at the "cold" end of a golden tube in which a temperature gradient is maintained. The growth of these crystals is fed by the diffusion of elements derived from the dissolution of parent minerals.

10.1.2
Clays Under Extreme Chemical Conditions

10.1.2.1
Extreme pH Values

Kaolinite
High pH conditions (pH > 12) have a very aggressive effect on clays. Bauer and Berger (1998) have shown that kaolinite like montmorillonite undergo a total dissolution while in contact with KOH 0.1–4 M solutions. However, these two minerals do not react identically. The activation energy of the dissolution reaction of kaolinite depends on the solution molarity (33 ± 8 kJ mol^{-1} 0.1 M and 51 ± 8 kJ mol^{-1} 3 M), contrary to smectite (52 ± 4 kJ mol^{-1}). These differences are related to the fact that both the tetrahedral and octahedral sheets of kaolinite (1:1 type) are exposed to the solution whereas only the tetrahedral sheets of smectite (2:1 type) are exposed.

The dissolution of kaolinite in presence of KOH solutions causes the precipitation of a series of potassic minerals: illite, KI-zeolite, K-phillipsite, K-feldspar (Bauer et al. 1998). K-feldspar is the stable phase whereas illite, KI-zeolite, and K-phillipsite are metastable. The series of precipitation-dissolution reactions can be explained by the Ostwald step rule, which enables the stable phase to form without going over a significant energy barrier (see Sect. 1.2.1.3). Illites crystallising during the first step exhibit different crystal habits depending on the silica oversaturation rate of solutions with respect to the stable phase (K-feldspar). Lath-shaped in the undersaturation domain, illites become isometric in the oversaturation domain. A very high oversaturation rate yields irregular particles while a very low oversaturation rate yields perfectly hexagonal particles (Bauer et al. 2000).

Smectites

Bauer and Velde (1999) have shown that smectites with different compositions (0% and 35% of tetrahedral charges) first dissolve congruently in the presence of KOH solutions with high pH values (13 to 14.5). Subsequently, the Si/Al ratio of solutions increases and causes the precipitation of zeolites. The major effect is the decrease in the coherent domain size, which is revealed by the intensity decrease of $d_{(001)}$ diffraction peaks. Montmorillonite is more sensitive to this destruction than the beidellitic smectite. The authors further suggest that I/S mixed layers are formed transiently.

The interactions between the cement waters and containment clays contemplated for the containment of nuclear waste storage have necessitated the study of the reactions of these minerals in the presence of solutions containing Ca^{2+}, Na^+ and K^+ ions under high pH conditions (13.5). When fluid/clay weight ratios are low (3:1), solutions are rapidly buffered and reactions stop in the early stages of the transformation of solids. Recent studies have shown from a Wyoming montmorillonite that the charge of octahedral sheets, the total CEC and the CEC after neutralisation of the octahedral charge remain unchanged (Rassineux et al. 2001). By contrast, the crystal structure of smectite is deeply modified. The total surface measured by ethylene glycol monoethyl (EGME) adsorption is reduced by about 40% whereas the expansion ratio after neutralisation of the octahedral charge (Hofmann and Klemen treatment) increases. It all takes place as if the number of interlayers bounded by two tetrahedral sheets increases (Fig. 10.3). Under more aggressive conditions (pH > 13.5), the montmorillonite is fully transformed into a beidellite having a higher layer charge.

10.1.2.2
Pillared Clays

Clays having interlayer spaces capable of expansion (mostly smectites) can acquire interesting catalytic properties from an industrial standpoint if three requirements are met (Bergaya 1990):

- interlayer spaces must retain their expansion at high temperature (over 500 °C);

- interlayer spaces must be accessible to the substances to be catalysed;

- interlayer spaces must have a high acidity.

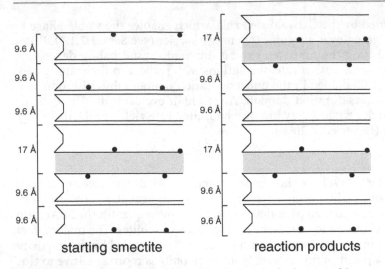

Fig. 10.3. Schematic representation of the changes in the layer stacking sequence caused by the reaction of a Wyoming montmorillonite with cement waters at pH = 13.5 (Rassineux et al. 2000)

"Pillared clays" are prepared so as to meet these requirements. Polymers are incorporated in the interlayer zone by cation exchange and are fixed in the structure by high-temperature deshydroxylation. The polymers are hydrated hydroxides of various metals: Zr, Cr, Ni and so on. The resulting interlayer spaces vary with the metal: 28 Å at 700 °C with Zr, 21 Å at 500 °C with Cr. The best-known polymer is the polymer of aluminium: $[Al_{13} O_4 (OH)_{24} (H_2O)_{12}]^{7+}$. After deshydroxylation, pillars are reduced to aluminium oxide (Al_2O_3) by release of H^+ protons that increase the acidity of the interlayer space (Fig. 10.4). Their size slightly decreases from 19 to 18.8 Å but their steady distribution remains unchanged (Van Damme and Fripiat 1985). Pillared clays retain a high specific surface (over $600 \, m^2 \, g^{-1}$) at 500 °C. They are preferred to zeolites whose performances are limited by a more rigid crystal structure.

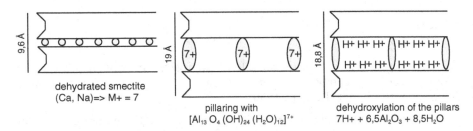

Fig. 10.4. A pillared clay is prepared by exchange between interlayer cations and a polymer (metal hydroxide). Acidity is obtained by deshydroxylation of the pillars (modified from Bergaya 1990)

10.1.3
Clays Under Irradiation Conditions

Natural or artificial irradiation causes the formation of two types of defects in crystal lattices:

- displacements of atoms related to direct elastic collisions (α articles, recoil nuclei). These are extended defects (fission tracks, collision cascades) that produce domains in which the structure of the mineral is disordered;

- ionisation of the atoms or ions of the crystal structure by inelastic interactions with their electrons (α, β, γ particles). The resulting excitation produces point defects, mainly vacancy/interstitial defects (Frenkel defects).

The point defects so created migrate under the effect of a thermal activation. They can agglomerate thus forming extended defects or appear on the surfaces of the mineral (healing). Some defects are paramagnetic and can be detected by electron paramagnetic resonance (EPR).

Kaolinite has been widely studied for its potentiality as a natural dosimeter that would make it usable as an indicator of the flow of mineralised fluids in uranium deposits (Muller and Calas 1993). Defects are due to oxygen ions that have lost one electron by radiation (trapped holes on oxygens bound to Si and Al atoms or radiation-induced defects, RID). The missing electric charge locally alters the intensity of the crystal field and magnetic field created by the electron clouds of neighbouring nuclei. This anomaly alters the shape of the EPR spectrum. Several types of trapped holes can be distinguished according to which orbital this single electron belongs to: Si–O (A and A′ centres), Al–O–Al (B centres).

The defects of naturally irradiated kaolinites are multiplied by artificial irradiation (Allard et al. 1994). Their number increases with the intensity of irradiation (Fig. 10.5a). The B defects are multiplied with smaller irradiation doses. The A defects are the more stable. The A-defect growth curves have an exponential pattern; they depend on the chemical purity (presence of Fe^{3+} ions) and on the order-disorder degree of crystals (Fig. 10.5b). The natural irradiation doses that crystals have received can be determined by extrapolation of exponential curves. Accordingly, kaolinites accompanying uranium deposits can be distinguished from those in wall rocks (Muller et al. 1990).

10.2
Natural Environments

Introduction

Many advances in the exploration of terrestrial and extraterrestrial environments have been made in the last decade. They show that clay minerals are much more widely occurring than initially thought. They may sometimes assume rather exotic appearances, like in meteorites for instance, but they subsist even under extreme conditions. Clay minerals are "tough".

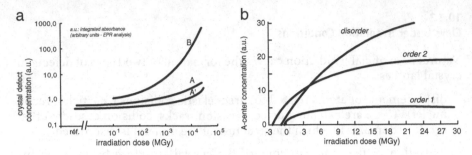

Fig. 10.5a,b. Radiation-induced defects. **a)** Increase in the concentration of A and A' defects (Si–O) and B defects (Al–O–Al) as a function of the radiation dose. **b)** Influence of the crystallinity state of kaolinite on the concentration of defects detected by irradiation: *disorder*: naturally irradiated disordered kaolinite; *order 2*: well crystallised but strongly naturally irradiated kaolinite; *order 1*: well crystallised and poorly naturally irradiated kaolinite (modified from Allard et al. 1994)

This section presents a few examples of their occurrence in magmatic and metamorphic environments on Earth where they were thought never to be found. One fascinating aspect is their presence in meteorites showing that molecular water is widely occurring in the Cosmos. Their catalytic properties might be at the origin of life.

10.2.1
High Temperatures: Post-Magmatic Crystallisation

The clays identified in basalts are classically considered to be either the products of a meteoric or hydrothermal alteration or the result of diagenetic reactions (Alt 1999). In both cases, they are to form at the expense of pre-existing phases (olivine, pyroxenes, plagioclases, glass) without any connection with the lava flow solidification processes. However, recent studies have shown that clays can form in basalts without alteration reactions (Mas 2000). Indeed, the growths of saponite over pyroxenes (Fig. 10.6a) or biotites (Fig. 10.6b) without any dissolution trace argue for a direct precipitation without destruction of the supporting mineral.

A basaltic lava flow cools down via its interfaces with the external environment, i. e. atmosphere and solid substratum (Fig. 10.7a). Accordingly, as regards gases and in particular water vapour, it behaves like a kind of naturally heated pressure vessel that seals from 980 °C on. Water remains in the interstices left available by anhydrous crystals (plagioclases, pyroxenes, olivines). These interstices have long been considered to be occupied by glass. In reality, they constitute totally crystallised micro-environments where quartz, K-feldspars and clay minerals (saponite, nontronite and chlorite) form by growth over anhydrous minerals and not by replacement of a pre-existing glass. Kinetics of the lava flow cooling is used as a temporal indicator of the variations in the mineralogical composition of clays from these micro-environments (Fig. 10.7b).

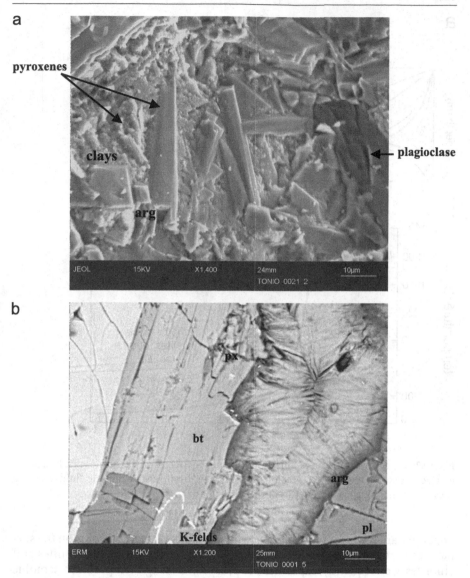

Fig. 10.6a,b. Post-magmatic crystallisation of clays in basaltic rock lava flows from the Moruroa atoll, French Polynesia (from Mas 2000). **a)** Saponite and nontronite crystals growing over automorphous pyroxenes and unaltered plagioclases (submarine lava flow, basalt, 1.70 m thick). **b)** Crystals of saponite growing on the intact (001) faces of a biotite (air lava flow, hawaite, 40 m thick)

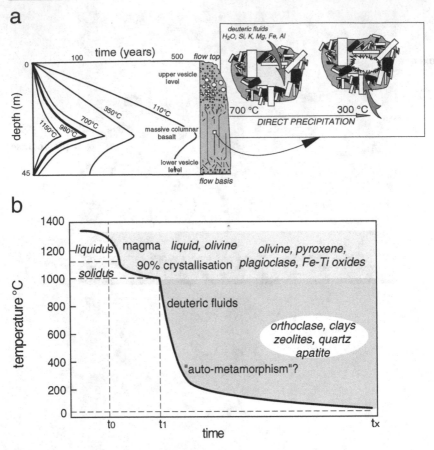

Fig. 10.7a,b. Post-magmatic crystallisations during the cooling process of basaltic lava flows. **a)** Cooling stages with direct precipitation of clays from post-magmatic fluids. **b)** The crystallisation sequence of minerals during the cooling process

Saponites are the first clay minerals to grow in the diktytaxitic voids on the surfaces of pyroxene, biotite or plagioclase crystals, followed by nontronites and chlorites. Consequently, saponites probably form at high temperature. Isotopic data indicate 300 °C (Mas 2000).

10.2.2
High Pressures: Subduction Zones

The mechanical strength of smectitic sediments under a confinement pressure greater than 100 MPa is at least twice as low as that of illitic sediments. This is due to the bonding weakness between the H_2O molecules saturating the interlayer spaces (Shinamoto and Logan 1981). In fact, the higher the pressure, the higher the temperature necessary to expel the water from the interlayer spaces (Koster van Groos and Guggenheim 1987). This explains that smectites "outlive" a deep burial in subduction zones and control, at least in part, the mechanical properties of the friction level (Vrolijk 1990).

The abundance of smectites in subduction-entrained sediments is closely related to the volcanic activity that scatters ashes in the atmosphere. Glass is rapidly altered in the water and forms bentonites (see Sect. 7.2.2.1) that can either constitute continuous beds or be scattered in turbidites (Fig. 10.8a). Smectite-rich sediments behave like a lubricant in the aseismic decollement zone. Burial necessarily causes a diagenesis leading to the transformation of

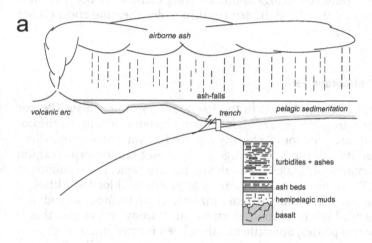

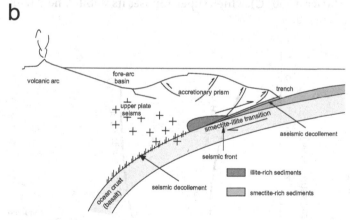

Fig. 10.8a,b. Clays and subduction zones (from Vrolijk 1990). a) The more abundant the volcanic ash input, the greater the smectite content in sediments. Scattered in turbiditic sediments (pelagic domain), smectites are more abundant at the level of the trench (semi-pelagic domain) as shown by the stratigraphic column. b) The mechanical strength of sediments is increased by the transformation of smectite into illite in the subduction zone. This transition corresponds to the seismic front

smectite into illite. In the case of subduction zones, this transition takes place at greater depth than in extension tectonic basins (see Sect. 8.1.2.1) for the following reasons:

- the burial rate is very high (25 km per million years in the Japan trench) and the geothermal gradient is very low (of the order of 10 °C/km);

- sediments are recent and the duration of diagenesis is short;

- dehydration is delayed by the rapid increase in pressure.

The smectite-to-illite transition takes place at depths greater than 10 km (Fig. 10.8b). This transition brings about a strong increase in the mechanical strength of sediments whose sudden rupture under shearing effect causes seisms (seismic front).

10.2.3
Metamorphism: Retrograde Path

The stability field of clays extends to the low-grade metamorphism (Peacor 1992). During the retrograde path of high-pressure metamorphism, magnesio-chloritoid crystals are pseudomorphically replaced by smectite-pyrophyllite-chlorite and smectite-chlorite-mica parageneses. These pseudomorphs exhibit intergranular contacts with talc and disthene that are typical of equilibrium micro-textures. Dioctahedral smectites are intergrown in chlorites. Although microprobe analyses yield compositions that are close to those of sudoites (Agard et al. 1999), high-resolution electron microscopy shows that this is a mixture of the two phases. Smectite has then been formed under high temperature conditions (at least 300 °C), which superimposes its stability field with that of pyrophyllite (Fig. 10.9).

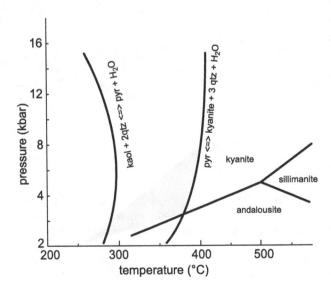

Fig. 10.9. *P-T-t* curve of high-pressure metamorphic rocks. The *shaded area* represents the domain in which dioctahedral smectites can crystallise during the retrograde path (modified from Agard et al. 1999)

10.2.4
Very Low Pressures: Extraterrestrial Objects

10.2.4.1
Meteorites

Phyllosilicates, particularly clay minerals, are sometimes very abundant in some meteorites or interstellar dusts (Rietmeijer 1998; Brearley and Jones 1998). They essentially form by alteration of anhydrous phases in the presence of water, and also in the presence of sulphur or CO_2. Numerous studies have been dedicated to these phyllosilicates that are the surviving evidence of the first moments of the Universe, comprising thermal metamorphism and aqueous alteration stages (Fig. 10.10a). The main species identified are:

– for interstellar dusts: Fe-rich saponite (10 to 15 Å), serpentine, talc, and more rarely, kaolinite, illite, pyrophyllite, mica;

– for chondritic meteorites whatever their type: serpentine-berthierine-cronstedtite, chrysotile, saponite more or less iron-rich, saponite/serpentine mixed layer and more rarely, chlorite, vermiculite and tochilinite/cronstedtite mixed layer.

The fact that the interlayer spaces of smectites still contain water molecules is noteworthy. Recall that this water has the thermodynamic status of ice (Mercury et al. 2001), and ice makes up most of the mass of comets.

The alteration that yielded these phyllosilicates could transform the primary minerals only if temperature conditions were high enough for water to be in a liquid or gaseous state. Serpentines and saponites are the dominant species. Their iron content is often higher than that of the olivines they replace. The chemical composition of the microsystems in which they were formed was probably controlled by the dissolution of several primary phases in the presence of sulphur and CO_2. This is attested by the formation of very special minerals such as tochilinite/cronstedtite mixed layers (Tomeoka and Busek 1985), tochlinite being itself a mixed-layer mineral comprising two sheets of Fe–Mg-hydroxides framing a sheet of Fe- Ni- Cu-sulphide (Fig. 10.10b). Saponite/serpentine mixed layers have been identified in several types of meteorites. They often contain ferrihydrite intergrowths ($5\,Fe_2O_3 \bullet 9\,H_2O$) whose formation is related to the dissolution of magnetite and Fe–Ni-sulphides (Tomeoka and Busek 1988) or to the oxidation of ferrous saponite (Brearley and Prinz 1992).

10.2.4.2
Soils of Mars

Mars missions (Viking) have enabled the Mars redness to be attributed to the dominant occurrence of hematite. Ferrihydrite and schwertmannite (hydroxylated ferric sulphate) are rarer. These phases are derived from the alteration of shergottitic basalts that form the planet's surface. The presence of palagonite

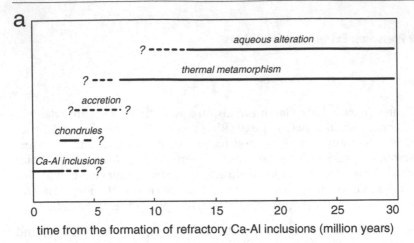

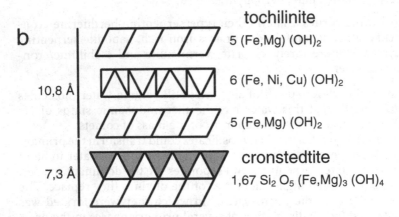

Fig. 10.10a,b. Clays in meteorites and interstellar dusts. **a)** Temporal sequence of events modifying the mineralogical composition of chondritic meteorites (Brearley and Jone 1998). *Full line*: measured times; *dotted line*: estimated times. **b)** Crystal structure of the tochilinite/cronstedtite mixed layer (modified from Mackinnon and Zolenshy 1984)

is highly probable (Banin et al. 1992). The rare Martian meteorites contain smectite- and illite-like clays combined with Fe-oxides and salts such as NaCl and $MgCl_2$ (Clark and van Hart 1981; Gooding 1992).

10.2.5
Clays and the Origin of Life

The effects of the catalytic properties of clays had a great importance in the early stages of the Earth's history when the surface conditions were very different from today's. Since Bernal (1951), numerous investigations have been undertaken to understand how the nuclides or their derivatives could be adsorbed on the internal surfaces (interlayers) or external surfaces of clay minerals and subsequently be concentrated so as to trigger their polymerisation, thus enabling clays to assume the role of screens protecting the "prebiological" molecules from degradation under the effect of ultraviolet radiation.

10.2.5.1
Adsorption of Nucleotides

Experiments of DNA and RNA molecule adsorption (Franchi et al. 1999) show that isotherms exhibit a classical pattern described by the Langmuir equation (Fig. 10.11):

$$S = S_{max} \frac{LC_e}{1 + LC_e} \tag{10.1}$$

where S and S_{max} are the adsorbed concentration and the maximum adsorption capacity ($\mu g\,g^{-1}$), respectively, L is the Langmuir affinity coefficient and C_e is the concentration of nucleic acid in the solution. Large-sized nucleotides are not adsorbed on the internal surfaces but overly the external edges of crystallites or particles. The smallest nucleotides can be adsorbed on the (001) faces. Adsorption slightly alters the molecular structure of nucleotides.

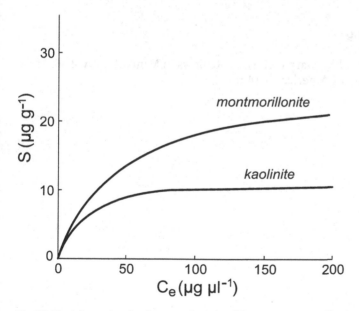

Fig. 10.11. Adsorption isotherms of nucleotides on montmorillonite and kaolinite (from Franchi et al. 1999)

10.2.5.2
Polymerisation of Nucleotides

The formation of peptide bonds from "prebiological" monomers classically takes place in two stages: (1) activation and (2) condensation of these molecules. Below 100 °C, without previous activation and without addition of condensation agents, the polymerisation of amino acids and dipeptides (Glycine, Alanine) has been achieved in two ways:

- by adsorption on clays subjected to drying-wetting cycles (Bujdak and Rode 1999);

- by catalytic reaction in aqueous solution at pH = 8 in the presence of Na-montmorillonite (Kawamura and Ferris 1999).

The bonds between activated nucleotides and montmorillonite are favoured by Mg^{2+} ions. However, the polymerisation rate is not dependent on the presence of these ions. This indicates that the greatest part of adsorption does not occur on the catalytic sites.

The kinetics of polymerisation depends on the species of nucleotide under consideration. For a given species, the rate constants of trimers are much higher than those of dimers but are not very different from those of bigger polymers. The variations in the rate constant value are dependent on the strength of the nucleotide-clay bonds: the rate constant is higher for pyrimidine than for purine because the former is less strongly bonded than the latter. These bond strengths are controlled by interactions of the Van der Waals type.

Suggested Reading

Papike JJ (ed) (1998) Planetary materials. Reviews in Mineralogy, vol 36. Mineralogical Society of America, 1014 pp

References

Aagaard P, Helgeson HC (1983) Activity/composition relations among silicates and aqueous solutions II. Chemical and thermodynamic consequences of ideal mixing of atoms among energetically equivalent sites in the montmorillonites, illites and mixed layer clays. Clays Clay Miner 31:207–217

Abercrombie HJ, Hutcheon IE, Bloch JD, Caritat P de (1994) Silica activity and the smectite-illite reaction. Geology 22:539–542

Agard P, Jullien M, Goffé B, Baronnet A, Bouybaouène M (1999) TEM evidence for high-temperature (300 °C) smectite in multistage clay-mineral pseudomorphs in pelitic rocks (Rif, Morocco). Eur J Mineral 11:655–668

Ahn JH, Peacor DR (1985) Transmission electron microscopic study of diagenetic chlorite in Gulf Coast argillaceous sediments. Clays Clay Miner 33:228–236

Ahn JH, Peacor DR, Coombs DS (1988) Formation mechanism of illite, chlorite and mixed-layer illite-chlorite in Triassic volcanogenic sediments from Southland Syncline, New Zealand. Contrib Mineral Petrol 99:82–89

Al Shaara M (1986) Etude géochimique et métallogénique des minéralisations (U-Ba) du nord di massif des Palanges (Aveyron, France). Thèse Univ. Paris VI, 174 pp

Allard T, Muller JP, Dran JC, Ménager MT (1994) Radiation-induced paramagnetic defects in natural kaolinites: alpha dosimetry with ion beam irradiation. Phys Chem Minerals 21:85–96

Allègre C (1985) De la pierre à l'étoile. Fayard, Paris, 300 pp

Allen PA (1997) Earth surface processes. Blackwell, Oxford, 404 pp

Alt JC (1999) Very-low grade hydrothermal metamorphism of basic igneous rocks. In: Frey M, Robinson D (eds) Low-grade metamorphism. Blackwell, Oxford, 169–201 pp

Alt JC, Honnorez J (1984) Alteration of the upper oceanic crust DSDP Site 417: mineralogy and chemistry. Contrib Mineral Petrol 87:149–169

Alt JC, Laverne C, Muehlenbachs K (1985) Alteration of the upper oceanic crust: mineralogy and processes in DSDP Hole 504B, Leg 83. In: Anderson RN, Honnorez J, Becker K (eds) Initial reports, DSDP vol 83. US Government Printing Office, Washington, DC, pp 217–247

Alt JC, Muehlenbach K, Honnorez J (1986) An oxygen isotopic profile through the upper kilometre of the oceanic crust, DSDP Hole 504B. Earth and Planet Sci Lett 80:217–299

Altaner SP, Hower J, Whitney G, Aronson JL (1984) Model for K-bentonite formation: evidence from zoned K-bentonites in the disturbed belt, Montana. Geology 12:412–415

Altaner SP, Ylagan RF (1997) Comparison of structural models of mixed-layer illite/smectite and reaction mechanisms of smectite illitisation. Clays Clay Miner 45:517–533

Ames LL, Sand LB (1958) Factors effecting maximum hydrothermal stability in montmorillonites. Am Mineral 43:641–648

Anceau A (1992) Sudoite in some Visean (Lower Carboniferous) K-bentonites from Belgium. Clay Miner 27:283–292

Anceau A (1993) Caractérisation des minéraux argileux des bentonites potassiques du Carbonifère Inférieur de la Belgique et des régions limitrophes. Thèse Univ. Liège, 165 pp

Anderson SJ, Sposito G (1991) Caesium-adsorption method for measuring accessible surface charge. Soil Sc Soc Am J 55:1569–1576

Andrews AJ (1977) Low temperature fluid alteration of oceanic layer 2 basalts. Can J Earth Sci 14:911–926

Andrews AJ (1980) Saponite and celadonite in layer 2 basalts, DSDP Leg 37. Contrib Mineral Petrol 73:323–340

April RH (1981) Trioctahedral smectite and interstratified chlorite/smectite in Jurassic strata of the Connecticut Valley. Clays and Clay Miner 29:31–39

Arch J, Maltman A (1990) Anisotropic permeability and tortuosity of wet sediments. J Geophys Res 95(B6):9035–9045

Arkai P (1991) Chlorite crystallinity: an empirical approach and correlation with illite crystallinity, coal rank and mineral facies as exemplified by Palaeozoic and Mesozoic rocks of north east Hungary. J Metam Geol 9:723–734

Aronson JL, Hower J (1976) Mechanism of burial metamorphism of argillaceous sediments. 2: Radiogenic argon evidence. Bull Geol Soc Am 87:738–744

Backer H, Richter H (1973) Die rezente hydrothermal-sedimentäre Lagerstätte Atlantis II – Tief im Roten Meer. Geol Rundschau 62:697–740

Badaut D (1988) Les argiles et les composés silico-ferriques des sédiments métallifères de la fosse Atlantis II (Mer Rouge). Thèse Univ. Paris-Sud, 217 pp

Badaut D, Besson G, Decarreau A, Rautureau R (1985) Occurrence of a ferrous trioctahedral smectite in recent sediments of Atlantis II Deep, Red Sea. Clay Miner 20:389–404

Badaut D, Blanc G, Decarreau A (1990) Variation des minéraux argileux ferrifères, en fonction du temps et de l'espace, dans les dépôts métallifères de la fosse Atlantis II en Mer Rouge. CR Acad Sci Paris 310:1069–1075

Badaut D, Decarreau A, Besson G (1992) Ferripyrophyllite and related Fe^{3+} – rich 2:1 clays in recent deposits of Atlantis II Deep, Red Sea. Clay Miner 27:227–244

Bai TB, Guggenheim S, Wang SJ, Rancourt DG, Koster van Groos AF (1993) Metastable phase relations in the chlorite-H2O system. Am Mineral 78:1208–1216

Bailey SW (1975) Chlorites. In: Gieseking JE (ed) Soil components, vol 2. Inorganic components. Springer, Berlin Heidelberg New York, pp 191–263

Bailey SW (1980) Structures of layer silicates. In: Brindley GW, Brown G (eds) Crystal structures of clay minerals and their X-Ray identification. Mineralogical Society Monographs 5, pp 1–123

Bailey SW (1982) Nomenclature for regular interstratifications. Am Mineral 67:394–398

Bailey SW (1988) Odinite, a new dioctahedral-trioctahedral Fe^{3+}-rich 1:1 clay mineral. Clay Miner 23:237–247

Bailey SW (1988) Hydrous phyllosilicates (exclusive of micas). Rev Mineral 19:725

Baker JC (1997) Green ferric clay in non-marine sandstones of the Rewan group, southern Bowen Basin, Eastern Australia. Clay Miner 32:499–506

Balan E, Allard T, Boizot B, Morin G, Muller JP (1999) Structural Fe^{3+} in natural kaolinites: new insights from electron paramagnetic resonance spectra fitting at X and Q-band frequencies. Clays Clay Miner 47:605–616

Banfield JF, Eggleton RA (1988) Transmission electron microscope study of biotite weathering. Clays Clay Miner 36:47–60

Banfield JF, Eggleton RA (1990) Analytical transmission electron microscope studies of plagioclase, muscovite, and K-feldspar weathering. Clays Clay Miner 38:77–89

Banin A, Clark BC, Wänke H (1992) Surface chemistry and mineralogy. In: Kieffer HH, Jakosky BM, Snyder CW, Matthews MS (eds) Mars. Univ Arizona Press, Tucson, AZ, pp 594–625

Barnes HL (1997) Geochemistry of hydrothermal ore deposits. Wiley, New York, 972 pp

Barnes I, O'Neil JR, Trescases JJ (1978) Present-day serpentinization in New Caledonia, Oman and Yugoslavia. Geochim Cosmochim Acta 42:144–145

Baronnet A (1976) Polytypisme et polymorphisme dans les micas. Contribution à l'étude du rôle de la croissance cristalline. Thèse Aix-Marseille III, 2 vol., 256 pp

Baronnet A (1988) Minéralogie. Dunod, 184 pp

Baronnet A (1994) Murissement d'Ostwald des minéraux: rappel des conditions limitantes. Bull Liaison Soc Fr Miner Cristall 2:28

Baronnet A (1997) Silicate microstructures at the sub-atomic scale. CR Acad Sci Paris 324(IIa):157–172

Baudracco J, Bel M, Perami R (1982) Effets de l'altération sur quelques propriétés mécaniques du granite du Sidobre (France). Bull Int Assoc Engin Geol 25:33–38

Bauer A, Berger G (1998) Kaolinite and smectite dissolution rate in high molar KOH solutions at 35 and 80 °C. Appl Geochem 13:905–916

Bauer A, Velde B, Berger G (1998) Kaolinite transformation in high molar KOH solutions. Appl Geochem 13:619–629

Bauer A, Velde B (1999) Smectite transformation in high molar KOH solutions. Clay Miner 34:259–273

Bauer A, Velde B, Gaupp R (2000) Experimental constraints on illite crystal morphology. Clay Miner 35:587–597

Beane RE (1982) Hydrothermal alteration in silicate rocks, Southwestern North America. In: Titley SR (ed)Advances in geology of the porphyry copper deposits, southwestern North America. University of Arizona Press, Tucson, AZ, pp 117–137

Beaufort D (1987) Interstratified chlorite/smectite ("metamorphic vermiculite") in the Upper Precambrian greywackes of Rouez, Sarthe, France. In: Schultz LG, Olphen H van, Mumpton FA (eds) Proc Inter Clay Conf, Denver 1985, pp 59–65

Beaufort D, Baronnet A, Lanson B, Meunier A (1997) Corrensite: a single phase or a mixed layered phyllosilicate of the saponite-chlorite conversion series? The case study of the Sancerre-Couy deep drill-hole (France). Am Mineral 82:109–124

Beaufort D, Berger G, Lacharpagne JC, Meunier A (2001) An experimental alteration of montmorillonite into a di + trioctahedral smectite assemblage at 100 and 200 °C. Clay Miner 36:211–225

Beaufort D, Cassagnabère A, Petit S, Lanson B, Berger G, Lacharpagne JC, Johansen H (1998) Kaolinite-to-dickite reaction in sandstone reservoirs. Clay Miner 33:297–316

Beaufort D, Meunier A (1983) Petrographic characterization of an argillic hydrothermal alteration containing illite, K-rectorite, K-beidellite, kaolinite and carbonates in a cupromolybdenic porphyry at Sibert, Rhône, France. Bull Min 106:535–551

Beaufort D, Meunier A (1994) Saponite, corrensite and chlorite-saponite mixed-layers in the Sancerre-Couy deep drill-hole (France). Clay Miner 29:47–61

Beaufort D, Papapanagiotou P, Patrier P, Fujimoto K, Kasai K (1995) High temperature smectites in active geothermal fields. In: Proc 8th Int Symp Water-Rock Interact, Vladivostok, pp 493–496

Beaufort D, Patrier P, Meunier A, Ottaviani MM (1992) Chemical variations in assemblages including epidote and/or chlorite in the fossil hydrothermal system of Saint Martin (Lesser Antilles). J Volcano Geotherm Res 51:95–114

Beaufort D, Westercamp D, Legendre O, Meunier A (1990) The fossil hydrothermal system of Saint Martin, Lesser Antilles: geology and lateral distribution of alterations. J Volcano Geotherm Res 40:219–243

Becker RH (1971) Carbon and oxygen isotopic ratios in iron formation and associated rocks from the Hamersley Range of Western Australia and their implications. PhD Dissertation, Univ of Chicago, Illinois, 258 pp

Belanteur N, Tacherifet S, Pakzad M (1997) Etude des comportements mécanique, thermo-mécanique et hydro-mécanique des argiles gonflantes fortement compactées. Rev Fr Géotechnique 78:31–50

Bergaya F (1990) Argiles à piliers. In: Matériaux argileux. Structure, propiétés et applications. Société Française de Minéralogie et de Cristallographie, pp 511–537

Berger G (1995) The dissolution rate of sanidine between 100 and 300 °C. In: Proc 8th Inter Symp Water-Rock Interaction, Vladivostok, pp 141–144

Berger G, Claparols C, Guy C, Daux V (1994) Dissolution rate of basalt glass in silica-rich solutions: implications for long-term alteration. Geochim Cosmochim Acta 58:4875–4886

Berger G, Turpault MP, Meunier A (1992) Dissolution-precipitation processes induced by hot water in a fractured granite. Part 2: Modelling of water-rock interaction. Eur J Mineral 4:1477–1488

Berger G, Velde B (1992) Chemical parameters controlling the propylitic and argillic alteration process. Eur J Mineral 4:1439–1454

Berg-Madsen V (1983) High-alumina glaucony from the Middle Cambrian of Öland and Bornholm, southern Baltoscandia. J Sediment Petrol 53:875–893

Bernal JD (1951) The physical basis of life. Routledge, London, 80 pp

Berndt ME, Seyfried WE, Beck JW (1988) Hydrothermal alteration processes at mid-ocean ridges: experimental and theoretical constraints from Ca and Sr exchange reactions and Sr isotopic ratios. J Geophys Res 93:4573–4583

Berry R (1999) Eocene and Oligocene Otay-type waxy bentonites of San Diego County and Baja California: chemistry, mineralogy, petrology and plate tectonic implications. Clays Clay Miner 47:70–83

Besson G, Drits VA, Daynyak LG, Smolliar BB (1987) Analysis of cation distribution in dioctahedral micaceous minerals on the basis of IR spectroscopy data. Clay Miner 22:465–478

Bethke CM, Altaner S (1986) Layer-by-layer mechanism of smectite illitisation and application to new rate law. Clays Clay Miner 34:136–145

Bettison LA, Schiffman P (1988) Compositional and structural variations of phyllosilicates from the Point Sal ophiolite, California. Am Mineral 73:62–76

Bettison-Varga L, MacKinnon IDR (1997) The role of randomly mixed-layer chlorite/smectite in the transformation of smectite to chlorite. Clays Clay Miner 45:506–516

Bevins RE, Rowbotham G, Robinson D (1991) Zeolite to prehnite-pumpellyite facies metamorphism of the Late Proterozoic Zig-Zag Dal Basalt Formation, eastern North Greenland. Lithos 27:155–165

Bigham JM, Fitzpatrick RW, Schulze DG (2002) Iron oxides. In: Dixon JB, Schulze DG (eds) Soil mineralogy with environmental applications. Soil Science Society of America Book Series 7:323–366

Billault V (2002) Texture, structure et propriétés cristallochimiques des chlorites ferreuses dans les réservoirs gréseux. Thèse Univ. Poitiers, 193 pp

Bird DK, Schiffman P, Elders WA, Williams AE, McDowell SD (1984) Calc-silicates mineralization in active geothermal systems. Econ Geol 79:671-695

Bischoff JL (1972) A ferroan nontronite from the Red Sea geothermal system. Clays Clay Miner 20:217-223

Bisdom EBA (1967) Micromorphology of weathered granite near the Ria de Arosa (NW Spain). Leiden Geol Med 37:33-67

Bish DL, Duffy CJ (1990) Thermogravimetric analyses of minerals. In: Stucki JW, Mumpton FA (eds) Thermal analysis in clay science. CMS Workshop Lectures 3, pp 96-157

Blanc P (1996) Organisation de l'empilement des minéraux interstratifiés illite/smectite: modélisation thermodynamique et application au domaine expérimental. Thèse Univ. Strasbourg, 205 pp

Blanc P, Bieber A, Frtitz B, Duplay J (1997) A short range interaction model applied to illite/smectite mixed-layer minerals. Phys Chem Minerals 24:574-581

Blatt H, Middleton GV, Murray RC (1980) Origin of sedimentary rocks. Prentice Hall, Englewood Cliffs, NJ

Blatter CL, Roberson HE, Thompson GR (1973) Regularly interstratified chlorite-dioctahedral smectite in dike-intruded shales, Montana. Clays Clay Miner 21:207-212

Bloch S, Bischoff JL (1979) The effect of low-temperature of basalt on the oceanic budget of potassium. Geology 7:193-196

Blum AE (1994) Determination of illite/smectite particle morphology using scanning force microscopy. In: Nagy KL, Blum AE (eds) Scanning probe microscopy of clay minerals. CMS Workshop Lectures 7, pp 172-202

Bocquier G, Muller JP, Boulangé B (1984) Les latérites. Connaissances et perspectives actuelles sur les mécanismes de leur différenciation. In: Livre Jubilaire du Cinquantenaire, A.F.E.S, Plaisir, pp 123-140

Bodine MW, Madsen BM (1987) Mixed-layer chlorite/smectites from a Pennsylvanian evaporite cycle, Grand County, Utah. In: Schultz LG, van Olphen H, Mumpton FA (eds) Proc Int Clay Conf, Denver, 1985. The Clay Minerals Society, pp 85-93

Boettcher AL (1966) Vermiculite, hydrobiotite, and biotite in the Rainy Creek igneous complex near Libby, Montana. Clay Miner 6:283-296

Bohor BF, Triplehorn DM (1993) Tonsteins: altered volcanic-ash layers in coal-bearing sequences. Geol Soc Am Spec Paper 285, 44 pp

Boles JR, Franks SG (1979) Clay diagenesis in Wilcox sandstones of southwest Texas: implications of smectite diagenesis on sandstone cementation. J Sediment Petrol 49:55-70

Bonhomme MG, Dos Santos RP, Renac C (1995) La datation potassium-argon des minéraux argileux. Etat des connaissances. Bull Centres Rech Explor-Prod Elf Aquitaine 19:199-223

Bonorino FG (1959) Hydrothermal alteration in the Front Range Mineral Belt, Montana. Bull Geol Soc Am 70:53-90

Bouchet A (1987) Minéralogie et géochimie des roches altérées du chapeau de fer de Rouez (Sarthe). Thèse Univ. Poitiers, 145 pp

Bouchet A, Lajudie A, Rassineux F, Meunier A, Atabek R (1992) Mineralogy and kinetics of alteration of a mixed layer kaolinite/smectite in nuclear waste disposal simulation experiment (Stripa site, Sweden). In: Meunier A (ed) Clays and hydrosilicate gels in nuclear fields, 113-123 pp

Bouchet A, Meunier A, Sardini P (2001) Minéraux argileux - Structures cristallines - Identification par diffraction de rayons X. Avec CD-ROM. Editions TotalFinaElf, 136 pp

Bouchet A, Meunier A, Velde B (1988) Hydrothermal mineral assemblages containing two discrete illite / smectite minerals. Bull Miner 111:587-599

Boulangé B (1984) Les formations bauxitiques latéritiques de Côte-d'Ivoire. Les faciès, leur transformation, leur distribution et l'évolution du modelé. Mem ORSTOM, 175 pp

Bradaoui M, Bloom PR (1990) Iron-rich high-charge beidellite in vertisols and mollisols of the high Chaouia region of Morocco. Soil Sci Am J 64:267–274

Brady NC, Weil RR (1999) The nature and properties of soils, 12th edn. Prentice Hall, Englewood Cliffs, New Jersey

Brearley AJ, Jones RH (1998) Chondritic meteorites. In: Papike JJ (ed) Planetary materials. Rev Mineral 36

Brearley AJ, Prinz M (1992) CI-like clasts in the Nilpena polymict ureilite: implications for aqueous alteration processes in CI chondrites. Geochim Cosmochim Acta 56:1373–1386

Brigatti MF (1983) Relationships between composition and structure in Fe-rich smectites. Clay Miner 18:177–186

Brindley GW (1961) Chlorite minerals. In: Brown G (ed) The X-ray identification and crystal structure of the clay Minerals. Mineralogical Society, London, UK, pp 242–296

Brindley GW, Brown G (1980) Crystal structures of clay minerals and their X-ray identification. Mineralogical Society Monograph 5, 485 pp

Brindley GW, Suzuki T, Thiry M (1983) Interstratified kaolinite/smectites from the Paris basin; correlations of layer proportions, chemical compositions and other data. Bull Mineral 106:403–410

Brindley GW, Chih-Chun Kau, Harrison JL, Lipsicas M, Raythatha R (1986) Relation between structural disorder and other characteristics of kaolinites and dickites. Clays Clay Miner 34:239–249

Browne PRL (1978) Hydrothermal alteration in active geothermal fields. Ann Rev Earth Planet Sci 6:229–250

Browne PRL (1998) Hydrothermal alteration in New Zealand geothermal systems. In: Arehart GB, Hulston JR (eds) Water-rock interaction 9. Balkemia, Rotterdam, pp 11–17

Browne PRL, Ellis AJ (1970) The Ohaaki-Broadland hydrothermal area, New Zealand. Mineralogy related to geochemistry. Am J Sci 269:97–131

Brusewitz AM (1986) Chemical and physical properties of Palaeozoic potassium bentonites from Kinnekulle, Sweden. Clays Clay Miner 34:442–454

Bujdak J, Rode BM (1999) Silica, alumina and clay catalysed peptide bond formation: enhanced efficiency of alumina catalyst. Origins Life Evol B 29:451–461

Burgess J (1978) Ions in solution: basic principles of chemical interactions. Wiley, New York

Burley SD, Flisch M (1989) K-Ar chronology and the timing of detrital I-S clay illitisation and authigenic illite precipitation in the Piper and Tartan Fields, Outer Moray Firth, UK North Sea. Clay Miner 24:285–315

Calarge LM, Meunier A, Formoso ML (2002) Chemical signature of two Permian volcanic ash deposits in a bentonite bed from Melo, Uruguay. Clays Clay Miner (in press)

Calle de la C (1977) Structure de vermiculites, facteurs conditionnant le mouvement des feuillets. Thèse Univ. Paris VI, 133 pp

Capuano RM (1992)The temperature dependence of hydrogen isotope fractionation between clay minerals and water: evidence for a geopressured system. Geochim Cosmochim Acta 56:2547–2554

Carslaw HS, Jaeger JC (1959) Conduction of heat in solids. Oxford University Press, New York, 510 pp

Caschetto S, Wollast R (1979) Dissolved aluminium in interstitial waters of recent marine sediments. Geochim Cosmochim Acta 43:425–428

Cases JM, Villiéras F, Michot L (2000) Les phénomènes d'adsorption, d'échange ou de rétention à l'interface solide-solution aqueuse. 1. Connaissance des propiétés structurales, texturales et superficielles des solides. CR. Acad Sci Paris, 331(IIa):763–773

Cassagnabère A, Iden IK, Johansen H, Lacharpagne JC, Beaufort D (1999) Kaolinite and dickite in Froy and Rind sandstone hydrocarbon reservoirs (Brent Formation from Norwegian continental shelf). In: Proc 11th Int Clay Conf, Ottawa, pp 97–102

Cathles LM (1977) An analysis of the cooling of intrusives by ground-water convection which includes boiling. Econ Geol 72:804–826

Cavaretta G, Gianelli G, Puxeddu M (1982) Formation of authigenic minerals and their use as indicators of the physicochemical parameters of the fluids in the Larderello-Travale geothermal field. Econ Geol 77:1071–1084

Chamley H (1971) Recherches sur la sédimentation argileuse en Méditerranée. Sci Géol, Strasbourg, Mém 35, 225 pp

Chamley H (1989) Clay sedimentology. Springer, Berlin Heidelberg New York, 623 pp

Chang HK, Mackenzie FT, Schoonmaker J (1986) Comparisons between the diagenesis of dioctahedral and trioctahedral smectite, Brazilian offshore basin. Clays Clay Miner 34:407–423

Chermarck JA (1993) Low temperature experimental investigation of the effect of high pH NaOH solutions on the Opalinus shale, Switzerland. Clays Clay Miner 40:650–658

Chermarck JA, Rimstidt JD (1990) The hydrothermal transformation rate of kaolinite to illite/muscovite. Geochim Cosmochim Acta 42:1–7

Chudaev OV (1978) Occurrence of clay minerals in flyschoid sediments of eastern Kamchatka (en russe). Lithol Polezn Iskop (1):105–115 (Traduction: Lithol Miner Resour 13:89–97)

Claret F (2001) Caractérisation structurale des transitions minéralogiques dans les formations argileuses: contrôle et imlications géochimiques des processus d'illitisation. Cas particulier d'une perturbation alcaline dans le Callovo-Oxfordien – Laboratoire souterrain Meuse – Haute Marne. Thèse Univ. Joseph Fourier, Grenoble, 174 pp

Clark BC, van Hart DC (1981) The salts of Mars. Icarus 45:370–378

Clauer N (1997) Un concept pour le comportement isotopique des minéraux argileux: une meilleure identification des époques, des mécanismes et des conditions de genèse et de diagenèse. CR Acad Sci 324(IIa):253–268

Clauer N, Chaudhuri S (1995) Clays in crustal environments. Isotope Dating and Tracing. Springer, Berlin Heidelberg New York, 359 pp

Clauer N, Keppens E, Stille P (1992) Sr isotopic constraints on the process of glauconitization. Geology 20:133–136

Clauer N, O'Neil JR, Bonnot-Courtois C (1982) The effect of natural weathering on the chemical and isotopic compositions of biotites. Geochim Cosmochim Acta 46:1755–1762

Clayton RN, O'Neil JR, Mayeda TK (1972) Oxygen isotope exchange between quartz and water. J Geophys Res 77:3057–3067

Clozel B, Allard T, Muller JP (1994) Nature and stability of radiation-induced defects in natural kaolinites: new results and a reappraisal of published works. Clays Clay Miner 42:657–666

Colman SM, Dethier DP (1986) Rates of chemical weathering of rocks and minerals. Academic Press, New York, 603 pp

Compton JS, Conrad ME, Vennemann TW (1999) Stable isotope evolution of volcanic ash layers during diagenesis of Miocene Monterey Formation, California. Clays Clay Miner 47:84–95

Coombs DS (1970) Present status of the zeolite facies. Am Chem Soc 2nd Int Zeolite Conf, pp 556–566

Courbe C, Velde B, Meunier A (1981) Weathering of glauconites: reversal of the glauconitization process in a soil profile in western France. Clay Miner 16:231–243

Couture R (1977) Composition and origin of palygorskite-rich and montmorillonite-rich zeolite-containing sediments from the Pacific Ocean. Chem Geol 19:113–130

Cradwick PDG, Farmer VC, Russell JD, Masson CR, Wada K, Yoshinaga N (1992) Imogolite, a hydrated aluminum silicate of tubular structure. Nature (London) Phys Sci 240:187–189

Craig H (1961) Isotopic variations in meteoric waters. Science 133:1702–1703

Creasey SC (1959) Some phase relations in hydrothermally altered rocks of porphyry copper deposits. Econ Geol 54:351–373

Cuadros J, Linares J (1995) Some evidence supporting the existence of polar layers in mixed-layer illite/smectite. Clays Clay Miner 43:467–473

Cuadros J, Altaner SP (1998) Compositional and structural features of the octahedral sheet in mixed-layer illite-smectite from bentonites. Eur J Miner 10:111–124

Cuadros J, Sainz-Diaz CI, Ramirez R, Hernandez-Laguna A (1999) Analysis of Fe segregation in the octahedral sheet of bentonitic illite-smectite by means of FTIR, ^{27}Al MAS NMR and reverse Monte Carlo simulations. Am J Sci 299:289–308

Dani N (1998) Petrologia das alteraçoes posmagmaticas e meteoricas das rochas alcalinas de Lages, SC-Brasil. Thèse Univ. Fed Rio Grande do Sul, Brasil, and Univ. Poitiers, France, 228 pp

Dasgupta S, Chaudhuri AK, Fukuoka M (1990) Compositional characteristics of glauconitic alteration of K-feldspar from India and their implications. J Sediment Petrol 60:277–281

Decarreau A (1990) Matériaux argileux. Structure, propiétés et applications. Soc Fr Mineral Cristall, Paris, 586 pp

Decarreau A, Bonnin D (1986) Synthesis and crystallogenesis at low temperature of Fe(III)-smectites by evolution of coprecipitated gels: experiments in partially reducing conditions. Clay Miner 21:861–877

Decarreau A, Petit S (1996) Clay synthesis and crystal growth. Short course on crystal growth in Earth Sciences. European Mineralogical Union, Lecture notes SC CG ES, pp 299–313

Decarreau A, Bonnin D, Badaut-Trauth D, Couty R, Kaiser P (1987) Synthesis and crystallogenesis of ferric smectite by evolution of Si-Fe coprecipitates in oxidising conditions. Clay Miner 22:207–223

Deconninck JF, Strasser A, Debrabant P (1988) Formation of illitic minerals at surface temperatures in Purbeckian sediments (Lower Berriasian, Swiss and French Jura). Clay Miner 23:91–103

Deer WA, Howie RA, Zussman J (1962) Rock-forming minerals, vol 3. Sheet silicates. Wiley, New York, 270 pp

Derjaguin B, Landau LD (1941) Theory of the stability of strongly charged lyophobic sols and the adhesion of strongly charged particles in solution of electrolytes. Acta Physichchim URSS 14:733–762

Delhoume JP (1996) Fonctionnement hydropédologique d'une toposéquence de sols en milieu aride (réserve de la Biosphère de Mapimi, Nord-Mexique). Thèse Doctorat Univ. Poitiers, 295 pp

Dixon JB, White GN (2002) Manganese oxides. In: Dixon JB, Schulze DG (eds) Soil mineralogy with environmental applications. Soil Science Society of America Book Series 7:367–388

Dong H, Peacor DR (1996) TEM observation of coherent stacking relations in smectite, I/S and illite of shales: evidence for MasEwan crystallites and dominance of the 2M1 polytypism. Clays Clay Miner 44:257–275

Douglas MC, MacEwan DMC, Wilson MJ (1980) Interlayer and intercalation complexes of clay minerals. In: Brindley GW, Brown G (eds) Crystal structures of clay minerals and their identification. Mineralogical Society Monograph 5, pp 197–248

Dove PM, Crerar DA (1990) Kinetics of quartz dissolution in electrolyte solutions using a hydrothermal mixed flow reactor. Geochim Cosmochim Acta 54:955–969

Drever JI (1982) The geochemistry of natural waters. Prentice Hall, Englewood Cliffs, NJ, 388 pp

Drits VA, Lindgreen H, Sakharov BA, Salyn AS (1997) Sequence structure transformation of illite-smectite-vermiculite during diagenesis of Upper Jurassic shales, North Sea. Clay Miner 33:351–371

Drits VA, Plançon A, Sakharov BA, Besson G, Tsipursky SI, Tchoubar C (1984) Diffraction effects calculated for structural models of K-saturated montmorillonite containing different types of defects. Clay Miner 19:541–561

Drits VA, McCarty DK (1996) The nature of diffraction effects from illite and illite-smectite consisting of interstratified trans-vacant and cis-vacant 2:1 layers: a semiquantitative technique for determination of layer-type content. Am Mineral 81:852–863

Drits VA, Varaxina TV, Sakharov BA, Plançon A (1994) A simple technique for identification of one-dimensional powder X-ray diffraction patterns for mixed-layer illite-smectites and other interstratified minerals. Clays Clay Miner 42:382–390

Drits VA, Salyn AL, Sucha V (1996) Structural transformations of interstratified illite-smectites from Dolna Ves hydrothermal deposits: dynamics and mechanisms. Clays Clay Miner 44:181–190

Drits VA, Tchoubar C (1990) X-ray diffraction by disordered lamellar structures. Theory and applications to microdivided silicates and carbons. Springer, Berlin Heidelberg New York, 371 pp

Ducloux J (1989) Fontenay-Le-Comte C-14. Carte pédologique de France à moyenne échelle. INRA, 17, 204 pp

Ducloux J, Meunier A, Velde B (1976) Smectite, chlorite and a regular interlayered chlorite-vermiculite in soils developed on a small serpentinite body, Massif Central, France. Clay Miner 11:121–135

Dudoignon P (1983) Altérations hydrothermales et supergène des granites. Etude des gisements de Montebras (Creuse), de Sourches (Deux-Sèvres) et des arènes granitiques de Parthenay. Thèse de Spécialité, Univ. Poitiers, 117 pp

Dudoignon P, Meunier A, Beaufort D, Gachon A, Buigues D (1989) Hydrothermal alteration at Muruoa Atoll (French Polynesia). Chem Geol 76:385–401

Dudoignon P, Pantet A (1998) Measurement and cartography of clay matrix orientations by image analysis and grey-level diagram decomposition. Clay Miner 33:629–642

Dudoignon P, Proust D, Gachon A (1997) Hydrothermal alteration associated with rift zones at Fangataufa Atoll (French Polynesia). Bull Volcanol 58:583–596

Eberl D (1978) Reaction series for dioctahedral smectites. Clays Clay Miner 26:327–340

Eberl D (1978) The reaction of montmorillonite to mixed layer clay: the effect of interlayer alkali and alkaline earth cations. Geochim Cosmochim Acta 42:1–7

Eberl DD, Blum A (1993) Illite crystallite thickness by X-ray diffraction. In: Reynods RC, Walker JR (eds) Computer applications to X-ray powder diffraction analysis of clay minerals. CMS Workshop Lectures 5, pp 124–153

Eberl DD, Hower J (1975) Kaolinite synthesis: the role of Si/al and (alkali)/(H^+) ratio in hydrothermal systems. Clays Clay Miner 23:301–309

Eberl DD, Hower J (1976) Kinetics of illite formation. Geol Soc Am Bull 87:1326–1330

Eberl DD, Srodon J (1988) Ostwald ripening and interparticle-diffraction effects for illite crystals. Am Mineral 73:1335–1345

Eberl DD, Srodon J, Kralik M, Taylor BE, Peterman ZE (1990) Ostwald ripening of clays and metamorphic minerals. Science 248:474–477

Eberl DD, Srodon J, Lee M, Nadeau PH, Northrop HR (1987) Sericite from the Silverton caldera, Colorado: correlation among structure, composition, origin and particle thickness. Am Mineral 72:914–934

Eberl DD, Srodon J, Northrop HR (1986) Potassium fixation in smectite by wetting and drying. In: Davis JA, Hayes KF (eds) Geochemical processes at mineral surfaces. Am Chem Soc Symp Ser 323:296–326

Edwards M (1986) Glacial environments. In: Reading HG (ed) Sedimentary environments and facies. Blackwell, Oxford, pp 445–470

Eggleton RA (1977) Nontronite: chemistry and X-ray diffraction. Clay Miner 12:181–194

Eggleton RA (1986) The relation between crystal structure and silicate weathering rate. In: Colman SM, Dethier DP (eds) Rates of chemical weathering of rocks and minerals. Academic Press, New York, pp 21–40

Eggleton RA, Busek PR (1980) High-resolution electron microscopy of feldspar weathering. Clays Clay Miner 28:173–178

Eggleton RA, Boland JN (1982) Weathering of enstatite to talc through a sequence of transitional phases. Clays Clay Miner 30:11–20

Ehrenberg SN, Aagaard P, Wilson MJ, Fraser AR, Duthie DML (1993) Depth-dependent transformation of kaolinite to dickite in sandstones of the Norwegian continental shelf. Clay Miner 28:325–352

Ehrmann WU, Melles M, Kuhn G, Grobe H (1992) Significance of clay mineral assemblages in the Antartic Ocean. Mar Geol 107:249–273

Elder JW (1981) Geothermal systems. Academic Press, New York, 508 pp

Elliott WC, Aronson JL, Matisoff G, Gautier DL (1991) Kinetics of the smectite to illite transformation in the Denver Basin: clay minerals, K-Ar data and mathematical results. Am Assoc Petrol Geol Bull 75:426–462

Ellis AJ, Mahon WAJ (1977) Chemistry and geothermal systems. Academic Press, New York, pp 392

Elprince AM, Mashhady AS, Aba-Hussayn MM (1979) The occurrence of pedogenic palygorskite (attapulgite) in Saudi Arabia. Soil Sci 128:211–218

Escande MA (1983) Echangeabilité et fractionnement isotopique de l'oxygène des smectites magnésiennes de synthèse. Etablissement d'un géothermomètre. Thèse Doct Univ. Paris-Sud, 150 pp

Eslinger E (1971) Mineralogy and oxygen isotope ratios of hydrothermal and low-grade metamorphic argillaceous rocks. PhD Thesis, Univ. Cleveland, 205 pp

Eslinger E, Pevear D (1988) Clay minerals for petroleum geologists and engineers. SEPM short course 22, Soc Econ Paleonto Mineral, 420 pp

Eslinger E, Savin SM (1973) Mineralogy and oxygen isotope geochemistry of the hydrothermally altered rocks of the Oaki-Broadlands, New-Zealand geothermal area. Am J Sci 273:240–267

Eslinger EV, Yeh HW (1981) Mineralogy of O^{18}/O^{16} and D/H ratios of clay-rich sediments from Deep Sea Drilling Project site 180, Aleutian Trench. Clays Clay Miner 29:309–315

Esposito KJ, Whitney G (1995) Thermal effects of thin igneous intrusions on diagenetic reactions in a Tertiary basin of Southwestern Washington. US Geol Survey Bull 2085-C, 36 pp

Essene EJ, Peacor DR (1995) Clay mineral thermometry: a critical perspective. Clays Clay Miner 43:540–553

Eswaran H, Bin WC (1978) A study of a deep weathering profile on granite in peninsular Malaysia. III Alteration of feldspars. Soil Soc Am J 42:154–158

Fagel N, Debrabant P, de Menocal P, Demoulin B (1992) Utilisation des minéraux sédimentaires argileux pour la reconstitution des variations paléoclimatiques à court terme en Mer d'Arabie. Oceanol Acta 15:125–136

Fang-Ru Chou Chang, Skipper NT, Sposito G (1998) Monte Carlo and molecular dynamics simulations of electrical double-layer structure in potassium-montmorillonite hydrates. Langmuir 14:1201–1207

Faure G (1986) Principles of isotope geology, 2nd edn. Wiley, New York, 589 pp

Feth TH, Bobertson CE, Polzer WE (1964) Sources of mineral constituents in waters from granitic rocks, Sierra Nevada, California and Nevada. U.S. Geol Surv Water Supply Paper 1535:70

Feuillet JP, Fleischer P (1980) Estuarine circulation: controlling factor of clay mineral distribution in James River estuary, Virginia. J Sediment Petrol 50:267–279

Fialips CI, Petit S, Decarreau A (2000) Hydrothermal formation of kaolinite from various metakaolins. Clay Miner 35:559–572

Fialips CI, Petit S, Decarreau A, Beaufort D (2000) Influence of synthesis pH on kaolinite "crystallinity" and surface properties. Clays Clay Miner 48:173–184

Fontanaud A, Meunier A (1983) Mineralogical facies of a serpentinized lherzolite from Pyrénées, France. Clay Miner 18:77–88

Fontes JC, Letolle R, Olive P, Blavoux B (1967) Oxygène 18 et tritium dans le bassin d'Evian. In: Isotopes in hydrology. IAEA, pp 401–415

Foscolos AE, Kodama H (1974) Diagenesis of clay minerals from lower Cretaceous shales of North Eastern British Columbia. Clays Clay Miner 22:319–335

Foster MD (1960) Interpretation of the composition of trioctahedral micas. Prof Pap U.S. Geol Survey 354-B:11–49

Foster MD (1962) Interpretation of the composition and a classification of the chlorites. U.S. Geol Survey Prof Paper 414-A:33

France-Lanord C, Derry L, Michard A (1993) Evolution of the Himalayas since Miocene time: isotopic and sedimentological evidence from the Bengal Fan. In Treloar PJ, Searle MP (eds) Himalayan Tectonics. Geol Soc London Special Publ 74, pp 603–621

Franchi M, Bramanti E, Morassi Bonzi L, Orioli PL, Vettori C, Gallori E (1999) Clay-nucleic acid complexes: characteristics and implications for the preservation of genetic material in primeval habitats. Origins Life Evo B 29:297–315

Frank-Kamenestkii VA, Kotov NV, Tomashenko AN (1973) The role of AlIV (tetrahedral) and AlVI (octahedral) in layer silicate synthesis and alteration. Geokhimiya 8:1153–1162 (English translation, Geochem Int 1973:867–874)

Fransolet AM, Schreyer W (1984) Sudoite, Di/trioctahedral chlorite: a stable low-temperature phase in the system $MgO-Al_2O_3-SiO_2-H_2O$. Contrib Mineral Petrol 86:409–417

Frey M (1969) A mixed layer Paragonite/Phengite of low grade metamorphism origin. Contrib Mineral Petrol 24:63–65

Frey M, Robinson G (1999) Low-grade metamorphism. Blackwell, Oxford, 313 pp

Frey M, Teichmüller M, Teichmüller R, Mullis J, Künzi B, Breitschmid A, Griner U, Schwizer B (1980) Very low grade metamorphism in external parts of the Central Alps: illite crystallinity, coal rank and fluid inclusion data. Eclogae Geol Helv 73:173–203

Friedman I, O'Neil JR (1977) Compilation of stable isotope fractionation factors of geochemical interest. In: Fleischer M (ed) Data of geochemistry, 6th edn. US Gov Printing Office, Washington, DC

Fritz B (1981) Etude thermodynamique et modélisation des réactions hydrothermales et diagénétiques. Sciences Géologiques Mém 65:197

Fritz P, Fontes JC (1980) Handbook of environmental isotope geochemistry, vol 1. Elsevier, Amsterdam, 328 pp

Fritz SJ, Toth TA (1997) An Fe-berthierine from a Cretaceous laterite: Part II. Estimation of Eh, pH and pCO_2 conditions of formation. Clays Clay Miner 45:580–586

Frost RL (1995) Fourier transform Raman spectrometry of kaolinite, dickite and halloysite. Clays Clay Miner 43:191–195

Frost RL, van der Gaast SJ (1997) Kaolinite hydroxyls-A raman microscopy study. Clay Miner 32:293–306

Frost RL, Thu Ha Tran, Kristof J (1997) FT-Raman spectroscopy of the lattice region of kaolinite and its intercalates. Vibration Spectros 13:175–186

Fyfe WS, Price N, Thompson AB (1978) Fluids in the Earth's crust. Elsevier, Amsterdam, 340 pp

Garrels RM (1984) Montmorillonite/illite stability diagrams. Clays Clay Miner 32:161–166

Garrels RM, Christ LL (1965) Solutions, minerals and equilibria. Freeman, Cooper, San Francisco, 450 pp. Traduction en français par Wollast R (1967) Equilibres des minéraux et de leurs solutions aqueuses. Gauthier-Villars, Paris, 335 pp

Garrels RM, Howard DF (1957) Reactions of feldspar and mica with water at low temperature and pressure. In: Proc 6th Conf Clays Clay Miner. Pergamon Press, New York, pp 68–88

Garrels RM, Mackenzie FT (1971) Evolution of sedimentary rocks. WW Norton, New York, 397 pp

Gautier DL (1986) Roles of organic matter in sediment diagenesis. SEPM Special Publication 38, Soc Econ Paleonto Mineral, 203 pp

Gautier DL, Kharaka YK, Surdam RC (1985) Relationships of organic matter and mineral diagenesis. SEPM short course 17, Soc Econ Paleonto Mineral, 279 pp

Ghent ED, Robbins DB, Stout MZ (1979) Geothermometry, geobarometry and fluid compositions of metamorphosed calc-silicates and pelites, Mica Creek, British Columbia. Am Mineralal 64:874–885

Gibbs R (1967) The geochemistry of the Amazon River Basin. Part I; The factors that control the salinity and the composition and concentration of suspended solids. Geol Soc Am Bull 78:1203–1232

Giese RF Jr (1988) Kaolin minerals: structures and stabilities. Chapter 3. In: Bailey SW (ed) Hydrous phyllosilicates (exclusive of micas). MSA Rev Mineral 19:29–66

Gilkes RJ, Scholz G, Dimmock GM (1973) Lateritic deep weathering of granite. J Soil Sci 24:523–536

Gillot F, Righi D, Räisänen ML (1999) Formation of smectites and their alteration in two chronosequences of podzols in Finland. Canadian J Soil Sci

Giorgetti G, Memmi I, Nieto F (1997) Microstructures of intergrown phyllosilicate grains from Verrucano metasediments (northern Apennines, Italy). Contrib Mineral Petrol 128:127–138

Giresse P, Odin GS (1973) Nature minéralogique et origine des glauconies du plateau continental du Gabon et du Congo. Sedimentology 20:457–488

Glassman JR, Lundegard PD, Clark RA, Penny BK, Collins ID (1989) Geochemical evidence for the history of diagenesis and fluid migration: Brent sandstone, Heather Field, North Sea. Clay Miner 24:255–284

Goldschmidt PM, Pfirman SL, Wollenburg I, Heinrich R (1992) Origin of sediment pellets from the Artic seafloor: sea ice or iceberg? Deep Sea Res 39:539–565

Gooding JL (1992) Soil mineralogy and chemistry on Mars: possible clues from salts and clays in SNC meteorites. Icarus 99:28–41

Goy-Eggenberger D (1998) Faible métamorphisme de la nappe de Morcles: Minéralogie et géochimie. Thèse Univ. Neuchâtel, Suisse, 192 pp

Grahame DC (1947) The electrical double layer and the theory of electrocapillarity. Chem Rev 41:441–501

Grauby O, Petit S, Decarreau A, Baronnet A (1993) The beidellite-saponite series: an experimental approach. Eur J Mineral 5:623–635

Grauby O, Petit S, Decarreau A, Baronnet A (1994) The nontronite-saponite series: an experimental approach. Eur J Mineral 6:99–112

Grim RE (1968) Clay Mineralogy, 2nd edn. McGraw Hill, New York, 596 pp

Gueddari M (1984) Géochimie et thermodynamique des évaporites continentales. Etude du lac Natron en Tanzanie et du Chott El Jerid en Tunisie. Sci Géol MéM 76, 143 pp

Gutsaenko GS, Samotoyin ND (1966) The decoration method applied to the study of clay minerals. In: Proc Int Clay Conf, Jerusalem, 1, pp 391–400

Güven N (1988) Smectites. In: Bailey W (ed) Hydrous phyllosilicates (exclusive of micas). Rev Mineral 19:497–559

Güven N (2001) Mica structure and fibrous growth of illite. Clays Clay Miner 49:189–196

Hannington MD, Jonasson IR, Herzig PM, Petersen S (1995) Physical and chemical processes of seafloor mineralization at mid-ocean ridges. In: Seafloor hydrothermal systems: physical, chemical, biological and geological interactions. Geophysical Monograph 91:115–175

Hanor JS (1988) Origin and migration of subsurface sedimentary brines. Soc Econ Paleont Mineral Tulsa, Short Course 24, 248 pp

Harder H (1978) Synthesis of iron layer silicate minerals under natural conditions. Clays Clay Miner 26:65–72

Hardie LA, Eugster HP (1970) The evolution of closed-basin brines. Min Soc Am Spec Paper 3:273–290

Harper CT (1970) Graphical solutions to the problem of $^{40}Ar^*$ loss from metamorphic minerals. Eclogae Geol Helv 63:1500–1507

Harris RC, Adams JAS (1966) Geochemical and mineralogical studies on the weathering of granitic rocks. Am J Sci 264:146–173

Hart R (1970) Chemical exchange between sea water and deep ocean basalts. EP SL 9:269–279

Hartmann M (1980) Atlantis II deep geothermal brine system. Hydrographic situation in 1977 and changes since 1965. Deep Sea Research 27A:161–171

Harward ME, Brindley GW (1966) Swelling properties of synthetic smectites in relation to lattice substitutions. Clays Clay Miner 13:209–222

Hazen RM, Wones DR (1972) The crystal structures of one-layer phlogopite and annite. Am Mineral 58:889–900

Hedenquist JW, Arribas A, Reynolds TJ (1998) Evolution of an intrusion-centered hydrothermal system: Far Southeast-Lepanto porphyry and epithermal Cu-Au deposits, Philippines. Econ Geol 93:373–404

Hein JR, Scholl DW (1978) Diagenesis and distribution of Late Cainozoic volcanic sediment in the Southern Bering Sea. Geol Soc Amer Bull 89:197–210

Helgeson HC, Delaney JM, Nesbitt HW, Bird DK (1978) Summary and critique of the thermodynamic properties of rock-forming minerals. Am J Sci 278A:1–229

Hellmuth KH, Siitari-Kauppi M, Lindberg A (1993) Study of the porosity and migration pathways in crystalline rock by impregnation with ^{14}C-polyméthylmethacrylate. J Contamin Hydrol 13:403–418

Helmold KP, van de Kamp PC (1984) Diagenetic mineralogy and controls on albitisation and laumontite formation in Paleogene arkoses, Santa Inez Mountains, California. In: McDonald DA, Surdam RC (eds) Clastic diagenesis. Am Assoc Petroleum Geol Mem 27, Tulsa, Oklahoma, pp 239–276

Henley RW, Ellis AJ (1983) Geothermal systems ancient and modern. Earth Sci Rev 19:1–50

Henley RW, Truesdell AH, Barton PB, Whitney JA (1984) Fluid-mineral equilibria in hydrothermal systems. Reviews in Economic Geology 1:267

Herbillon AJ, Makumbi MN (1975) Weathering of chlorite in a soil derived from chlorite schist under humid tropical conditions. Geoderma 13:89–104

Herrero CP, Gregorkiewitz M, Sanz J, Serratosa JM (1987) ^{29}Si MAS-NMR spectroscopy of mica-type silicates: observed and predicted distribution of tetrahedral Al-Si. Phys Chem Minerals 15:84–90

Herrero CP, Sanz J, Serratosa JM (1985) Si, Al distribution in micas: analysis of high-resolution ^{29}Si NMR spectroscopy. J Phys C Solid State 18:13–22

Herrero CP, Sanz J, Serratosa JM (1989) Dispersion of charge deficits in the tetrahedral sheet of phyllosilicates. Analysis from ^{29}Si NMR spectra. J Phys Chem 93:4311–4315

Hess PC (1966) Phase equilibria of some minerals in the K_2O-Na_2O-Al_2O_3-SiO_2-H_2O system at 25 °C and 1 atmosphere. Am J Sci 264:289–309

Heuvel RC van den (1966) The occurrence of sepiolite and attapulgite in the calcareous zone of a soil near Las Cruces, N. Mexico. Clays Clay Miner 13:193–207

Hillier S (1993) Origin, diagenesis and mineralogy of chlorite minerals in Devonian lacustrine mudrocks, Orcadian Basin, Scotland. Clays Clay Miner 41:240–259

Hillier S (1994) Pore-lining chlorites in siliclastic reservoir sandstones: electron microprobe, SEM and XRD data, and implications for their origin. Clay Miner 29:665–679

Hillier S (1995) Erosion, sedimentation and sedimentary origin of clays. In: Velde B (ed) Origin and mineralogy of clays. clays and the environment. Springer, Berlin Heidelberg New York, pp 162–219

Hillier S, Velde B (1991) Octahedral occupancy and the chemical composition of diagenetic (low-temperature chlorites) Clay Miner 26:149–168

Hillier S, Velde B (1992) Chlorite interstratified with a 7 Å mineral: an example from offshore Norway and possible implications for the interpretation of the composition of diagenetic chlorites. Clay Miner 27:475–486

Hodder APW, Green BE, Lowe DJ (1990) A two-stage model for the formation of clay minerals from tephra-derived volcanic glass. Clay Miner 25:313–327

Hoefs J (1980) Stable isotope geochemistry, 2nd edn. Springer, Berlin Heidelberg New York, 208 pp

Hofmann U, Klemen R (1950) Verlust der Austauschfähigkeit von Lithiumionen an Bentonit durch Erhitzung. Zeitsch Anorg Chem 262:95–99

Holdren GR Jr, Speyer P (1986) Stoichiometry of alkali feldspar dissolution at room temperature and various pH values. In: Colman SM, Dethier DP (eds) Rates of chemical weathering of rocks and minerals. Academic Press, New York, pp 61–81

Holtzapffel T, Chamley H (1986) Les smectites lattées du domaine Atlantique depuis le Jurassique supérieur: gisement et signification. Clay Miner 21:133–148

Hornibrook ERC, Longstaffe FJ (1996) Berthierine from the lower Cretaceous Clearwater Formation, Alberta, Canada. Clays Clay Miner 44:1–21

Horton DG (1985) Mixed layer illite/smectite as a paleotemperature indicator in the Amethyst vein system. Creede district, Colorado, USA. Contrib Mineral Petrol 91:171–179

Howard JJ (1981) Lithium and potassium saturation of illite/smectite clays from interlaminated shales and sandstones. Clays Clay Miner 29:136–142

Howard JJ, Roy DM (1985) Development of layer charge and kinetics of experimental smectite alteration. Clays Clay Miner 33:81–88

Hower J (1961) Some factors concerning the nature and origin of glauconite. Am Mineral 47:886–896

Hower J (1981) Shale diagenesis. In: Longstaff FJ (ed) Short course in clays and the resource geologist. Miner Assoc Canada, pp 60–80

Hower J, Eslinger EV, Hower ME, Perry EA (1976) Mechanism of burial and metamorphism of argillaceous sediments: 1. Mineralogical and chemical evidence. Geol Soc Am Bull 87:725–737

Hower J, Mowatt TC (1966) The mineralogy of illites and mixed-layer illite/montmorillonites. Am Mineral 51:825–854

Huang WL, Longo JM, Pevear DR (1993) An experimentally derived kinetic model for smectite-to-illite conversion and its use as a geothermometer. Clays Clay Miner 41:162–177

Huff WD, Türkmenoglu AG (1981) Chemical characteristics and origin of Ordovician K-bentonites along the Cincinnati Arch. Clays Clay Miner 29:113–123

Huff WD, Anderson TB, Rundle CC, Odin GS (1991) Chemostratigraphy, K-Ar ages and illitization of Silurian K-bentonites from the Central Belt of the Southern Uplands-Down-Longford terrane, British Isles. J Geol Soc 148:861–868

Humphris S, Thompson RN, Marriner GF (1980) The mineralogy and chemistry of basalt weathering, Holes 417A and 418B. DSDP Initial Reports, vol. 51–52–53, pp 1201–1217

Hunziker JC (1986) The evolution of illite to muscovite: an example of the behaviour of isotopes in low-grade metamorphic terrains. In: Deutsch S, Hofmann AW (eds) Isotopes in geology. Chem Geol 57:34–40

Hutcheon I, Oldershaw A, Ghent ED (1980) Diagenesis of Cretaceous sandstones of the Kootenay Formation at Elk Valley (southeast British Columbia) and Mt. Allan (southwestern Alberta). Geochim Cosmochim Acta 44:1425–1435

Iijima A, Matsumoto R (1982) Berthierine and chamosite in coal measures of Japan. Clays Clay Miner 30:264–274

Iijima A, Utada M (1966) Zeolites in sedimentary rocks, with reference to the depositional environments and zonal distribution. Sedimentology 28:185–200

Iijima A, Utada M (1971) Present-day zeolitic diagenesis of the Neogene geosynclinal deposits in the Niigata oil field, Japan. In: Molecular sieve zeolites-I. Adv Chem Ser 101:342–348

Iiyama JT, Roy R (1963) Controlled synthesis of heteropolytypic (mixed layer) clay minerals. Clays Clay Miner 10:4–22

Iiyama JT, Roy R (1963) Unusually stable saponite in the system Na_2O-MgO-Al_2O_3-SiO_2. Clay Miner Bull 5:161–171

Ildefonse P (1980) Mineral facies developed by weathering of a meta-gabbro, Loire-Atlantique (France). Geoderma 24:257–273

Ildefonse P (1987) Analyse pétrologique des altérations prémétéoriques et météoriques de deux roches basaltiques (basalte alcalin de Belbex, Cantal et hawaïte de M'Bouda, Cameroun). Thèse doctorat, Univ. Paris 7, 323 pp

Imbert T, Desprairies A (1987) Neoformation of halloysite on volcanic glass in a marine environment. Clay Miner 31:81–91

Inoue A (1985) Chemistry of corrensite: a trend in composition of trioctahedral chlorite/smectite during diagenesis: J Coll Arts Sci Chiba Univ. B-18:69–82

Inoue A, Aoki M (1999) In situ growing kaolinite under hydrothermal conditions of Ohyunuma explosion crater lake, Hokkaido, Japan. In: Proc 11th Int Clay Conf, Ottawa, pp 673–680

Inoue A, Aoki M, Ito H (2000) Mineralogy of Ohunuma explosion crater lake, Hokkaido, Japan. Part 2: dynamics of kaolinite formation. Clay Sci (in press)

Inoue A, Kitagawa R (1994) Morphological characteristics of illitic clay minerals from a hydrothermal system. Am Mineral 79:700–711

Inoue A, Utada M (1991) Smectite to chlorite transformation in thermally metamorphosed volcanoclastic rocks in the Kamikita area, northern Honshu, Japan. Am Mineral 76:628–640

Inoue A, Utada M, Nagata H, Watanabe T (1984) Conversion of trioctahedral smectite to interstratified chlorite/smectite in Pliocene acidic pyroclastic sediments of the Ohyu district, Akita Prefecture, Japan. Clay Sci 6:103–106

Inoue A, Utada M, Wakita K (1992) Smectite to illite conversion in natural hydrothermal systems. Appl Clay Sci 7:131–145

Inoue A, Velde B, Meunier A, Touchard G (1988) Mechanism of illite formation during smectite-to-illite conversion in a hydrothermal system. Am Mineral 73:1325–1334

Iriarte Lecumberri I (2003) Synthèse de minéraux argileux dans le système SiO_2 – Al_2O_3 – Fe_2O_3 – MgO – Na_2O – H_2O. PhD Thesis (cotutella) Univ. Poitiers and Granada, 241 pp

Jaboyedoff M (1999) Transformation des interstratifiés illite-smectite vers l'illite et la phengite: un exemple dans la série carbonatée du domaine Briançonnais des Alpes suisses romandes. Thèse Univ. Lausanne, 452 pp

Jaboyedoff M, Kübler B, Thélin P (1999) An empirical equation for weakly swelling mixed-layer minerals, especially illite-smectite. Clay Miner 34:601–617

Jaeger JC (1964)Thermal effects of intrusions. Reviews of Geophysics 2:443–466

Jaeger JC (1968) Cooling and solidification of igneous rocks. In: Hess H (ed) Basalts, vol 2. Wiley, New York, pp 504–535

Jakobsen HJ, Nielsen NC, Lindgreen H (1995) Sequences of charged sheets in rectorite. Am Mineral 80:247–252

Jahren JS, Aagaard P (1989) Compositional variations in diagenetic chlorites and illites and relationships with formation-water chemistry. Clay Miner 24:157–170

Jeans CV, Mitchell JG, Scherer M, Fisher MJ (1994) Origin of the Permo-Triassic clay mica assemblage. Clay Miner 29:575–589

Jennings S, Thompson GR (1987) Diagenesis in the Plio-Pleistocene sediments of the Colorado River Delta, southern California. J Sediment Petrol 56:89–98

Jeong GY, Kim HB (2003) Mineralogy, chemistry, and formation of oxidized biotite in the weathering profile of granitic rocks. Am Mineral 88:352–364

Jiang WT, Peacor DR, Buseck PR (1994) Chlorite geothermometry? Contamination and apparent octahedral vacancies. Clays Clay Miner 42:593–605

Jones BF, Galan E (1988) Sepiolite and palygorskite. In: Bailey SW (ed) Hydrous phyllosilicates (exclusive of micas). Rev Mineral 19:631–674

Juteau T, Noack Y, Whitechurch H (1979) Mineralogy and geochemistry of alteration products in holes 417A and 417D basement samples (Deep Sea Drilling Project LEG 51). In: Donnelly T, Francheteau J, Bryan W, Robinson P, Flower M, Salisbury M (eds) Initial reports of the deep sea drilling project. Vol. Ll, Lll, Lll, pp 1273–1297

Juteau T, Maury R (1999) Géologie de la croûteocéanique. Pétrologie et dynamique endogènes. Dunod, Paris, 367 pp

Kaleda KG, Cherkes ID (1991) Alteration of glauconite minerals in contact with heated (100 °C) water. Int Geol Rev 33:203–208

Karlin R (1980) Sediment sources and clay mineral distributions off the Oregon coast. J Sediment Petrol 50:543–560

Kawamura K, Ferris J (1999) Clay catalysis of oligonucleotide formation: kinetics of the reaction of the 5'-phosphorimidazolides of nucleotides with the non-basic heterocycles uracil and hypoxanthine. Origins Life Evo B 29:563–591

Keller J, Ryan WBF, Ninkovich D, Altherr R (1978) Explosive volcanic activity in the Mediterranean over the past 200,000 years as recorded in deep-sea sediments. Geol Soc Am Bull 89:591–564

Kharpoff AM (1984) Miocene red clays of the South Atlantic: dissolution facies of calcareous oozes at Deep Sea Drilling Project sites 519 to 523, Leg 73. In: Hsü KJ, Labrecque JL et al. (eds) Initial report of deep sea drilling project 73, Washington, DC, pp 515–535

Kisch HJ (1981) Coal rank and illite crystallinity associated with the zeolite facies of Southland and the pumpellyite-bearing facies of Otago, southern New Zealand. N Zeal J Geol Geophys 24:349–360

Kitagawa R (1992) Surface microtopographies of pyrophyllite from the Shozokan area, Chugoku Province, Southwest Japan. Clay Sci 8:285–295

Kitagawa R (1995) Coarsening process of a hydrothermal sericite sample using surface microtopography and transmission electron microscopy techniques. In: Churchman GJ, Fitzpatrick RW, Eggleton RA (eds) Clays controlling the environment. In: Proc Int Clay Conf, Adelaide, pp 249–252

Kitagawa R (1997) Surface microtopographies of clay minerals by Au-decoration method and its technical problems. Nendo Kagaku 4:215–220

Kitagawa R, Matsuda T (1992) Microtopography of irregularly-interstratified mica and smectite. Clays Clay Miner 40:114–121

Klein G de V (1991) Basin – forming processes. In: Force ER, Eidel JJ, Maynard JB (eds) Sedimentary and diagenetic mineral deposits: a basin analysis approach to exploration. Soc Econ Geol Reviews in Economic Geology 5:25–41

Kloprogge JT, Jansen JBH, Geus JW (1990) Characterization of synthetic Na-beidellite. Clays Clay Miner 38:409–414

Kloprogge JT, Komarneni S, Amonette J (1999) Synthesis of smectite clay minerals: a critical review. Clays Clay Miner 47:529–554

Kloprogge JT, van der Eerden AMJ, Jansen JBH, Geus JW, Schuiling RD (1993) Synthesis and paragenesis of Na-beidellite as function of temperature, water pressure and sodium activity. Clays Clay Miner 41:423–430

Koizumi M, Roy R (1959) Synthetic montmorillonoids with variable exchange capacity. Am Mineral 44:788–805

Kolla V, Kostecki JA, Robinson F, Biscaye PE, Ray PK (1981) Distributions and origins of clay minerals and quartz in surface sediments of the Arabian Sea. J Sediment Petrol 51:563–569

Korzhinskii DS (1959) Physicochemical basis of the analysis of the paragenesis of minerals (Trans.). Consultant Bureau, New York, 143 pp

Kossovskaya AG, Drits VA (1970) The variability of micaceous minerals in sedimentary rocks. Sedimentology 15:83–101

Köster HM, Ehrlicher U, Gilg HA, Jordan R, Murad E, Onnich K (1999) Mineralogical and chemical characteristics of five nontronites and Fe-rich smectites. Clay Miner 34:579–599

Koster van Groos AF, Guggenheim S (1987) Dehydration of a Ca- and a Mg-exchanged montmorillonite (Swy-1) at elevated pressures. Am Mineral 72:292–298

Koster van Groos AF, Guggenheim S (1990) High-pressure differential thermal analysis: application to clay minerals. In: Stucki JW, Mumpton FA (eds) Thermal analysis in clay science. CMS Workshop Lectures 3, pp 50–94

Kounestron O, Robert M, Berrier J (1977) Nouvel aspect de la formation des smectites dans les vertisols. CR Acad Sci Paris 284:733–736

Kretz R (1994) Metamorphic crystallization. Wiley, New York, 507 pp

Kristmannsdottir H (1975)Hydrothermal alteration of basaltic rocks in Icelandic geothermal areas. In: Proc 2nd UN Symposium on the Development and Use of Geothermal Resources. Lawrence Berkeley Laboratory, CA, pp 441–445

Kristmannsdottir H (1977) Types of clay minerals in hydrothermally altered basaltic rocks, Reykjanes, Iceland. Jokull 26:30–39

Kristmannsdottir H (1978) Alteration of basaltic rocks by hyrothermal activity at 100–300 °C. Developments in Sedimentology 27:359–367

Kübler B (1964) Les argiles, indicateurs de métamorphisme. Rev Inst Franç Pétrol 19:1093–1112

Kübler B (1969) Crystallinity of illite. Detection of metamorphism in some frontal part of the Alps. Referate der Vorträge auf der 47. Jahrestagung der Deutschen Mineral Ges 29–40

Kübler B (1973) La corrensite, indicateur possible de milieux de sédimentation et du degré de transformation d'un sédiment. Bull Centre Rech Pau-S.N.P.A. 7:543–556

Kübler B (1984) Les indicateurs des transformations physiques et chimiques dans la diagenèse, température et calorimétrie. In: Lagache M (ed) Thermométrie et barométrie géologiques. Soc Fr Mineral Cristall 2:489–596

Kuhlemann J, Lange H, Paetsch H (1993) Implication of a connection between clay mineral variations and coarse grained debris and lithology in the central Norwegian-Grennland Sea. Mar Geol 114:1–11

Kyser TK (1987) Equilibrium fractionation factors for stable isotopes. In: Kyser TK (ed) Stable isotope geochemistry of low temperature processes. Short course handbook 13, Miner Assoc Can, Toronto, pp 1–84

Laffon B, Meunier A (1982) Les réactions minérales des micas hérités et de la matrice argileuse au cours de l'ltération supergène d'une marne (Roumazières, Charente). Sci Géol 35:225–236

Lagache M (1984) Thermométrie et barométrie géologique. Soc Fr Minéral Cristall:663 pp

Lagaly G, Weiss A (1969) Determination of the layer charge in mica-type layer silicates. In: Proc Int Clay Conf, Tokyo, Japan, 1, pp 61–80

Laird DA (1994) Evaluation of the structural formula and alkylammonium methods of determining layer charge. In: Mermut AR (ed) Layer charge characteristics of 2:1 silicate clay minerals, pp 80–103

Lambert SJ, Epstein S (1980) Stable isotope investigations of active geothermal system in Valles Caldera, Jemez Mountains, New Mexico. J Volcanol Geotherm Res 8:111–129

Land LS, Dutton SP (1978) Cementation of a Pennsylvanian deltaic sandstone: isotopic data. J Sediment Petrol 48:1167–1176

Lanson B (1990) Mise en évidence des mécanismes de transformation des interstratifiés illite/smectite au cours de la diagenèse. Thèse Univ. Paris VI, 366 pp

Lanson B, Champion D (1991) The I/S to illite reaction in the late stage diagenesis. Am J Sci 291:473–506

Lanson B, Beaufort D, Berger G, Baradat J, Lacharpagne JC (1996) Illitization of diagenetic kaolinite-to-dickite conversion series: late stage diagenesis of the Lower Permian Rotliegend sandstone reservoir. Offshore of the Netherlands. J Sediment Res 66:501–518

Lanson B, Meunier A (1995) La transformation des interstratifiés ordonnés (S≥1) illite-smectite en illite dans les séries diagénétiques. Etat des connaissances et perspective. Bull Centres Rech Explor.-Prod. Elf Aquitaine 19:149–165

Laperche V (1991) Etude de l'état et de la localisation des cations compensateurs dans les phyllosilicates par les méthodes spectrométriques. Thèse Univ. Paris VII, 104 pp

Larsen G, Chilingar GV (1983) Diagenesis in sediments and sedimentary rocks 1–2. Developments in Sedimentology 25(A-B):579

Lasaga AC, Kirkpatrick RJ (1981) Kinetics of geochemical processes. Rev Mineral 8:398

Lasaga AC (1981) Rate laws of chemical reactions. In: Lasaga AC, Kirkpatrick RJ (eds) Kinetics of geochemical processes. Rev Mineral 8:1–68

Laverne C (1987) Les altérations des basaltes en domaine océanique. Minéralogie, Pétrologie et Géochimie d'un système hydrothermal: le puits 504B, Pacifique oriental. Thèse Univ. Aix-Marseille

Lawrence JR (1979) $^{18}O/^{16}O$ of the silicate fraction of recent sediments used as a provenance indicator of the South Atlantic. Mar Geol 33(M1–M7)

Lawrence JR, Drever JI (1981) Evidence for cold water circulation at DSDP sire 395: isotopes and chemistry of alteration products. J Geophys Res 86:5125–5133

Lawrence JR, Drever JI, Anderson TF, Brueckner HK (1979) Importance of alteration of volcanic material in the sediments of Deep Sea Drilling site 323: chemistry. Geochim Cosmochim Acta 43:573–588

Lawrence JR, Taylor HP (1971) Deuterium and oxygen-18 correlation: Clay minerals and hydroxides in Quaternary soil compared to meteoric waters. Geochim Cosmochim Acta 35:993–1003

Lawrence JR, Taylor HP (1972) Hydrogen and oxygen isotope systematics in weathering profiles. Geochim Cosmochim Acta 36:1377–1393

Lee M (1984) Diagenesis of the Permian Rotliegendes sandstone, North Sea: K/Ar, O^{18}/O^{16} and petrologic evidence. PhD Thesis, Case Western Reserve University, Cleveland, 205 pp

Lee JH, Ahn JH, Peacor DR (1985) Textures in layered silicates: progressive changes through diagenesis and low-temperature metamorphism. J Sediment Petrol 55:532–540

Lee JH, Peacor DR, Lewis DD, Wintsch RP (1984) Chlorite-illite/muscovite interlayered and interstratified crystals: a TEM/STEM study. Contrib Mineral Petrol 88:372–385

Léger CL (1997) Etude expérimentale des solutions solides octaédriques ternaires dans les smectites. Thèse Univ. Poitiers, 156 pp

Levi B, Aguirre L, Nyström JO (1982) Metamorphic gradients in buriam metamorphosed vesicular lavas: comparison of basalt and spilite in Cretaceous basic flows from central Chile. Contrib Mineral Petrol 80:49–58

Longstaffe FJ, Ayalon A (1987) Oxygen-isotope studies of clastic diagenesis in the Lower Cretaceous Viking Formation, Alberta: implications for the role of meteoric water. In: Marshall JD (ed) The diagenesis of sedimentary sequences. Geol Soc Spec 36:277–296

Longstaffe FJ, Racki MA, Ayalon A (1992) Stable isotope studies of diagenesis in berthierine-bearing oil sands, Clearwater Formation Alberta, Canada. In: Kharaka YK, Maest AS (eds) Water rock interaction (WRI-7). Proc 7th Int Symp, Park City, USA, pp 955–958

Loveland PJ (1981) Weathering of a soil glauconite in southern England. Geoderma 25:35–54

Lovering TS (1949) Rock alteration as a guide to ore – East Tintic district, Utah. Econ Geol Mon 1:65

Lovering TS (1950) The geochemistry of argillic and related types of alteration. Colorado School of Mines Quart 45–1B:231–260

Lowe DJ (1986) Controls and rates of weathering and clay mineral genesis in airfall tephras: a review and New Zealand case study. In: Colman SM, Dutie DP (eds) Rates of chemical weathering of rocks and minerals. Academic Press, New York, pp 265–230

Lowell JD, Guilbert JM (1970) Lateral and vertical alteration and mineralization zoning in porphyry ore dposits. Econ Geol 65:373–408

Luo X, Brigaud F, Vasseur G (1992) Compaction coefficient of argillaceous sediments: their implication, significance and determination. Norw Petrol Soc Elsevier, pp 321–332

Lynn WC, Arhens RJ, Smith AL (2002) Soil minerals, their geographic distribution, and soil taxonomy. In: Dixon JB, Schulze DG (eds) Soil mineralogy with environmental applications. Soil Science Society of Am Book Series 7:691–709

Mackinnon IDR, Zolensky M (1984) Proposed structures for poorly characterized phases in the C2 M carbonaceous chondrite meteorites. Nature 309:240–242

Madejovà J, Komadel P, Cicel B (1994) Infrared study of octahedral site populations in smectites. Clay Miner 29:319–326

Maes A, Verheyden D, Cremers A (1985) Formation of highly selective cesium-exchange sites in montmorillonites. Clays Clay Miner 33:251–257

Makumbi L, Herbillon AJ (1972) Vermiculitisation expérimentale d'une chlorite. Bull Groupe franç Argiles 24:153–164

Malla PB, Douglas LA (1987) Identification of expanding layer silicates: layer charge vs. expanding properties. Proc Int Clay Conf, Denver 1985, pp 277–283

Mamy J, Gautier JP (1976) Les phénomènes de diffraction des rayons X et électroniques par les réseaux atomiques. Application à l'étude de l'ordre cristallin dans les minéraux argileux. II: Evolution structurale de la montmorillonite associée au phénomène de fixation irréversible du potassium. Ann Agron 27:1–16

Manceau A, Calas G (1985) Heterogeneous distribution of nickel in hydrous silicates from New Caledonia ore deposits. Am Mineral 70:549–558

Manceau A, Calas G (1986) Ni-bearing clay minerals. 2. X-ray absorption study of Ni-Mg distribution. Clay Miner 21:341–360

Manceau A, Calas G, Petiau J (1985) cation ordering in Ni-Mg phyllosilicates of geological interest. In: EXAFS and near edge structure III. Springer, Berlin Heidelberg New York, pp 358–361

Mas A (2000) Etude pétrographique et minéralogique des mésostases d'unités basaltiques et hawaïtiques de l'Atoll de Moruroa (Polynésie Française). Origine des phyllosilicates en présence. Thèse Univ. Poitiers, 358 pp

Matsuda T, Kodama H, Yang AF (1997) Ca-rectorite from Sano mine, Nagano prefecture, Japan. Clays Clay Miner 45:773–780

Matsuhisa Y, Goldsmith JR, Clayton RN (1979) Oxygen isotopic fractionation in the system quartz-albite-anorthite-water. Geochim Cosmochim Acta 43:1131–1140

McBride MB (1994) Environmental chemistry of soils. Oxford University Press, New York, 406 pp

McDowell SD, Elders WA (1980) Authigenic layer silicate minerals in borehole Elmore 1, Salton sea geothermal field, California, USA. Contrib Mineral Petrol 74:293–310

McEwan DMC (1961) Montmorillonite minerals. In: Brown G (ed) The X-ray identification and crystal structures of clay minerals. Mineralogical Society, London, pp 143–207

McIlrath IA, Morrow DW (1990) Diagenesis. Geoscience Canada, reprint series 4, Geol Assoc Canada, 338 pp

McMurtry GM, Wang Chung-Ho, Yeh Hsueh-Wen (1983) Chemical and isotopic investigations into the origin of clay minerals from the Galapagos hydrothermal mounds field. Geochim Cosmochim Acta 47:475–489

Mélières F (1973) Les minéraux argileux de l'estuaire du Guadalquivir (Espagne). Bull Gr Fr Argiles 25:161–172

Merceron T, Vieillard P, Fouillac AM, Meunier A (1992) Hydrothermal alterations in the Echassières granitic cupola (Massif Central, France). Contrib Mineral Petrol 112:279–292

Mercury L (1998) Equilibres minéraux-solutions en domaine non saturé des sols. Thermodynamique de l'eau non capillaire et modélisation physicochimique. Doc BRGM, 283

Mercury L, Vieillard P, Tardy Y (2001) Thermodynamics of ice polymorphs and "ice-like" water in hydrates and hydroxides. Appl Geochem 16:161–181

Méring J (1949) L'interférence des rayons X dans les systèmes à stratification désordonnée. Acat Crystallogr 2:371–377

Méring J (1975) Smectites. In: Gieseking JE (ed) Soil components, vol 2: Inorganic components. Springer, Berlin Heidelberg New York, pp 97–119

Mermut AR (1994) Problems associated with layer charge characterization of 2:1 phyllosilicates. In: Mermut AR (ed) Layer charge characteristics of 2:1 silicate clay minerals, pp 106–122

Merriman R, Frey M (1999) Patterns of very low-grade metamorphism in metapelitic rocks. In: Frey M, Robinson D (eds) Low-grade metamorphism. Blackwell, Oxford, pp 61–107

Meunier A (1983) Micromorphological advances in rock weathering studies. In: Bullock P, Murphy CP (eds) Int Working Meeting Soil Micromorphology, pp 467–483

Meunier A (1980) Les mécanismes de l'altération des granites et le rôle des microsystèmes. Etude des arènes du massif granitique de Parthenay. Mémoires SGF, 1980, 140, 80 pp

Meunier A (1995) Hydrothermal alteration by veins. In: Velde B (ed) Origin and mineralogy of clays. Clays and the environment. Springer, Berlin Heidelberg New York, pp 247–267

Meunier A, Beaufort D, Parneix JC (1987) Dépôts minéraux et altérations liées aux microfracturations des roches: un moyen pour caractériser les circulations hydrothermales. Bull Soc Géol France 8, t. III, 5:163–171

Meunier A, Lanson B, Beaufort D (2000) Vermiculitization of smectite interfaces and illite layer growth as a possible dual model for illite-smectite illitization in diagenetic environments: a synthesis. Clay Miner 35:573–586

Meunier A, Velde B (1976) Mineral reactions at grain contact in early stages of granite weathering. Clay Miner 11:235–250

Meunier A, Velde B (1986) Construction of potential-composition and potential-potential phase diagrams for solid solution-type phases: graphical considerations (a graphical method based on Korzhinskii' equipotential theory. Bull Mineral 109:657–666

Meunier A, Velde B (1989) Solid solutions in illite/smectite mixed layer minerals and illite. Am Mineral 74:1106–1112

Meunier A, Velde B, Griffault L (1998) The reactivity of bentonites: a review. An application to clay barrier stability for nuclear waste storage. Clay Miner 33:187–196

Meybeck M (1987) Global chemical weathering of surficial rocks estimated from the river dissolved loads. Am J Sci 287:401–428

Millot G (1964) Géologie des argiles. Masson, Paris, 510 pp

Monnier F (1982) Thermal diagenesis in the Swiss molasse basin: implications for oil generation. Can J Earth Sci 19:328–342

Montanez IP, Gregg JM, Shelton KL (1997) Basin-wide diagenesic patterns: integrated petrologic, geochemical and hydrologic considerations. SEPM special publ 57, Soc Sedim Geol Tulsa, 302 pp

Montoya JW, Hemley JJ (1975) Activity relations and stabilities in alkali feldspar and mica alteration reactions. Econ Geol 70:577–594

Morton JP (1985) Rb-Sr dating of diagenesis and source age of clays in Upper Devonian black shales of Texas. Geol Soc Am Bull 96:1043–1049

Moore DM, Reynolds RC (1989) X-ray diffraction and the identification and analysis of clay minerals. Oxford University Press, New York, 332 pp

Mossman JR, Clauer N, Liewig N (1992) Dating thermal anomalies in sedimentary basins: the diagenetic history of clay minerals in the Triassic sandstones of the Paris Basin (France). Clay Miner 27:211–226

Mukarami T, Utsunomyia S, Yokoyama T, Kasama T (2003) Biotite dissolution process and mechanisms in the laboratory and in nature: early stage weathering environment and vermiculitization. Am Mineral 88:377–386

Muller JP, Bocquier G (1986) Dissolution of kaolinite and accumulation of iron oxides in lateritic-ferruginous nodules: mineralogical and microstructural transformations. Geoderma 37:113–136

Muller JP, Ildefonse P, Calas G (1990) Paramagnetic defect centers in hydrothermal kaolinite from an altered tuff in the Nopal uranium deposit, Chihuahua, Mexico. Clays Clay Miner 38:600–608

Muller JP, Manceau A, Callas G, Allard T, Ildefonse P, Hazemann JL (1995) Crystal chemistry of kaolinite and Fe-Mn oxides: relation with formation conditions in low-temperature systems. Am J Sci 297:393–417

Murakami T, Sato T, Watanabe T (1993) Microstructure of interstratified illte/smectite at 123 K: a new method for HERTM examination. Am Mineral 78:465–468

Nadeau PH, Wilson J, McHardy WJ, Tait JM (1984) Interstratified clays as fundamental particles. Science 225:923–925

Nagasawa K (1978) Kaolin minerals. In: Sudo T, Shimoda S (eds) Clays and clay minerals in Japan. Development in Sedimentology 26. Elsevier, Amsterdam, pp 189–219

Nagy KL (1994) Application of morphological data obtained using scanning force microscopy to quantification of fibrous illite growth rates. In: Nagy KL, Blum AE (eds) Scanning probe microscopy of clay minerals. CMS Workshop Lectures 7, pp 204–239

Nagy KL, Blum AE (1994) Scanning Probe Microscopy of Clay Minerals. CMS Workshop Lectures, The Clay Mineralogical Society, Boulder, 239 pp

Nahon D (1991) Introduction to the petrology of soils and chemical weathering. Wiley, New York, 334 pp

Nakasawa H, Yamada H, Fujita T (1992) Crystal synthesis of smectite applying very high pressure and temperature. Appl Clay Sci 6:395–401

Nespolo M (2001) Perturbative theory of mica polytypism. Role of the M2 layer in the formation of inhomogeneous polytypes. Clays Clay Miner 49:1–23

Neuhoff PS, Fridriksson T, Anorsson S, Bird DK (1999) Porosity evolution and mineral paragenesis during low-grade metamorphism of basaltic lavas at Teigarhorn, eastern Iceland. Am J Sci 299:467–501

Newman ACD (1987) Chemistry of clays and clay minerals. Mineralogical Society Monograph 6, Mineralogical Society, 480 pp

Newman ACD, Brown G (1966) Chemical changes during the alteration of micas. Clay Miner 23:337–342

Newman ACD, Brown G (1987) The chemical constitution of Clays. In: Newman ACD (ed) Chemistry of clays and clay minerals. Mineral Soc Monograph 6, Mineralogical Society, pp 1–128

Nguyen Kha, Rouiller J, Souchier B (1976) Premiers résultats concernant une étude expérimentale du phénomène d'appauvrissement dans les Pélosols. Sci Sol 4:269–268

Nickkling WG (1994) Aeolian sediment transport and deposition. In: Pye K (ed) Sediment transport and depositional processes. Blackwell, Oxford, pp 293–350

Norrish K, Pikering JG (1983) Clay minerals. In: Soils: an Australian viewpoint. Academic Press, London, pp 281–308

Northrop DA, Clayton RN (1966) Oxygen isotope fractionations in systems containing dolomite. J Geol 74:174–195

Norton D, Knight J (1977) Transport phenomena in hydrothermal systems: cooling plutons. Am J Sci 277:937–981

Odin GS, Matter A (1981) De glauconarium origine. Sedimentology 28:611–624

Odin GS (1988) Green marine clays. Development in Sedimentology 45, Amsterdam, Elsevier, 445 pp

Odin GS, Fullagar PD (1988) Geological significance of the glaucony facies. In: Odin GS (ed) Green marine clays. Oolithic ironstone facies, Verdine facies, Glaucony facies and Celadonite bearing facies – A comparative study. Development in Sedimentology 5, Elsevier, Amsterdam, pp 295–332

Odin GS, Bailey SW, Amouric M, Fröhlich F, Waychunas G (1988) Mineralogy of the verdine facies. In: Odin GS (ed) Green marine clays. Oolitic Ironstone facies, Verdine facies, Glaucony facies and Celadonite-bearing facies – A comparative study. Development in Sedimentology 5, Elsevier, Amsterdam, pp 159–219

Odin GS, Desprairies A, Fullagar PD, Bellon H, Decarreau A, Frölich F, Zelvelder M (1988) Nature and geological significance of celadonite. In: Odin GS (ed) Green marine clays. Oolithic ironstone facies, Verdine facies, Glaucony facies and Celadonite bearing facies – A comparative study. Development in Sedimentology 5, Elsevier, Amsterdam, pp 337–398

Odom E (1984) Glauconite and celadonite minerals. In: Micas, Rev Mineral 13:545–572

Olis AC, Malla PB, Douglas LA (1990) The rapid estimation of the layer charges of 2:1 expanding clays from a single alkylammonium ion expansion. Clay Miner 25:39–50

Ollier C, Pain C (1996) Regolith, soils and landforms. John Wiley & Sons, LTD, Chichester, 316 pp

O'Neil JR, Clayton RN, Mayeda TK (1969) Oxygen isotope fractionation in divalent metal carbonates. J Chem Phys 51:5547–5558

O'Neil JR, Taylor HP (1967) Oxygen isotope equilibration between muscovite and water. J Geophys Res 74:6012–6022

Page R, Wenk HR (1979) Phyllosilicate alteration of plagioclase studied by transmission electron microscopy. Geology 7:393–397

Papike JJ (1998) Planetary materials. Rev Mineral 36:1014

Papapanagiotou P, Beaufort D, Patrier P, Traineau H (1992) Clay mineralogy of the < 0.2 μm rock fraction in the MI-1 drill hole of the geothermal field of Milos (Greece). Bull Geol Soc Greece, XXVIII(2):575–586

Paquet H (1983) Stability, instability and significance of attapulgite in the calcretes of Mediterranean and tropical areas with marked dry seasons. Sci Géol 72:131–140

Paquet H, Clauer N (1997) Soils and sediments. Mineralogy and geochemistry. Springer, Berlin Heidelberg New York, 369 pp

Paradis S, Velde B, Nicot E (1983) Chloritoïd-pyrophyllite-rectorite facies rocks from Brittany, France. Contrib Mineral Petrol 83:342–347

Patrier P, Beaufort D, Meunier A, Eymeri JP, Petit S (1991) Determination of the non-equilibrium ordering state in epidotes from the ancient geothermal field of Saint-Martin: application of the Mossbaüer spectroscopy. Am Mineral 76:602–610

Patrier P, Beaufort D, Touchard G, Fouillac AM (1990) Crystal size of epidote: a potentially exploitable geothermometer in geothermal fileds? Geology 18:1126–1129

Peacor DR (1992) Diagenesis and low-grade metamorphism of shales and slates. In: Busek PR (ed) Minerals and reactions at the atomic scale: transmission electron microscopy. Rev Mineral 27:335–380

Pédro G (1989) Geochemistry, mineralogy and microfabric of soils in soils and their management. Elsevier, Amsterdam, pp 59–90

Pédro G (1993) Argiles des altérations et des sols. Colloque Georges Millot: Sédimentologie et Géochimie de Surface. Académie des Sciences, pp 1–17

Perry E, Hower J (1970) Burial diagenesis in Gulf Coast pelitic sediments. Clays Clay Miner 18:165–178

Petit S, Decarreau A (1990) Hydrothermal (200 °C) synthesis and crystal chemistry of iron-rich kaolinites. Clay Miner 25:181–196

Pevear DR (1992) Illite age analysis, a new tool for basin thermal history analysis. In: Kharaka YK, Maest AS (eds) Proc 7th Int Symp on water-rock interaction. Balkema, Rotterdam, pp 1251–1254

Pye K (1987) Aeolian dust and dust deposits. Academic Press, New York, 334 pp

Pignon F, Magnin A, Piau JM (1996) Thixotropic colloidal suspensions and flow curves with minimum: identification of flow regimes and rheometric consequences. J Rheology 40:573–587

Pirajno F (1992) Hydrothermal mineral deposits, principles and fundamental concepts for the exploration geologist. Springer, Berlin Heidelberg New York, 709 pp

Plançon A, Drits VA (2000) Phase analysis of clays unsing an expert system and calculation programs for X-ray diffraction by two- and three-component mixed layer minerals. Clays Clay Miner 48:57–62

Plançon A (2001) Order-disorder in clay mineral structures. Clay Miner 36:1–14

Pollastro RM, Barker CE (1986) Application of clay minerals, vitrinite reflectance and fluid inclusion studies to the thermal and burial history of the Pinedale anticline. Green River Basin, Wyoming. In: Gautier DL (ed) Roles of organic matter in sediment diagenesis. Soc Econ Paleontol Mineral Spec Publ 38:73–83

Porrenga DH (1967) Glauconite and chamosite as depth indicators in the marine environment. Mar Geol 5:495–501

Porrenga DH (1968) Non-marine glauconitic illite in the lower Oligocene of Aardebrug, Belgium. Clay Miner 7:421–430

Prêt D, Sardini P, Beaufort D, Zellagui R, Sammartino S (2004) Porosity distribution in a clay gouge by image processing of ^{14}C-PolyMethylMetacrylate (^{14}C-PMMA) autoradiographs: case study of the fault of St. Julien (Basin of Lodève, France). Applied Clay Sci 27, 107–118 pp

Proust D, Eymeri JP, Beaufort D (1986) Supergene vermiculitization of a magnesian chlorite: iron and magnesium removal processes. Clays and Clay Miner 34:572–580

Proust D, Lechelle J, Meunier A, Lajudie A (1990) Hydrothermal reactivity of mixed-layer kaolinite/smectite and implications for radioactive waste disposal: I. High-charge to low-charge smectite conversion. Eur J Miner 2:313–325

Proust D, Meunier A, Fouillac AM, Dudoignon P, Sturz A, Charvet J, Scott SD (1992) Preliminary results on the mineralogy and geochemistry of basalt alteration, hole 794D. In: Proceedings of the Ocean Drilling Program, Scientific Results, 127/128, 2, pp 883–889

Proust D, Velde B (1978) Beidellite crystallization from plagioclase and amphibole precursors: local and long-range equilibrium during weathering. Clay Miner 13:199–209

Pruissen DJ, Capkova P, Driessen RAJ, Schenk H (2000) Structure analysis of intercalated smectites using molecular simulations. Applied Catalysis A: General 193:103–112

Purdy JW, Jäger E (1976) K-Ar ages of rock forming minerals from the central Alps. Mem Inst Geol Min Univ. Padova, 30 pp

Pusch R, Madsen FT (1995) Aspects on the illitization of the Kinnekulle bentonites. Clays Clay Miner 43:261–270

Putnis A (2003) Introduction to mineral sciences. Cambridge University Press, 457 pp

Putnis A, McConnell JDC (1980) Principles of mineral behaviour. Blackwell Scientific Publications, Oxford

Pytte AM (1982) The kinetics of the smectite to illite reaction in contact metamorphic shales. MS Thesis, Dartmouth College, Hanover, NH, 78 pp

Pytte AM, Reynolds RC (1989) The thermal transformation of smectite to illite. In: Naesser ND, MCCulloh TH (eds) The thermal history of a sedimentary basin: methods and case history. Springer, Berlin Heidelberg New York, pp 133–140

Radoslovich EW, Norrish K (1962) The cell dimensions and symmetry of layer lattice silicates. I. Some structural considerations. Am Mineral 47:599–636

Renac C (1994)Diagenèse des minéraux argileux dans la marge passive cévenole (Forage G.P.F. Balazuc 1, France): structure cristalline, morphologie, composition isotopique, datation K/Ar et inclusions fluides. Thèse Doctorat Univ. Poitiers, 185 pp

Ransom B, Helgeson HC (1993) Compositional end members and thermodynamic components of illite and dioctahedral aluminous smectite solid solutions. Clays Clay Miner 41:537–550

Rassineux F, Beaufort D, Merceron T, Bouchet A, Meunier A (1987) Diffraction sur lame mince pétrographique avec un détecteur à localisation linéaire. Analusis 15:333–336

Rassineux F, Griffault L, Meunier A, Berger G, Petit S, Vieillard P, Zellagui R, Munoz M (2001) Expandability-layer stacking relation during experimental alteration of a Wyoming bentonite in pH 13.5 solutions at 35 and 60 °C. Clay Miner 36:197–210

Reyes AG (1990) Petrology of Philippine geothermal systems and the application of alteration mineralogy to their assessment. J Volcanol Geotherm Res 43:279–309

Reyes AG, Cardile CM (1989) Characterization of clay scales forming in Philippine geothermal wells. Geothermics 3:429–446

Reynolds RC Jr (1980) Interstratified clay minerals. In: Brindley GW, Brown G (eds) Crystal structures of the clay minerals and their X-ray identification. Mineralogical Society, London, pp 249–303

Reynolds RC Jr (1985) NEWMOD a computer program for the calculation of one-dimensional diffraction patterns of mixed-layered clays. RC Reynolds, 8 Brook Rd., Hanover, NH, USA

Reynolds RC Jr (1988) Mixed layer chlorite minerals. Rev Mineral 19:601–630

Reynolds RC Jr (1992) X-ray diffraction studies of illite/smectite from rocks, < 1 μm randomly oriented powders, and < 1 μm oriented powder aggregates: the absence of laboratory-induced artifacts. Clays Clay Miner 40:387–396

Rice RM (1973) Chemical weathering on the Carnmenellis granite. Min Mag 39:429–447

Rietmeijer FJM (1998) Interplanetary dust particles. In: Papike JJ (ed) Planetary materials. Rev Mineral 36(2–1):2–95

Righi D, Huber K, Keller C (1999) Clay formation and podzol development from postglacial morains in Switzerland. Clay Miner 34:319–332

Righi D, Meunier A (1991) Characterization and genetic interpretation of clays in an acid brown soil (Dystrochrept) developed in a granite saprolite. Clays Clay Miner 39:519–530

Righi D, Meunier A (1995) Origin of clays by rock weathering and soil formation. In: Velde B (ed) Origin and mineralogy of clays. Clays and the environment. Springer, Berlin Heidelberg New York, pp 43–161

Righi D, Ranger J, Robert M (1988) Clay minerals as indicators of some soil forming processes in the temperate zone. Bull Miner 111:326–632

Righi D, Terribile F, Petit S (1998) Pedogenic formation of high-charge beidellite in a vertisol of Sardinia (Italy). Clays Clay Miner 46:167–177

Robert JL, Kodama H (1988) Generalization of the correlations between OH-stretching wavenumbers and composition of micas in the system K_2O-MgO-Al_2O_3-SiO_2-H_2O: a single model for trioctahedral and dioctahedral micas. Am J Sci 288-A:196–212

Robert M, Pédro G (1972) Etablissement d'un schéma de l'évolution expérimentale des micas trioctaédriques en fonction des conditions du milieu (pH concentration). In: Proc Int Clay Conf Madrid, pp 433–447

Roberson HE (1988) Random mixed-layer chlorite-smectite: does it exist? Abstract for Clay Mineral Society, 25th annual meeting, Grand Rapids, MI, USA

Roberson HE, Lahann RW (1981) Smectite to illite conversion rates: effects of solution chemistry. Clays Clay Miner 29:129–135

Roberson HE, Reynods RC, Jenkins DM (1999) Hydrothermal synthesis of corrensite: a study of the transformation of saponite to corrensite. Clays Clay Miner 47:212–218

Robinson D, Bevins RE (1994) Mafic phyllosilicates in low-grade metabasites. Characterization using deconvolution analysis. Clay Miner 29:223–237

Roddick JC, Farrar E (1971) High initial argon ratios in hornblendes. Earth Planet Sci Lett 12:208–214

Ross GJ, Kodama H (1976) Experimental alteration of a chlorite into a regularly interstratified chlorite-vermiculite by chemical oxidation. Clays Clay Miner 24:183–190

Routson RC, Kittrick JA (1971) Illite solubility. Soil Sci Soc Am Proc 35:714–718

Russell JD, Clarck DR (1978) The effect of Fe for Si substitution on the b-dimension of nontronite. Clay Miner 13:133–138

Ryan PC, Reynolds RC Jr (1996) The origin and diagenesis of grain-coating serpentine-chlorite in Tuscloosa Formation sandstone, U.S. Gulf Coast. Am Mineral 81:213–225

Sainz-Diaz CI, Hernandez-Laguna A, Dove MT (2001) Modelling of dioctahedral 2:1 phyllosilicates by means of transferable empirical potentials. Phys Chem Minerals 28:130–141

Sakai H, Tutsumi M (1978) D/H fractionation factors between serpentine and water at 100 to 500 °C and 2,000 bar water pressure and the D/H ratio of natural serpentine. Earth Planet Sci Lett 40:231–242

Salomons W, Hofman P, Boelens R, Mook WG (1975) The oxygen isotopic composition of the fraction less than 2 microns (clay fraction) in recent sediments of Western Europe. Mar Geol 18:M23-M28

Sanz J, Robert JL (1992) Influence of structural factors on 29Si and 27Al NMR chemical shifts of phyllosilicate 2:1. Phys Chem Minerals 19:39–45

Sardini P, Meunier A, Siitatri-Kauppi M (2001) Porosity distribution of minerals forming crystalline rocks. In: Cidu R (ed) Water-rock interaction, vol. 2. Balkema, Rotterdam, pp 1375–1378

Sato M (1960) Oxidation of sulfide ore bodies. I. Geochemical environments in terms of Eh and pH. Econ Geol 55:928–961

Savin SM, Epstein S (1970) The oxygen and hydrogen isotope geochemistry of clay minerals. Geochim Cosmochim Acta 34:25–42

Savin SM, Lee ML (1988) Isotopic studies of phyllosilicates. Rev Mineral 19:189–223

Sayles FL (1979) The composition and diagenesis of interstitial solutions: I. Fluxes across the seawater-sediment interface in the Atlantic Ocean. Geochim Cosmochim Acta 43:527–545

Sayles FL, Mangelsdorf PC (1977) The equilibration of clay minerals with sea water: exchange reactions. Geochim Cosmochim Acta 41:951–960

Shafiqullah M, Damon E (1974) Evaluation of K-Ar isochron methods. Geochim Cosmochim Acta 38:1341–1358

Schiffman P, Fridleifsson GO (1991) The smectite-chlorite transition in drillhole NJ-15, Nesjavellir geothermal field, Iceland: XRD, BSE and electron microprobe investigations. J Metam Geol 9:679–696

Schmidt ST, Robinson D (1997) Metamorphic grade and porosity/permeability controls on mafic phyllosilicate distributions in a regional zeolite to greenschist facies transition in the North Shore Volcanic Group, Minnesota. Bull Geol Soc Am 109:683–697

Schultz LG, van Olphen H, Mumpton FA (eds) (1985) Proc Int Clay Conf 1985. The Clay Minerals Society, Bloomington, Indiana, pp 85–93

Scott RB, Hajash A (1976) Initial submarine alteration of basalt pillow lavas: a microprobe study. Am J Sci 276:480–501

Scott AD, Smith SJ (1966) Susceptibility of interlayer potassium in mica to exchange with sodium. 14th Nat Conf Clay Clay Miner, New York. Pergamon Press, Oxford, pp 69–81

Shannon RD, Prewitt CT (1976) Revised effectice ionic radii and systematic studies of interatomic distances in halides and chalcogenides. Acta Crystallographica A32:751–767

Shau YH, Peacor DR, Essene EJ (1990) Corrensite and mixed-layer chlorite/corrensite in metabasalt from nothern Taiwan: TEM/AEM, EMPA, XRD, and optical studies. Contrib Mineral Petrol 105:123–142

Sheppard SMF, Gilg HA (1996) Stable isotope geochemistry of clay minerals. The story of sloppy, sticky, lumpy and tough. Clay Miner 31:1–24

Shinamoto T, Logan JM (1981) Effects of simulated clay gouges on the sliding behavior of Tennessee sandstone. Tectonophysics 75:243–255

Shutov VD, Drits VA, Sakharov BA (1969) On the mechanism of a postsedimentary transformation of montmorillonite to hydromica. In: Heller L (ed) Proc Int Clay Conf, Tokyo. Israel University Press, Jerusalem, pp 523–531

Siffert B (1962) Quelques réactions de la silice en solution: la formation des argiles. Mém Serv Carte Géol Als Lorr 21, pp 86

Sikora W, Stoch L (1972) Mineral forming processes in weathering crusts of acidic magmatic and metamorphic rocks of lower Silesia. Miner Pol 3:39–52

Sillén LG (1961) The physical chemistry of sea water. In: Sears M (ed) Oceanography. Am Assoc for the Advancement of Science, Washington, DC, pp 549–581

Singer A, Stoffers P, Heller-Kallai L, Safranek D (1984) Nontronite in a deep-sea core from the South Pacific. Clays Clay Miner 32:375–383

Singh B (1996) Why does halloysite roll? – A new model. Clays Clay Miner 44:191–196

Singh B, Mackinnon IDR (1996) Experimental transformation of kaolinite to halloysite. Clays Clay Miner 44:825–834

Small JS (1993) Experimental determination of the rates of precipitation of authigenic illite and kaolinite in the presence of aqueous oxalate and comparison of the K/Ar ages of authigenic illite in the reservoir sandstones. Clays Clay Miner 41:191–208

Smith DE (1998) Molecular computer simulations of the swelling properties and interlayer structure of cesium montmorillonite. Langmuir 14:5959–5967

Smith JV, Yoder HS (1956) Experimental and theoretical studies of the mica polymorphs. Mineral Mag 31:209–235

Soil Survey Staff (1998) Keys to soil taxonomy, 8th edn. US Department of Agriculture, Natural Resources Conservation Service, Washington, DC, 326 pp

Soil Survey Staff (1999) Soil taxonomy. A basic system of classification for making and interpreting soil surveys, 2nd edn. US Department of Agriculture, Natural Resources Conservation Service. Agriculture Handbook 436, 869 pp

Somerton WH (1992) Thermal properties and temperature-related behavior of rock/fluid systems. Developments in Petroleum Science 37, Elsevier, Amsterdam, 257 pp

Spear FS, Ferry JM, Rumble III D (1982) Analytical formulation of phase equilibria: the Gibbs' method. In: Ferry JM (ed) Characterization of metamorphism through mineral equilibria. Rev Mineral 10:105–122

Spiers GA, Dudas MJ, Muehlenbachs K, Pawluck S (1986) Isotopic evidence for clay mineral weathering and authigenesis in Cryoboralfs. Soil Sci Soc Am J 49:467–474

Sposito G (1984) The surface chemistry of soils. Oxford University Press, New York, 234 pp

Sposito G (1989) The chemistry of soils. Oxford University Press, New York, 277 pp

Srodon J (1980) Precise identification of illite/smectite interstratifications by X-ray powder diffraction. Clays Clay Miner 28:401–411

Srodon J (1981) X-ray identification of randomly interstratified illite/smectite in mixtures with discrete illite. Clay Miner 16:297–304

Srodon J (1984) X-ray powder diffraction identification of illitic materials. Clays Clay Miner 32:337–349

Srodon J, Eberl DD (1987) Illite. In: Bailey SW (ed) Micas. Rev Miner 13:495–544

Stakes DS, O'Neil JR (1982) Mineralogy and stable isotope geochemistry of hydrothermally altered oceanic rocks. Earth Planet Sci Lett 57:285–304

Stalder P (1979) Organic and inorganic metamorphism in the Staveyannaz sandstone of the Swiss Alps and equivalent sandstones in France and Italy. J Sediment Petrol 49:463–482

Staudigel H, Muelenbachs K, Richardson SH, Hart SR (1981) Agents of low temperature ocean crust alteration. Contrib Mineral Petrol 77:150–157

Steinberg M, Holtzapffel T, Raurureau M (1987) Characterization of overgrowth structures formed around individual clay particles during early diagenesis. Clays Clay Miner 35:189–195

Steiner A (1968) Clay minerals in hydrothermally altered rocks in Wairakei, New Zealand. Clays Clay Miner 16:193–213

Stern O (1924) Zur theorie der electrolytischen doppelschicht. Z Elektrochem 30:508–516

Stumm W (1992) Chemistry of the solid-water interface. Processes at the mineral-water and particle-water interface in natural systems. Wiley, New York, 428 pp

Sucha V, Kraus I, Gerthofferova H, Petes J, Serekova M (1993) Smectite to illite conversion in bentonites and shales of the East Slovak Basin. Clay Miner 28:243–253

Suquet H, De la Calle C, Pezerat H (1975) Swelling and structural organization of saponite. Clays Clay Miner 23:1–9

Surdam RC, Boles JR (1979) Diagenesis of volcaniclastic sandstones. Soc Econ Paleont Mineral Special publication 26:227–242

Talibudeen O, Goulding KWT (1983) Charge heterogeneities in smectites. Clays Clay Miner 31:37–42

Tardy Y (1993) Pétrologie des latérites et des sols tropicaux. Masson, 535 pp

Tardy Y, Garrels RM (1974) Method for estimating the Gibbs energies of formation of layer silicates. Geochim Cosmochim Acta 38:1101–1116

Tardy Y, Garrels RM (1976) Prediction of Gibbs energies of formation. I – Relationships among Gibbs energies of formation of hydroxides, oxides and aqueous ions. Geochim Cosmochim Acta 40:1051–1056

Tardy Y, Garrels RM (1977) Prediction of Gibbs energies of formation of compounds from the elements. II – Monovalent and divalent metal silicates. Geochim Cosmochim Acta 41:87–92

Tardy Y, Roquin C (1992) Geochemistry and evolution of lateritic landscapes. In: Martini IP, Chesworth W (eds) Weathering, soils and paleosoils. Elsevier, Amsterdam, pp 407–443

Taylor G, Eggleton RA (2001) Regolith geology and morphology. John Wiley & Sons, LTD, Chichester, 375 pp

Tessier D (1984) Etude expérimentale de l'organisation des matériaux argiles. Hydratation, gonflement et structuration au cours de la dessiccation et de la réhumectation. Thèse Univ. Paris VII, INRA Versailles PuB, 360 pp

Tessier D (1990) Organisation des matériaux argileux en relation avec leur comportement hydrique. In: Decarreau A (ed) Matériaux argileux. Structure, propriétés et applications. Soc Fr Minér Cristall Paris, pp 389–445

Thiry M (2000) Paleoclimatic interpretation of clay minerals in marine deposits: an outlook from the continental origin. Earth Sci Rev 49:201–221

Thiry M, Simon-Coinçon R, Schmidt JM (1999) Paléoaltérations kaoliniques: signification climatique et signature dans la colonne sédimentaire. CR Acad Sci Paris, 329(IIa):853–863

Thompson G (1973) A geochemical study of low-temperature interaction of sea water and oceanic igneous rocks. Am Geophys Union Trans 54:1015–1019

Thompson GR, Hower J (1973) An explanation for the low radiometric ages from glauconite. Geochim Cosmochim Acta 37:1473–1491

Thompson GR, Hower J (1975) The mineralogy of glauconite. Clays and Clay Miner 23:289–300

Thornber MR (1975) Supergene alteration of sulfides. A chemical model based on massive nickel deposit at Kambalda, Western Australia. Chem Geol 15:1–14

Thornber MR (1983) Weathering of sulfide orebodies. The chemical processes of gossan formation. Chemical aspects of gossan assessment. In: Smith RE (ed) Geochemical exploration of deeply weathered terrains. CSIRO, Wembley, Australia, pp 67–72; 92–103

Thornber MR, Allchurch PD, Nickel EH (1981) Variation in a gossan geochemistry at the Perseverance nickel sulfide dposit, Western Australia: a descriptive and experimental study. Econ Geol 76:1764–1774

Tissot BP, Welte DH (1978) Petroleum formation and occurrence. A new approach to oil and gas exploration. Springer, Berlin Heidelberg New York, 638 pp

Titley SR (1982) The style and progress of mineralization and alteration in porphyry copper systems. In: Titley SR (ed) Advances in geology of the porphyry copper deposits, southwestern North America. University Arizona Press, Tucson, AZ, pp 93–116

Titley SR, Thompson RC, Haynes FM, Manske SL, Robinson LC, White JL (1986) Evolution of the fractures and alteration in the Sierra-esperanza hydrothermal system, Pima County, Arizona. Econ Geol 81:343–370

Tomasson J, Kristmannsdottir H (1972) High temperature alteration minerals and thermal brines, Reykjanes, Iceland. Contrib Mineral Petrol 36:123–134

Tomeoka K, Busek PR (1985) Indicators of aqueous alteration in CM carbonaceous chondrites: microtextures of a layered mineral containing Fe, S, O and Ni. Geochim Cosmochim Acta 49:2149–2163

Tomeoka K, Busek PR (1988) Matrix mineralogy of the Orgueil CI carbonaceous chondrite. Geochim Cosmochim Acta 52:1627–1640

Trauth N (1977) Argiles évaporitiques dans la sédimentation carbonatée continentale et épicontinentale tertiaire. Sci Géol 49:195

Trescases JJ (1997) The lateritic nickel-ore deposits. In: Paquet H, Clauer N (eds) Soils and sediments. Mineralogy and geochemistry. Springer, Berlin Heidelberg New York, pp 125–138

Trolard F, Tardy Y (1989) A model for Fe^{3+} -kaolinite, Al^{3+} -goethite, Al^{3+} -hematite equilibria in laterites. Clay Miner 24:1–21

Tsoar H, Pye K (1987) Dust transport and the question of desert loess formation. Sedimentology 34:139–153

Turpault MP, Berger G, Meunier A (1992a) Dissolution-precipitation processes induced by hot water in a fractured granite. Part 1: Wall-rock alteration and vein deposition processes. Eur J Mineral 4:1457–1475

Turpault MP, Meunier A, Guilhaumou N, Touchard G (1992b) Analysis of hot fluid infiltration in fractured granite by fluid inclusion study. Appl Geochim Suppl 1:269–276

Urey HC (1947) The thermodynamic properties of isotopic substances. J Chem Soc 562–581

Utada M (1980) Hydrothermal alteration related to igneous acidity in Cretaceous and Neogene formations of Japan. Mining Geol Jpn Spec Issue 12:79–92

Van Damme H (1995) Scale invariance and hydric behaviour of soils and clays. CR Acad Sci Paris 320(IIa):665–681

Van Damme H, Fripiat JJ (1985) A fractal analysis of adsorption processes by pillaring swelling clays. J Chem Phys 82:2785–2789

Van den Heuvel RC (1966) The occurrence of sepiolite and attapulgite in the calcareous zone of a soil near Las Cruces, New Mexico. Clays Clay Miner 13:193–207

Van Olphen H (1977) An introduction to clay colloid chemistry, 2nd edn. Wiley, New York, 187 pp

Vantelon D, Pelletier M, Michot LJ, Barres O, Thomas F (2001) Fe, Mg and Al distribution in the octahedral sheet of montmorillonites. An infrared study in the OH-bending region. Clay Miner 36:369–379

Varajao A, Meunier A (1995) Particle morphological evolution during the conversion of I/S to illite in lower Cretaceous shales from Sergipe-Alagoas basin, Brazil. Clays Clay Miner 43(1):35–59

Vasseur G (1988) Propagation de la chaleur dans la terre et flux géothermique. In: Berest P, Weber P (eds) La thermomécanique des roches. Manuels et méthodes, BRGM, 16, pp 102–130

Veblen DR, Guthrie GD, Livi KJT, Reynolds RC Jr (1990) High-resolution transmission electron microscopy and electron diffraction of mixed-layer illite/smectite: experimental results. Clays Clay Miner 38:1–13

Veblen RR, Buseck PR (1980) Microstructures and reaction mechanisms in biopyriboles. Am Mineral 65:599–623

Vedder W (1964) Correlations between infrared spectrum and chemical composition of mica. Am Mineral 49:736–768

Vegard L (1928) Die Röntgenstrahlen im Dienste der Erforschung der Materie. Z Kristallogr 42:239–259

Velde B (1972) Celadonite mica: solid solution and stability. Contr Miner Petrol 37:235–247

Velde B (1976) The chemical evolution of glauconite pellets as seen by microprobe determinations. Mineral Mag 30:753–760

Velde B (1985) Clay minerals. A physico-chemical explanation of their occurrence. Developments in Sedimentology 40, Elsevier, Amsterdam, 427 pp

Velde B (1995) Origin and mineralogy of clays. Clays and the environment. Springer, Berlin Heidelberg New York, 334 pp

Velde B (1995) Compaction and diagenesis. In: Velde B (ed) Origin and mineralogy of clays. Clays and the environment. Springer, Berlin Heidelberg New York, pp 220–246

Velde B (1999) Structure of surface cracks in soils and muds. Geoderma 93:101–124

Velde B, Odin GS (1975) Further imformation related to the origin of glauconite. Clays Clay Miner 23:376–381

Velde B, Vasseur G (1992) Estimation of the diagenetic smectite to illite transformation in time-temperature space. Am Mineral 77:697–709

Velde B, Lanson B (1993) Comparison of I/S transformation and maturity of organic matter at elevated temperatures. Clays Clay Miner 41:178–183

Velde B, Renac C (1996) Smectite-to-illite conversion and K-Ar ages. Clay Miner 31:25–32

Verwey EJW, Overbeek JTG (1948) Theory of the lyophobic colloids. Elsevier, Amsterdam

Vidal O (1997) Experimental study of the thermal stability of pyrophyllite, paragonite, and clays in a thermal gradient. Eur J Miner 9:1123–140

Vieillard P (1994) Prediction of enthalpy of formation based on refined crystal structures of multisite compounds. 1. Theories and examples. Geochim Cosmochim Acta 58:4049–4063

Vieillard P (2000) A new method for the prediction of Gibbs free energies of formation of hydrated clay minerals based on the electronegativity scale. Clays Clay Miner 48:459–473

Villiéras F, Micot L, Cases J, Berend I, Bardot F, François M, Gerard G, Yvon J (1996) Static and dynamic studies of the energetic surface heterogeneity of clay minerals. In: Rudzinsky W, Steele WA, Zgrablich G (eds) Equilibria and dynamics of gas adsorption on heterogeneous solid surfaces. Elsevier, Amsterdam, pp 1–47

Vrolijk P (1990) On the mechanical role of smectite in subduction zones. Geology 18:703–707

Wada K (1989) Allophane and imogolite. In: Dixon JB, Weed SB (eds) Minerals in soil environments, 2nd edn. Soil Sci Soc Am Madison, pp 1051–1087

Walker GG (1961) Vermiculite minerals. In: Brown G (ed) The X-ray identification and crystal structures of clay minerals. Mineralogical Society, London, pp 297–324

Walker GF (1975) Vermiculites. In: Gieseking JE (ed) Soil components, vol 2: Inorganic components. Springer, Berlin Heidelberg New York, pp 155–189

Walker JR, Thompson GR (1990) Structural variations in illite and chlorite in a diagenetic sequence from the imperial valley, California. Clays Clay Miner 38:315–321

Watanabe T (1988) The structural model of illite/smectite interstratifed minerals and the diagram for their identification. Clay Science 7:97–114

Weaver CE, Pollard LD (1975) The chemistry of clay minerals. In: Developments in sedimentology 15, Elsevier, Amsterdam, 213 pp

Weaver CE (1989) Clays, muds and shales. Developments in sedimentology 44. Elsevier, Amsterdam, 819 pp

Wenner DB, Taylor HP (1971) Temperatures of serpentinization of ultramafic rocks based on O^{18}/O^{16} fractionation between coexisting serpentine and magnetite. Contr Mineral Petrol 32:165–168

White AF, Brantley SL (1995) Chemical weathering rates of silicate minerals. Reviews in Mineralogy 31, 583 pp

Whitney G (1983) Hydrothermal reactivity of saponite. Clays Clay Miner 31:1–8

Whitney G, Northrop HR (1987) Diagenesis and fluid flow in the San Juan Basin, New Mexico. Regional zonation in the mineralogy and stable isotope composition of clay minerals in sandstones. Am J Sci 287:353–382

Whitney G, Northrop HR (1988) Experimental investigation of the smectite to illite reaction: dual reaction mechanisms and oxygen isotope systematics. Am Mineral 73:77–90

Wiewiora A (1990a) Crystallochemical classifications of phyllosilicates based on the unified system of projection of chemical composition: III. The mica group. Clay Miner 25:73–81

Wiewiora A (1990b) Crystallochemical classifications of phyllosilicates based on the unified system of projection of chemical composition: III. The serpentine-kaolin group. Clay Miner 25:93–98

Wiewiora A, Weiss Z (1990) Crystallochemical classifications of phyllosilicates based on the unified system of projection of chemical composition: III. The chlorite group. Clay Miner 25:83–92

Wildman WE, Whittig LD (1971) Serpentine stability in relation to formation of iron-rich montmorillonite in some California soils. Am J Sci 56:587–602

Wilkinson M, Haszeldine RS (2002) Fibrous illite in oilfield sandstones – a nucleation kinetic theory of growth. Terra Nova 14:56–60

Wilson MJ (1987) Soil smectites and related interstratified minerals: recent developments. In: Schultz G, van Olphen H, Mumpton FA (eds) Proc Int Clay Conf, Denver. The Clay Minerals Society, Bloomington, pp 167–173

Wilson MJ (1994) Clay mineralogy: spectroscopic and chemical determinative methods. Chapman and Hall, London, 366 pp

Wise WS, Eugster HP (1964) Celadonite: synthesis, thermal stability and occurrence. Am Mineral 1031–1083

Woff RG (1967) Weathering of Woodstock granite near Baltimore, Maryland. Am J Sci 265:106–117

Wood JR, Hewitt TA (1984) Reservoir diagenesis and convective fluid flow. In: MacDonald DA, Surdam RC (eds) Clastic diagenesis. AAPG Memoir 37, pp 99–110

Worden RH, Morad S (2003) Clay mineral cements in sandstones. Int Assoc Sediment Special Publ 34, Blackwell, Oxford, 509 pp

Xu H, Veblen DR (1996) Interstratification and other reaction microstructures in the chlorite-berthierine series. Contrib Mineral Petrol 124:291–301

Yamada H, Nakasawa H (1993) Isothermal treaments of regularly interstratified montmorillonite-beidellite at hydrothermal conditions. Clays Clay Miner 41:726–730

Yamada H, Nakasawa H, Yoshioka K, Fujita T (1991) Smectites in the montmorillonite-beidellite series. Clay Miner 26:359–369

Yeh HW (1974) Oxygen isotope studies of the ocean sediments during sedimentation and burial diagenesis. PhD Thesis, Univ. Cleveland, 136 pp

Yeh HW (1980) D/H ratio and late-stage dehydration of shales during burial. Geochim Cosmochim Acta 44:341–352

Yeh HW, Savin SM (1977) Mechanism of burial metamorphism of argillaceous sediments, 3. O-isotope evidence. Geol Soc Am Bull 88:1321–1330

Yeh HW, Eslinger E (1986) Oxygen isotopes and the extent of diagenesis of clay minerals during sedimentation and burial in the sea. Clays Clay Miner 34:403–406

Young DA, Smith DE (2000) Simulations of clay mineral swelling and hydration: dependence upon interlayer ion size and charge. J Phys Chem B 104:9163–9170

Zane A, Sassi FP, Sassi R (1996) New data on chlorites as petrogenic indicator mineral in metamorphic rocks. Plinius 16:212–213

Zan L, Gianelli G, Passerini P, Toisi C, Hagas AO (1990) Geothermal exploration in the Republic of Djibouti: thermal and geological data of the Hanlé and Asa areas. Geothermics 19:561–582

Zellagui R (2001) Mécanismes d'argilisation des gouges argileuses associées aux failles dans les bassins silicoclastiques: etude du faisceau de failles du Mas d'Alary-St Jean (bassin de Lodève) et de la faille de Caire Brun (bassin d'Anot). Thèse Univ. Poitiers, 249 pp

Zen EA (1959) Clay mineral carbonate relations in sedimentary rocks. Am J Sci 257:29–43

Zierenberg RA, Shanks III WC (1983) Mineralogy and geochemistry of epigenetic features in metalliferrous sediment, Atlantis II Deep, Red Sea. Econ Geol 78:57–72

Index